Advanced Sampling Methods

Raosaheb Latpate · Jayant Kshirsagar ·
Vinod Kumar Gupta · Girish Chandra

Advanced Sampling Methods

 Springer

Raosaheb Latpate
Savitribai Phule Pune University
Pune, Maharashtra, India

Vinod Kumar Gupta
Former ICAR National
Professor at IASRI
Delhi, India

Jayant Kshirsagar
Ekta Shikshan Prasarak Mandal's Arts,
Commerce and Science College
Ahmednagar, India

Girish Chandra
Indian Council of Forestry
Research and Education
Dehradun, Uttarakhand, India

ISBN 978-981-16-0624-3   ISBN 978-981-16-0622-9   (eBook)
https://doi.org/10.1007/978-981-16-0622-9

This Springer imprint is published by the registered company Springer Nature Singapore Pte Ltd.
The registered company address is: 152 Beach Road, #21-01/04 Gateway East, Singapore 189721,
Singapore

Foreword

Sampling is so deep-rooted in our habits that it usually takes place unconsciously. The authors have undertaken the large task of surveying the sampling literature of the past few decades to provide a reference book for researchers in the area. They have done an excellent job. Starting with the elementary level, the authors very clearly explain subsequent developments. In fact, even the most modern innovations of survey sampling, both methodological and theoretical, have found a place in this concise volume. The book contains various types of sampling schemes along with exercises at the end of each chapter. In this connection, I would like to emphasize that the book presents a very lucid account of various types of adaptive cluster sampling, ranked set sampling, and resampling techniques. With its own distinctiveness, this book is indeed a very welcome addition to the already existing rich literature on survey sampling. The authors of the book are well-established researchers in the area of theory and applications of survey sampling. The present book further highlights their potential and expression in this subject. I know Dr. Latpate as a colleague since 2009 when he started teaching sampling in the Department of Statistics, Savitribai Phule Pune University, India. He guided a Ph.D. student in adaptive sampling, and now he has a command over sampling theory and its recent developments. I thank all the book authors for their efforts in making the new developments of the sampling theory available to readers.

I am sure readers will appreciate the quality of the subtle alchemy that transmutes arduous demonstrations into simple explanations. I wish the book a very great success.

David D. Hanagal
Department of Statistics
Savitribai Phule Pune University
Pune, India

Preface

Sample survey techniques emerged from attempts which aim to measure the characteristics of the population by the use of scientific knowledge of this population which is difficult or cost and time prohibitive to enumerate. The roots of sample survey techniques and thinking can be traced back to our epics. One such oldest and strongest application of survey sampling can be witnessed in the *Nala–Damyanti Akhyan* in the Indian epic the Mahabharata. It states that Nala and King Bhangasuri were moving in a chariot through a forest. Bhangasuri told Nala that if he can count the number of fallen leaves and fruits, he (Bhangasuri) can state the number of fruits and leaves on two strongest branches of the Vibhitak tree. One hundred one are the number of leaves and one fruit informed Nala after counting the fallen leaves and fruit. Bhangasuri avers 2095 fruits and fifty million leaves on the two strongest branches of the tree (actually, it is 5 *koti* leaves and 1 koti is 10 million). Nala counts all night and is duly amazed by morning. Bhangasuri accepts his due; "I of dice possess the science and in numbers thus am skilled", said Bhangasuri. Vahuca replied, "That science if to me thou will impart, in return, O king, receive thou my surpassing skill in steeds".

Early empires often collated the census of population or recorded the trade in various commodities. The Roman Empire was one of the first states to extensively gather data on the size of the empire's population, geographical area, and wealth. Rulers and kings needed information about lands, agriculture, commerce, and population of their states to assess their military potential, wealth, taxation, and other aspects of government. Babylonians and Egyptians carried censuses of agriculture. A population census was conducted in ancient China to determine the revenue and military strength of its provinces. Census of people and registration of property was conducted in ancient Rome. Roman Emperor Augustus conducted surveys on births and deaths of the citizens of the empire as well as the amount of livestock each owned and the crops each harvested. Romans also developed methods of collecting, organizing, and summarizing data. During the middle ages, censuses of population, household goods, and land were conducted. People began keeping records on births, deaths, and accidents in order to determine insurance rates during the fourteenth century CE. Many other such instances can be traced in the history of development as the needs and requirements changed with time. In the modern era, many developments in survey techniques have been realized through renowned sampling practitioners including A. L. Bowley, L. H. C. Tippett, F. Yates, J. Neyman, E. S. Pearson, etc. From purposive sampling to probability sampling, a judicious mixture of these two to design- and model-based sampling, inferential aspects of survey sampling, etc. have been the landmarks of the theory of survey sampling and estimation, and this has totally changed the perception of survey sampling and estimation.

However, we strongly feel that no contribution to the theory of survey sampling can be complete without the mention of four stalwarts of the yester years from India, namely Prof. P. C. Mahalanobis, Prof. P. V. Sukhatme, Prof. D. Basu, and Prof. J. N. K. Rao from the recent times. These names have always fascinated authors and every researcher engaged in survey sampling research. In the words of Harold Hotelling, "No technique of random sample has, so far as I can find, been developed in the United States or elsewhere, which can compare in accuracy with that described by Prof. Mahalanobis", and Sir R. A. Fisher commented, "Indian Statistical Institute has taken the lead in the original development of the technique of sample surveys, the most potent fact-finding process available to the administration".

The motivation to write this book came from teaching the theory of survey sampling to graduates, postgraduates, and doctoral students of various universities and research organizations. The experience of the authors in handling sample surveys and survey data, which blends judiciously with the theory of survey sampling, has also been a motivating factor to write a book on survey sampling. A couple of brain-storming sessions of Raosaheb with Vinod Gupta about this project and then the suggestions of Vinod Gupta about how to manage the contents for better and purposeful presentation and also for the inclusion of some topics, especially randomized response technique and also sharing some lecture notes that helped in the preparation of the manuscript, enabled Vinod Gupta to join the team as an author at a very later stage. The authors also realized that research organizations like ICAR—Indian Agricultural Statistics Research Institute, New Delhi, and Forest Research Institute, Dehradun, are institutions that help in producing theories of survey sampling and also developing methodologies for the estimation of population parameters related to agriculture and forestry. The knowledge of advanced sampling methods and estimation is essential for achieving gains within limited resources. In forestry experiments, throughout the country, more than 200 research projects are implemented every year, which deal directly with forest surveys including the household, ecological, biodiversity assessment, insect, pest, and disease surveys, spatial surveys, etc.

The contents and coverage of this book have been presented through a total of 22 chapters. The theories of different sampling methods and estimation thereof have been described. The first seven chapters discuss the basic fundamentals of survey sampling and describe conventional sampling methods. Chapter 1 describes the introduction to the sampling techniques. Chapter 2 provides the detail about simple random sampling. The ratio and regression methods of estimation are also included in this chapter. The methods of stratified and systematic sampling are given in Chaps. 3 and 4, respectively. The cluster sampling with equal and unequal cluster sizes and multistage sampling are given in Chap. 5. Chapter 6 covers double sampling methods. The probability proportional to size sampling has been described in Chap. 7. The knowledge of these conventional methods is essential to help the reader follow the latter chapters, which deal with advanced and more recent sampling methodologies. In some spatial populations, the performance of conventional methods is poor. In many situations, the surveyor deals with sensitive issues like the estimation of illegal timber felling from the forest which are difficult to obtain through conventional methods. Therefore, the randomized response technique has been introduced in Chap. 8. Resampling methods which include bootstrapping and jackknifing are given in Chap. 9.

Thompson in 1990 introduced the adaptive cluster sampling (ACS) method for the estimation of rare plant and animal species and developed the estimators for the population mean/total. This method is useful when the population of interest is hidden or elusive and conventional sampling methods are difficult to use. These types of problems have been observed in various areas including forestry, ecological and environmental sciences, and agriculture experiments. ACS is not cost-effective in the case when the sampling size goes very high. To control the sample size, we have introduced a new method which uses the domain knowledge about the negative correlation between the variable of interest and the auxiliary variable.

The major portion of this book (Chaps. 10–18) covers different types of ACS. Chapter 10 covers the introduction and important estimators on ACS. Since Thompson's introduction on ACS, sufficient developments have taken place. Chapter 11 elaborates two-stage ACS along with the different estimators under this design. The adaptive cluster double sampling has been introduced in Chap. 12. The absence of sample size in ACS has also been discussed in the form of inverse ACS and placed in Chap. 13. In Chap. 14, the generalization of inverse ACS for two stages has been described. In some situations of stratified sampling, the population is rare and clumped within the strata. For such situations, stratified ACS has been given in Chap. 15.

Chapter 16 deals with Negative ACS. Negative ACS double sampling and two-stage negative ACS have been introduced in Chaps. 17 and 18, respectively. This is one of those few books that discuss these recent developments in ACS and are not available on the issues of providing real-life applications. The idea of McIntyre (1952), popularly known as ranked set sampling (RSS), has become one of the important sampling methods nowadays not only as an alternative to classical methods but also in the reduction of measurement cost. The balanced and unbalanced RSSs have been introduced in Chap. 19. Chapter 20 covers the use of RSS in parametric and non-parametric inferences. Some useful versions of RSS are given in Chap. 21. The last chapter of the book describes the non-sampling errors. The real-world applications of some of the above methods are given and illustrated with R software to understand the procedure in detail.

We would like to thank the research students, teaching staff, and postgraduate students for making concrete suggestions. Also, we are thankful to our family members for their support and encouragement during the writing of this book.

The present book can be used as a reference by researchers and practitioners of the theory of survey sampling and its applications. This book will also serve as a useful textbook for graduate and postgraduate students with keen interest in the theory of survey sampling and estimation. We expect that readers would benefit immensely from the content and the faculty would find this book useful for teaching students. We welcome suggestions from readers so that the same can be incorporated in the subsequent editions of the book so as to make it further purposeful and useful.

Pune, India   Raosaheb Latpate
Ahmednagar, India   Jayant Kshirsagar
New Delhi, India   Vinod Kumar Gupta
Dehradun, India   Girish Chandra

Contents

1	**Introduction**		1
	1.1	Introduction	1
	1.2	Sample Survey	2
	1.3	Questionnaire Construction	3
	1.4	Population, Sample, and Estimator	4
	1.5	Advanced Sampling Methods	6
	1.6	Exercises	8
	References		8
2	**Simple Random Sampling**		11
	2.1	Introduction	11
		2.1.1 Some Population Parameters	11
		2.1.2 Some Sample Values or Statistics	12
	2.2	Simple Random Sampling with Replacement	13
	2.3	Simple Random Sampling Without Replacement	16
	2.4	Estimation of Population Proportions	22
	2.5	Determination of Sample Size	24
	2.6	Use of Auxiliary Information	25
		2.6.1 Ratio Method of Estimation	26
		2.6.2 Regression Method of Estimation	31
	2.7	Exercises	34
3	**Stratified Random Sampling**		37
	3.1	Introduction	37
	3.2	Stratified Random Sampling with Proportional Allocation	40
	3.3	Stratified Random Sampling with Optimal Allocation	43
	3.4	Post-Stratified Random Sampling	47
	3.5	Stratified Random Sampling for Proportions	49
	3.6	Exercises	51
	Reference		53

4   Systematic Sampling   55
  4.1   Introduction   55
  4.2   Comparison of Systematic Sampling with Simple Random
         Sampling   58
  4.3   Comparison of Systematic Sampling with Stratified
         Random Sampling....................................   59
  4.4   Circular Systematic Sampling   60
  4.5   Exercises   60

5   Cluster Sampling   61
  5.1   Introduction   61
  5.2   Single-Stage Cluster Sampling   62
         5.2.1   Equal Cluster Sizes.......................   62
         5.2.2   Unequal Cluster Sizes.....................   66
  5.3   Multistage Cluster Sampling   66
         5.3.1   Two-stage Cluster Sampling   68
  5.4   Exercises   72
  References   75

6   Double Sampling...   77
  6.1   Introduction   77
  6.2   Ratio Estimator in Double Sampling   77
  6.3   Regression Estimator in Double Sampling   80
  6.4   Exercises   83

7   Probability Proportional to Size Sampling   85
  7.1   Introduction   85
  7.2   Probability Proportional to Size Sampling
         with Replacement   85
  7.3   Probability Proportional to Size Sampling Without
         Replacement.......................................   89
  7.4   Inclusion Probability Proportional to Size Sampling
         Designs   94
         7.4.1   Sampford's (1967) IPPS Scheme.............   94
         7.4.2   Midzuno–Sen (1952) Sampling Strategy   95
         7.4.3   Rao–Hartley–Cochran (RHC) Sampling Strategy ...   95
  7.5   Classes of Linear Estimators   96
  7.6   Exercises   97
  References   97

8   Randomized Response Techniques   99
  8.1   Introduction   99
  8.2   Warner's Randomized Response Design................   101
  8.3   Three Versions of Warner's Design   103
         .3.1   Unrelated Question Design (Greenberg et al.
                 1969, 1971)   103

	8.3.2	Forced Response Design (Boruch 1971)	105
	8.3.3	Disguised Response Design (Kuk 1990)	106
8.4	Some Case Studies		107
8.5	Exercises		108
References			109

9 Resampling Techniques 113
9.1	Introduction		113
9.2	Monte Carlo Methods		114
	9.2.1	Evaluating an Integral	114
	9.2.2	Variance of Monte Carlo Estimators	116
9.3	Bootstrapping Method		116
9.4	Jackknife Method		117
9.5	Importance Sampling		119
9.6	Examples		121
9.7	Exercises		124
References			124

10 Adaptive Cluster Sampling 125
10.1	Introduction		125
10.2	Notations and Terminology		127
10.3	Illustration of Methodology		128
10.4	Sampling Units and their Inclusion Probabilities		131
10.5	Illustration of Computing Inclusion Probabilities		132
10.6	Different Approaches of Estimation		133
	10.6.1	Model-Based Approach	134
	10.6.2	Design-Based Approach	134
	10.6.3	Model Assisted-Cum-Design-Based Approach	135
10.7	Variable Probability Estimation		135
10.8	Estimation of Population Mean/Total without Using the Edge Units		136
	10.8.1	Different Estimators in ACS	137
	10.8.2	Improved Unbiased Estimators in ACS	140
10.9	Estimation of Population Mean in ACS by using Edge Units		143
10.10	Restricted Adaptive Sampling Design		146
	10.10.1	Estimation of an Upper Bound for the Population Total	146
	10.10.2	Estimation of a Lower Bound for the Population Total	148
	10.10.3	Illustrative example of RAS	149
10.11	Estimation of the Final Sample Size in ACS		150
10.12	Pilot Study for Estimating the Final Sample Size		153

10.13   Exercises   154
References   155

11   Two-Stage Adaptive Cluster Sampling   157
11.1   Introduction   157
11.2   Notations and Selection Procedure   158
    11.2.1   Notations   158
    11.2.2   Selection Procedure   158
11.3   Overlapping Clusters Scheme   159
    11.3.1   Modified HT Estimator   159
    11.3.2   Modified HH Estimator   161
11.4   Non-overlapping Clusters Scheme   163
    11.4.1   Choice of a Scheme   164
11.5   Illustrative Example   164
11.6   Exercises   168
References   169

12   Adaptive Cluster Double Sampling   171
12.1   Introduction   171
12.2   Notations and Sampling Design   172
12.3   Estimators Under ACDS   173
    12.3.1   Regression Type Estimator   173
    12.3.2   HT Type Estimator   176
12.4   Exercises   176
References   177

13   Inverse Adaptive Cluster Sampling   179
13.1   Introduction   179
13.2   Inverse Sampling   180
    13.2.1   Sampling Until k Units of P_M are Observed   180
    13.2.2   Mixed Design Stopping Rule   181
13.3   Inverse Sampling with Unequal Selection Probabilities   182
13.4   General Inverse Sampling   185
13.5   Regression Estimator Under Inverse Sampling   186
    13.5.1   General Inverse Sampling   186
    13.5.2   Inverse Sampling with Unequal Selection
             Probabilities   188
13.6   Inverse Adaptive Cluster Sampling   188
13.7   General Unequal Probability Inverse ACS   189
13.8   Exercises   190
References   192

14   Two-Stage Inverse Adaptive Cluster Sampling   193
14.1   Introduction   193
14.2   Two- Stage Inverse Adaptive Cluster Sampling   194

          14.2.1   New Design   194
          14.2.2   Estimators of Population Total..............   195
   14.3   Simulation Study and Discussion   199
   14.4   Exercises   202
   References   202

15 **Stratified Inverse Adaptive Cluster Sampling**   203
   15.1   Introduction   203
   15.2   Stratified Inverse ACS Design   203
   15.3   Simulation Study.................................   206
   15.4   Exercises   208
   References   208

16 **Negative Adaptive Cluster Sampling**.......................   209
   16.1   Introduction   209
   16.2   Negative Adaptive Cluster Sampling   210
          16.2.1   Procedure of NACS   211
          16.2.2   Difference in ACS, ACDS, and NACS   212
   16.3   Different Estimators in NACS   213
          16.3.1   Modified Ratio Type Estimator   213
          16.3.2   Modified Regression Estimator   215
          16.3.3   Product Estimator   216
   16.4   Sample Survey   217
   16.5   Exercises   226
   References   227

17 **Negative Adaptive Cluster Double Sampling**.................   229
   17.1   Introduction   229
   17.2   Negative Adaptive Cluster Double Sampling (NACDS)......   230
          17.2.1   Sampling Design and Notations...............   230
   17.3   Different Estimators.................................   232
          17.3.1   Regression Estimator   232
          17.3.2   Ratio Estimator..........................   233
   17.4   Sample Survey   234
   17.5   Exercises   239
   References   240

18 **Two-Stage Negative Adaptive Cluster Sampling**   241
   18.1   Introduction   241
   18.2   Two-Stage NACS   242
   18.3   Different Estimators.................................   243
          18.3.1   Composite HT Estimator....................   243
          18.3.2   Two-Stage Regression Estimator   246
   18.4   Sample Survey   249

18.5   Exercises   255
    References   255

19   **Balanced and Unbalanced Ranked Set Sampling**   257
    19.1   Introduction   257
    19.2   Balanced RSS   260
    19.3   Unbalanced RSS   262
          19.3.1   Skewed Underlying Distributions   262
          19.3.2   Symmetric Underlying Distributions   266
    19.4   Some Important Applications   270
    19.5   Imperfect Ranking in RSS   271
    19.6   Exercises   271
    References   273

20   **RSS in Other Parameteric and Non-parametric Inference**   275
    20.1   Introduction   275
    20.2   Estimation of Location and Scale Parameters   275
    20.3   Estimating Population Proportion   279
    20.4   Quantile Estimation   280
    20.5   RSS in Non-parametric Inference   281
          20.5.1   One Sample Case   282
          20.5.2   Two Sample Case   284
    20.6   Exercises   285
    References   286

21   **Important Versions of RSS**   289
    21.1   Introduction   289
    21.2   Extreme RSS (ERSS)   289
    21.3   Median Ranked Set Sampling (MRSS)   291
    21.4   Double-Ranked Set Sampling (DRSS)   292
    21.5   Some Other Versions   293
          21.5.1   Double ERSS (Samawi 2002)   293
          21.5.2   Quartile RSS (Muttlak 2003a) and Percentile
                    RSS (Muttlak 2003b)   293
          21.5.3   LRSS (Al-Nasser 2007)   293
          21.5.4   Balanced Groups RSS (Jemain et al. 2008)   293
          21.5.5   Selective Order RSS (Al-Subh et al. 2009)   294
          21.5.6   Robust ERSS (Al-Nasser and Bani-Mustafa
                    2009)   294
          21.5.7   Quartile DRSS (Al-Saleh and Al-Omari 2002)   294
          21.5.8   Record RSS (Salehi and Ahmadi 2014)   294
          21.5.9   Mixed RSS (Haq et al. 2014)   294
          21.5.10   Neoteric RSS (Zamanzade and Al-Omari 2016)   295
          21.5.11   Robust Extreme-DRSS (Majd and Saba 2018)   295
          21.5.12   Extreme-cum-Median RSS (Ahmed and Shabbir
                    2019)   295

    21.6   Exercises 295

    References 296

22 Non-sampling Errors 297

    22.1   Introduction 297

    22.2   Types of Non-sampling Errors 297

            22.2.1   Measurement Error 298

            22.2.2   Absence of Pilot Study 298

            22.2.3   Inadequate Sample Size 298

            22.2.4   Response Errors 299

            22.2.5   Non-response Error 299

    22.3   Exercises 301

    Reference 301

Chapter 1
Introduction

1.1 Introduction

Solving scientific problems requires data. It can be obtained by conducting experiments or sample surveys. The purpose of every sample survey is to obtain the information about populations by selecting the sample. Population is a group of units defined according to the objective of the survey. Here, the term 'unit' means the sampling unit which is the smallest element upon which the measurements are to be made for drawing the inferences. Examples of populations are all fields under a specified crop as in area, all agricultural holdings above specified size as in agricultural surveys, or the number of forest fringe villages in the country. The sample surveys are conducted by government and non-government organizations, researchers, sociologists, and business firms to get answers to certain specific questions which cannot be obtained directly through mathematical and statistical formulations. The sample surveys are conducted on demography (sex ratio), labor force (employment), health and living conditions, political opinion poll, marketing, etc. Surveys have thus grown into a universally accepted approach for information gathering. While conducting a sample survey, it should be conducted in such a way that inference related to the population should have a valid statistical background. A sampling method is a scientific and objective-oriented procedure of selecting units from the population. It provides a sample which may be truly representative of the population. Examples of samples are provided by a handful of grains taken from a sack, or a spoonful of rice taken from a pressure cooker.

The purpose of sampling theory is to develop the methods of sample selection from a finite or infinite population and of estimation that provides estimates of the unknown population parameters, generally population total, population mean, or population proportion which are precise enough for our purpose. Survey samples can broadly be categorized into two types: probability samples and nonprobability samples. Surveys based on probability samples are capable of providing mathemat-

© The Author(s), under exclusive license to Springer Nature Singapore Pte Ltd. 2021
R. Latpate et al., *Advanced Sampling Methods*,
https://doi.org/10.1007/978-981-16-0622-9_1

ically sound statistical inferences about a larger target population. Inferences from probability-based surveys may, however, suffer from many types of bias.

There is no way of measuring the bias or sampling errors of estimators in surveys which are not based on probability sampling. Surveys based on nonprobability samples are not externally valid. They can only be said to be representative of sampling units that have actually been observed. Henceforth, a sample survey would always mean one wherein sampling units have been selected by the probability sampling mechanism, unless otherwise stated. The use of nonprobability sampling is only considered in certain extreme situations, where probability sampling is not possible. For example, in surveying and analyzing the chemical contents of extinct plant species, some areas of natural forests are not accessible/approachable, therefore, the approachable areas based on the expert's advice may be chosen.

Sample survey methods, based on probability sampling, have more or less replaced complete enumeration (or census) methods on account of several well-known advantages. It is well recognized that the introduction of the probability sampling approach has played an important role in the development of survey techniques. The concept of representativeness through probability sampling techniques introduced by Neyman (1934) provided a sound base to the survey approach of data collection. One of the salient features of probability sampling is that besides providing an estimate of the population parameter; it also provides an idea about the precision of the estimate (sampling error). Throughout this book, the attention would be restricted to sample surveys and not the complete survey. For a detailed exposition of the concepts of sample surveys, reference may be made to the texts of Cochran (1977), Des and Chandhok (1998), Murthy (1977), Sukhatme et al. (1984), and Mukhopadhyay (1998a, b).

In this book, we present a basic and advanced body of knowledge concerning survey sampling theory, methods and its applications in agriculture, forestry, environmental sciences, medical sciences, industry, economics, government, biological sciences, humanities, and science and technology.

1.2 Sample Survey

A method of gathering information from a secured subgroup/sample aim to infer the whole population characteristics as closely as possible. Sample surveys reduce the time, cost and enhance the accuracy of results because of the trained surveyors and appropriate sampling methods. The following are the examples of sample surveys:

1. TV networks to estimate how many and what type of people watch their programs.
2. Political opinion poll to estimate the total votes obtained by a particular candidate/party.
3. Household budget survey to estimate the average and source of income level of households of a specific geographical domain.
4. Rare species estimation as per the IUCN categorization.

A sample must be chosen so that it is representative of the population. It differs from census. In census, each and every individual's characteristics are measured and in a sample survey, the characteristics of selected individuals are measured. Census is too costly and time consuming. Also, it gives a less precise estimate due to the less-trained enumerators which leads to serious non-sampling errors.

Steps in a Sample Survey
The following steps are used while conducting the sample surveys.

1. State the objective(s).
2. Define the population.
3. Decide on data required to fulfill the objectives.
4. Decide on the sample design.
5. Decide on the method of measurement.
6. Conduct the pilot study.
7. Revise questionnaire, determine the sample size, and choose appropriate sampling design by a pilot study.
8. Collection of the data.
9. Analyze the result.
10. Present the final conclusion in the form of a report.

1.3 Questionnaire Construction

Many of the surveys depend on the questionnaire to be filled by the respondents. The questionnaire should be constructed in such a way that it contains precise questions to be answered so as to get more accurate information and to reduce the poor or non-responses. Some of the important points that need to be taken care of are as listed below.

1. Make the respondents aware about the purpose of the research and benefits. Ensure that their names are not disclosed for any purposes. The survey should be voluntary and anonymous as far as possible.
2. It should be short, simple, and specific to the objective of the sample survey. The already available information should be avoided.
3. Keep logical question progression.
4. The flow of questions should be maintained.
5. The arrangement of questions should be like, for example, the first key questions are followed by the sensitive questions.
6. Keep allied questions together.
7. It should be reasonable when requesting forecasts.
8. Mix fully structured and open-ended questions in the questionnaire.
9. Every question should be relevant to the objective of the sample survey.

1.4   Population, Sample, and Estimator

A finite population is a collection of known number N of distinct and identifiable sampling units. If Us denote the sampling units, the population of size N
may be represented by the set $U = \{U_1, U_2, \ldots, U_i, \ldots, U_N\}$. The study variable
is denoted by y having value Y_i on unit i; $i = 1, 2, \ldots, N$. We may represent by
$\mathbf{Y} = (Y_1, Y_2, \ldots, Y_i, \ldots, Y_N)'$ an N-component vector of the values of the study
variable Y for the N population units. The $\mathbf{Y}$ is assumed fixed though unknown.
Sometimes, the auxiliary information is also available on some other characteristic
x related to the study variable Y. The auxiliary information is generally available for
all the population units. We may represent by $\mathbf{X} = (X_1, X_2, \ldots, X_i, \ldots, X_N)'$ an
N-component vector of the values of the auxiliary variable X for the N population
units. The total $X = X_1 + X_2 + \cdots + X_i + \cdots + X_N$ is generally known.

A list of all the sampling units in the population along with their identity is known
as a sampling frame. The sampling frame is a basic requirement for sampling from
finite populations. It is assumed that the sampling frame is available and it is perfect
in the sense that it is free from under or over coverage and duplication. However, in
most of the cases, it is not readily available.

The probability selection procedure selects the units from U with probability
$P_i, i \in U$. We shall denote by $\mathbf{P} = (P_1, P_2, \ldots, P_i, \ldots, P_N)'$ an N-component vector of the initial selection probabilities of the units such that $\mathbf{P}'\mathbf{1} = 1$. Generally,
$P \sim g(n, N, X)$; e.g. $P_i = \frac{1}{N} \ \forall \ i \in U$; or $P_i = \frac{n}{N} \ \forall \ i = 1, 2, \ldots, k, k = N/n$; or
$P_i = \frac{X_i}{X_1 + X_2 + \cdots + X_N}, \ \forall \ i \in U$.

A non-empty set $\{s : s \in U\}$, obtained by using probability selection procedure
$\mathbf{P}$, is called an unordered sample. The cardinality of s is n, which is also known as
the (fixed) sample size. However, there shall be occasions wherein we shall discuss
the equal probability with replacement and unequal probability with replacement
sampling schemes. In such cases, the sample size is not fixed. A set of all possible
samples is called sample space S. While using probability selection procedures, the
sample s may be drawn either with or without replacement of units. In the case of
with replacement sampling, the cardinality of S is $\eta = N^n$, and the probability is very
strong that the sample selected may contain a unit more than once. In without replacement sampling, the cardinality of S is $\nu = \binom{N}{n}$, and the probability that the sample
selected may contain a unit more than once is zero. Throughout, it is assumed that
the probability sampling is without replacement of units unless specified otherwise.

Given a probability selection procedure $\mathbf{P}$ which describes the probability of selection of units one by one, we define the probability of selection of a sample s as $P(s) =$
$g(P_i) : i \in s, s \in S$. We also denote by $\mathbf{p} = \{p(1), p(2), \ldots, p(s), \ldots, p(\nu)\}'$ a ν-
component vector of selection probabilities of the samples. Obviously, $p(s) \geq 0$ and
$\mathbf{p}'\mathbf{1} = 1$. It is well known that given a unit-by-unit selection procedure, there exists
a unique mass selection procedure; the converse is also true.

After the sample is selected, data are collected from the sampled units. Let y_i be the value of study variable on the ith unit selected in the sample s, $i \in s$ and $s \in S$. We shall denote by $\mathbf{y} = (y_1, y_2, \ldots, y_i, \ldots, y_n)'$ an n-component vector of the sampled observations. It is assumed here that the observation vector $\mathbf{y}$ is measured without error and its elements are the true values of the sampled units.

The problem in sample surveys is to estimate some unknown population parameter $\theta = f(\mathbf{Y})$ or $\theta = f_1(\mathbf{Y}, \mathbf{X})$. We shall focus on the estimation of population total, $\theta = \mathbf{Y}'\mathbf{1} = \sum_{i \in U} Y_i$, or population mean $\theta = \bar{Y}_N = N^{-1}\mathbf{Y}'\mathbf{1} = N^{-1}\sum_{i \in U} Y_i$. An estimator e for a given sample s is a function such that its value depends on $y_i, i \in s$. In general, $e = h(\mathbf{y}, \mathbf{X})$ and the functional form $h(., .)$ would also depend upon the functional form of θ, besides being a function of the sampling design. We can also write $e_s = h\{\mathbf{y}, p(s)\}$.

A *sampling design* is defined as
$d = [\{s, p(s)\} : s \in S]$.
Further $\sum_{s \in S} p(s) = 1$.
We shall denote by $D = \{d\}$, a class of sampling designs.
S is called the *support* of the sampling plan and $v = \binom{N}{n}$ is called the support size. A sampling plan is said to be a fixed-size sampling plan if whenever $p(s) \geq 0$, the corresponding subsets of units are composed of the same number of units. We shall restrict the discussion to fixed-size sample only and sample size would always mean fixed-size sample.

The triplet (S, p, e_s) is called the sampling strategy.

A familiarity with the expectation and variance operators is assumed in the sequel. An estimator e_s is said to be unbiased for the estimation of population parameter θ if $E_d(e_s) = \theta$ with respect to a sampling design d, where E_d denotes the expectation operator. The bias of an estimator e_s for estimating θ, with respect to a sampling design d, is $B_d(e_s) = E_d(e_s) - \theta$. Variance of an unbiased estimator e_s for θ, with respect to sampling design d, is $V_d(e_s) = E_d\{e_s - E_d(e_s)\}^2 = E_d(e_s - \theta)^2 = E_d(e_s)^2 - \theta^2$. The mean squared error of a biased estimator e_s for θ is given by $MSE_d(e_s) = E_d(e_s - \theta)^2 = E_d\{e_s - E_d(e_s) + E_d(e_s) - \theta\}^2 = V_d(e_s) + \{B_d(e_s)\}^2$.

Sampling Distribution

Consider again a finite population of N distinct and identifiable units. The problem is to estimate some population parameter θ using a sample of size n drawn without replacement of units using some predefined sampling plan $\mathbf{p} = \{p(s); s \in S\}$. There are v samples in the sample space S. The sampling distribution is given in Table 1.1.

Now, draw a random number between 0 and 1 (i.e. a uniform variate). Let the random number drawn be R. Then the kth sample would be selected if $p(s_1) + p(s_2) + \cdots + p(s_{k-1}) < R \leq p(s_1) + p(s_2) + \cdots + p(s_k)$.

Table 1.1 Sampling distribution of the sampling plan $\mathbf{p} = \{p(s); s \in S\}$

Sr. No.	Sample(s)	Probability of selection p(s)	Cumulative sum of p(s)
1	s_1	$p(s_1)$	$p(s_1)$
2	s_2	$p(s_2)$	$p(s_1) + p(s_2)$
3	s_3	$p(s_3)$	$p(s_1) + p(s_2) + p(s_3)$
.	.	.	.
.	.	.	.
.	.	.	.
$k-1$	s_{k-1}	$p(s_{k-1})$	$p(s_1) + p(s_2) + p(s_3) + \cdots + p(s_{k-1})$
k	s_k	$p(s_k)$	$p(s_1) + p(s_2) + p(s_3) + \cdots + p(s_k)$
.	.	.	.
.	.	.	.
.	.	.	.
v	s_v	$p(s_v)$	$p(s_1) + p(s_2) + p(s_3) + \cdots + p(s_v) = 1$

1.5 Advanced Sampling Methods

Sampling designs for a sample survey are one of the most applicable statistical techniques used in various fields including agriculture, ecology, environmental science, health science, forestry, biosciences, humanities, etc. Several sampling designs are already developed. In forestry, the assessment of rare, endangered, and threatened flora and fauna are a challenging task as they do not have homogeneous distribution and their assessment is necessarily required from time to time by IUCN and other international organizations. For such a situation, the method of adaptive cluster sampling (ACS) (Thompson 1990) is found suitable, however, it needs to be improved for specific situations in nature. The ACS differs from the classical sampling designs in a significant way. In classical sampling designs, the sample acquisition does not begin unless the sample selection is complete. The measurements are obtained on selected sampling units only after the entire sample is acquired. In ACS, the initial sample is used as a pilot survey. Here, after selecting the initial sample, the measurements on the selected sampling units are taken. This information is then used to decide which units should be included to get the final sample.

In ACS, the final sample size is a variable quantity. Considering it as a random variable, we have obtained its expected value in this book. The newly proposed negative ACS (NACS) method has also been included in this book as a book chapter. In this method, the process of adding the units to the initial sample is different than that of ACS. In this sampling design, two highly negatively correlated variables X and Y have been considered. X is highly abundant in the population whereas Y is rare. Taking observations on X is rather economical and easy as compared to Y.

Further, the different estimators of population total have been proposed and their variances also have been derived. A sample survey is presented in Western Ghats of Maharashtra to estimate the number of evergreen plants.

The other new negative adaptive cluster double sampling (NACDS) is the combination of double sampling and NACS. It is an efficient method for the estimation of the parameters of a rare and clustered population in the presence of an auxiliary variable. If an auxiliary variable is highly positively correlated with the study variable, then the auxiliary information is also rare. In such situations, the maximum utilization of the auxiliary information at the design stage and estimation stage is not possible. Hence, the auxiliary variable which is highly negatively correlated with the study variable and the population related to the study variable is highly clustered and patchy has been considered. According to the nature of such kind of population, the auxiliary information is abundant and we can exploit it at the design stage and estimation stage. The adaptation of units is based on the auxiliary information instead of the study variable. Regression and ratio estimators based on Särndal's estimators and the product type ratio Horvitz–Thompson (HT) estimator are proposed. The estimators of the variances of these estimators are also obtained. A sample survey in Tamhini Ghat, Maharashtra, India, is conducted using this new NACDS methodology.

Salehi and Seber (1997) used a two-stage sampling procedure in ACS to overcome the problems associated with ACS. We have proposed a two-stage negative adaptive cluster sampling design. It is a new design, which is a combination of two-stage sampling and NACS. In this design, we consider an auxiliary variable which is highly negatively correlated with the variable of interest, and the auxiliary information is completely known. In the first stage of this design, an initial random sample is drawn by using the auxiliary information. Further, using Thompson's (1990) adaptive procedure, networks in the population are discovered. These networks serve as the primary-stage units (PSUs). In the second stage, random samples of unequal sizes are drawn from the PSUs to get the secondary-stage units (SSUs). The values of the auxiliary variable and the variable of interest are recorded for these SSUs. A regression estimator is proposed to estimate the population total of the variable of interest. A new estimator, composite Horvitz–Thompson (CHT)-type estimator, is also proposed. It is based on only the information on the variable of interest. Variances of the above two estimators along with their unbiased estimators are derived. Using this proposed methodology, a sample survey was conducted at Western Ghats of Maharashtra, India.

General Inverse ACS was introduced by Christman and Lan (2001). We have proposed a new sampling design which is a combination of two-stage inverse cluster sampling and ACS designs. At the first stage, the population is divided into non-overlapping clusters and a random sample of pre-fixed number of clusters is selected from these clusters. At the second stage, an initial sample of a fixed size is selected from each of these selected clusters. Further, the number of units satisfying some predetermined condition (number of successes) is decided for each cluster. The estimator of population total at each stage is proposed by using the Rao–Blackwellization procedure. All these methods are highly useful for assessing density and distribution of any rare and endangered species.

An idea of McIntyre (1952), popularly known as ranked set sampling (RSS), has become one of the important sampling methods nowadays not only as an alternative of classical methods but also in reduction of measurement cost. It is considered one of the sections of environmental sampling because of its practical applications in forestry, ecological and environmental sciences, agriculture, etc. After the mathematical foundation of RSS laid by Takahasi and Wakimoto in the year 1968, many theoretical developments in parametric and non-parametric estimation and various variants of the original idea of RSS along with the effect of imperfect judgment ranking have been made. Some of the applications successfully concluded are estimating shrub phytomass in Appalachian Oak forests, herbage mass in pure grass swards/mixed grass–clover swards, mean Reid vapor pressure, mean pool area for 20 of 21 streams, milk yield based on 402 sheep, weights of browse and herbage in a pine–hardwood forest of east Texas, response of developmental programs, and regenerations of longleaf pine.

The method randomized response technique (RRT), initially proposed by Warner (1965), for the structured survey interviews which consist of sensitive questions and reduce the potential bias, is also included in this book. Three important variants of RRT are discussed.

The real-world applications of all the above methods are given and some of them are illustrated with R software to understand the methods and procedure.

1.6   Exercises

1. Define the terms population and sample.
2. Write a short note on mean squared error.
3. Requisites of good questionnaire construction.
4. What is need for sampling as compared to census?
5. Enlist the importance of sampling?
6. What is meant by sampling frame? Give an example?
7. What is the difference between an estimator and an estimate?
8. Suppose a population consists of 5 units having y-values 3, 2, 5, 6, and 9. Find all possible samples of size 2 (without replacement) and show that sample mean is an unbiased estimator of a population mean of 5.

References

Christman, M.C., Lan, F.: Inverse adaptive cluster sampling. Biometrics **57**(4), 1096–1105 (2001)

Cochran, W.G.: Sampling Techniques, 3rd edn. Wiley, New York (1977)

Des, R., Chandhok, P.: Sample Survey Theory. Narosa Publishing House, New Delhi (1998)

McIntyre, G.A.: A method for unbiased selective sampling using ranked sets. Aust. J. Agric. Res. **3**, 385–390 (1952)

Mukhopadhyay, P.: Theory and Methods of Survey Sampling. Prentice Hall of India Private Limited, New Delhi (1998a)

Mukhopadhyay, P.: Small Area Estimation in Survey Sampling. Narosa Publishers, New Delhi (1998b)

Murthy, M.N.: Sampling: Theory and Methods. Statistical Publishing Society, Calcutta, India (1977)

Neyman, J.: On the two different aspects of the representative method—The method of stratified sampling and the method of purposive selection. J. R. Stat. Soc. **97**, 558–625 (1934)

Salehi, M.M., Seber, G.A.F.: Two stage adaptive cluster sampling. Biometrics, **53**, 959–970 (1997)

Sukhatme, P.V., Sukhatme, B.V., Sukhatme, S., Asok, C.: Theory of Sample Surveys with Applications. Indian Society of Agricultural Statistics, Indian Agricultural Statistics Research Institute, New Delhi (1984)

Thompson, S.K.: Adaptive cluster sampling. J. Am. Stat. Assoc. **85**(412), 1050–1058 (1990)

Warner, S.L.: Randomized response: a survey technique for eliminating evasive answer bias. J. Am. Stat. Assoc. **60**(309), 63–69 (1965)

Chapter 2
Simple Random Sampling

2.1 Introduction

Simple random sampling (SRS) is the simplest method of selecting a sample of n units out of N units by drawing units one by one with or without replacement. Every unit in the population has an equal probability of selection. This sampling method is useful whenever the underlined population is homogeneous.

2.1.1 Some Population Parameters

A finite population is a collection of known number N of distinct and identifiable sampling units. If U's denote the sampling units, the population of size N may be represented by the set

$$U = \{U_1, U_2, \ldots, U_i, \ldots, U_N\},$$

where $N \sim$ is the population size or the number of elements in the population.
$y \sim$ is the study variable (or the characteristic of interest).
$y_i \sim$ is the value of the study variable y for the ith element of the population.
Represent by $\mathbf{Y} = (y_1, y_2, \ldots, y_i, \ldots, y_N)'$ an N-component vector of the values of the study variable Y for the N population units.

$Y = \sum\limits_{i=1}^{N} y_i \sim$ is the population total of the study variable y, generally unknown.

$\bar{Y} = \frac{Y}{N} = \frac{1}{N} \sum\limits_{i=1}^{N} y_i \sim$ is the population mean of the study variable y, generally unknown.

$\sigma_y^2 = \frac{1}{N} \sum\limits_{i=1}^{N} (y_i - \bar{Y})^2 \sim$ is the population variance of the study variable y, generally unknown.

© The Author(s), under exclusive license to Springer Nature Singapore Pte Ltd. 2021
R. Latpate et al., *Advanced Sampling Methods*,
https://doi.org/10.1007/978-981-16-0622-9_2

$S_y^2 = \frac{N}{N-1}\sigma_y^2 = \frac{1}{N-1}\sum_{i=1}^{N}(y_i - \bar{Y})^2 \sim$ is the mean square of the population elements for the study variable y, generally unknown.

$x \sim$ is the Auxiliary variable related to the study variable y.

$x_i \sim$ is the value of the auxiliary variable x for the ith element of the population.

Represent by $\mathbf{X} = (x_1, x_2, \ldots, x_i, \ldots, x_N)'$ an N-component vector of the values of the auxiliary variable X for the N population units.

$X = \sum_{i=1}^{N} x_i \sim$ is the population total of the auxiliary variable x, generally known.

$\bar{X} = \frac{X}{N} = \frac{1}{N}\sum_{i=1}^{N} x_i \sim$ is the population mean of the auxiliary variable x, generally known.

$\sigma_x^2 = \frac{1}{N}\sum_{i=1}^{N}(x_i - \bar{X})^2 \sim$ is the population variance of the auxiliary variable x.

$S_x^2 = \frac{N}{N-1}\sigma_x^2 = \frac{1}{N-1}\sum_{i=1}^{N}(x_i - \bar{X})^2 \sim$ is the mean square of the population elements for the auxiliary variable x.

$S_{xy} = \frac{1}{N-1}\sum_{i=1}^{N}(x_i - \bar{X})(y_i - \bar{Y}) \sim$ is the covariance between the study variable y and the auxiliary variable x for the population elements.

$\rho = \frac{S_{xy}}{S_x S_y} \sim$ is the correlation coefficient between the study variable y and the auxiliary variable x for the population units.

$C_y = \frac{S_y}{\bar{Y}} \sim$ is the coefficient of variation of the study variable y for the population elements.

$C_x = \frac{S_x}{\bar{X}} \sim$ is the coefficient of variation of the auxiliary variable x for the population elements.

$\rho_t = \frac{\sum_{i=1}^{N}(y_i - \bar{Y})(y_{i+t} - \bar{Y})}{N\sigma_y^2}, \quad t = 1, 2, \ldots, m(<N) \sim$ is the tth-order circular serial correlation coefficient for the study variable y between the population elements.

$\rho_t = \frac{\sum_{i=1}^{N-m}(y_i - \bar{Y})(y_{i+t} - \bar{Y})}{N\sigma_y^2}, \quad t = 1, 2, \ldots, m(<N) \sim$ is the tth-order linear serial correlation coefficient for the study variable y between the population elements.

2.1.2   Some Sample Values or Statistics

$n \sim$ is the fixed sample size or the number of elements in the sample.

$y_i \sim$ is the value of the study variable y for the ith element of the sample.

Denote by $\mathbf{y} = (y_1, y_2, \ldots, y_i, \ldots, y_n)'$ an n-component vector of the sampled observations. It is assumed here that the observation vector $\mathbf{y}$ is measured without error, and its elements are the true values of the sampled units.

$\bar{y} = \frac{1}{n}\sum_{i=1}^{n} y_i \sim$ is the sample mean of the study variable y.

$s_y^2 = \frac{1}{n-1} \sum\limits_{i=1}^{n} (y_i - \bar{y})^2 \sim$ is the sample mean square of the sampled elements of the study variable y.

$x_i \sim$ is the value of the auxiliary variable x for the ith element of the sample.

$\bar{x} = \frac{1}{n} \sum\limits_{i=1}^{n} x_i \sim$ is the sample mean of the auxiliary variable x.

$s_x^2 = \frac{1}{n-1} \sum\limits_{i=1}^{n} (x_i - \bar{x})^2 \sim$ is the sample mean square of the sampled elements of the auxiliary variable x.

$s_{xy} = \frac{1}{n-1} \sum\limits_{i=1}^{n} (x_i - \bar{x})(y_i - \bar{y}) \sim$ is the sample covariance between the study variable y and the auxiliary variable x for the sampled elements.

$\hat{\rho} = \frac{s_{xy}}{s_x s_y} \sim$ is the sample correlation coefficient between the study variable y and the auxiliary variable x for the sampled units.

2.2 Simple Random Sampling with Replacement

Simple random sampling with replacement (SRSWR) is a method of drawing n units from N units such that at every draw, each unit in the population has the same chance of being selected, and draws are made with replacement of selected units. For the SRSWR scheme, $P_i = \frac{1}{N}$, $\forall i = 1, 2, \ldots, N$. Further, the probability of selecting unit i at the rth draw $P_i^{(r)} = P_i^{(1)} = \frac{1}{N}$, $\forall i \in U; r = 1, 2, \ldots, n; s \in S$. The sample space S under SRSWR consists of N^n possible samples. The probability of selecting a particular sample of size n under SRSWR is $p(s) = \frac{1}{N} \cdot \frac{1}{N} \cdots \frac{1}{N} = \frac{1}{N^n}$.

An unbiased estimator of population mean under SRSWR is given by

$$\bar{y}_n = \frac{1}{n} \sum_{i=1}^{n} y_i. \tag{2.1}$$

The variance of the estimator (2.1) under SRSWR is given by

$$V(\bar{y}_n) = \frac{\sigma_y^2}{n} \tag{2.2}$$

where $\sigma_y^2 = \frac{1}{N} \sum\limits_{i=1}^{N} (y_i - \bar{Y})^2$.

An unbiased estimator of (2.2) under SRSWR is given by

$\hat{V}(\bar{y}_n) = \frac{s_y^2}{n}$

where $s_y^2 = \frac{1}{n-1} \sum\limits_{i=1}^{n} (y_i - \bar{y}_n)^2$.

Theorem 2.1 *In SRSWR, sample mean* $\bar{y}_n = \frac{1}{n} \sum_{i=1}^{n} y_i$ *is an unbiased estimator of population mean* $\bar{Y} = \frac{1}{N} \sum_{i=1}^{N} y_i$.

Proof The sample mean for SRSWR is

$$\bar{y}_n = \frac{1}{n} \sum_{i=1}^{n} y_i .$$

Taking expectation on both sides, $E(\bar{y}_n) = \frac{1}{n} \sum_{i=1}^{n} E(y_i)$.

Since each unit has an equal probability of selection, i.e. $\frac{1}{N}$, and by using definition of expectation

$$E(\bar{y}_n) = \frac{1}{n} \sum_{i=1}^{n} \frac{1}{N} \sum_{i=1}^{N} y_i$$

$$= \frac{n}{n} \frac{1}{N} \sum_{i=1}^{N} y_i$$

$$= \frac{1}{N} \sum_{i=1}^{N} y_i$$

$$= \bar{Y}.$$

Hence the theorem.   □

Theorem 2.2 *The variance of the sample mean under SRSWR is* $V(\bar{y}_n) = \frac{\sigma_y^2}{n}$.

Proof Taking variance on both sides of (2.1),

$$V(\bar{y}_n) = \frac{1}{n^2} \sum_{i=1}^{n} V(y_i). \tag{2.3}$$

Since the selection of units is independent,

also, $V(y_i) = \frac{1}{N} \sum_{i=1}^{N} (y_i - \bar{Y})^2$.

Using this result, Eq. (2.3) becomes

$$V(\bar{y}_n) = \frac{1}{n^2} \sum_{i=1}^{n} \frac{1}{N} \sum_{i=1}^{N} (y_i - \bar{Y})^2 = \frac{n}{n^2} \frac{1}{N} \sum_{i=1}^{N} (y_i - \bar{Y})^2$$

$$V(\bar{y}_n) = \frac{\sigma_y^2}{n}.$$

Hence the theorem.   □

Theorem 2.3 *The estimator of variance under SRSWR* $\hat{V}(\bar{y}_n)$ *is an unbiased estimator of* $V(\bar{y}_n)$.

Proof The unbiased estimator of variance under SRSWR is $\hat{V}(\bar{y}_n) = \frac{s_y^2}{n}$.
Applying expectation on both sides,

$$E\big(\hat{V}(\bar{y}_n)\big) = \frac{1}{n} E(s_y^2). \tag{2.4}$$

Let

$$E(s_y^2) = E\Big[\frac{1}{n-1}\sum_{i=1}^{n}(y_i - \bar{y}_n)^2\Big]$$

$$= \frac{1}{n-1}E\Big[\sum_{i=1}^{n} y_i^2 - n\bar{y}_n^2\Big].$$

$$E(s_y^2) = \frac{1}{n-1}\Big[\sum_{i=1}^{n} E(y_i^2) - nE(\bar{y}_n^2)\Big]. \tag{2.5}$$

By using the definition of variance,

$$V(\bar{y}_n) = E(\bar{y}_n - E(\bar{y}_n))^2.$$
$$= E(\bar{y}_n^2) - (E(\bar{y}_n))^2.$$

Using Theorem 2.1, $E(\bar{y}_n) = \bar{Y}$. After rearranging the terms,

$$E(\bar{y}_n^2) = V(\bar{y}_n) + \bar{Y}^2,$$
$$E(\bar{y}_n^2) = \frac{\sigma_y^2}{n} + \bar{Y}^2.$$

Using these results, Eq. (2.5) becomes

$$E(s_y^2) = \frac{1}{n-1}\Big[\sum_{i=1}^{n} E(y_i^2) - n\Big(\frac{\sigma_y^2}{n} + \bar{Y}^2\Big)\Big]$$

$$= \frac{1}{n-1}\Big[\sum_{i=1}^{n} \frac{1}{N}\sum_{i=1}^{N} y_i^2 - n\Big(\frac{\sigma_y^2}{n} + \bar{Y}^2\Big)\Big]$$

$$= \frac{1}{n-1}\Big[\frac{n}{N}\sum_{i=1}^{N} y_i^2 - n\Big(\frac{\sigma_y^2}{n} + \bar{Y}^2\Big)\Big]$$

$$= \frac{n}{n-1}\Big[\frac{1}{N}\sum_{i=1}^{N} y_i^2 - \bar{Y}^2 - \frac{\sigma_y^2}{n}\Big]$$

$$= \frac{n}{n-1}\Big[\sigma_y^2 - \frac{\sigma_y^2}{n}\Big].$$

$$E(s_y^2) = \sigma_y^2. \tag{2.6}$$

Using Eqs. (2.6) in (2.4),

$$E\big(\hat{V}(\bar{y}_n)\big) = \frac{1}{n}\sigma_y^2,$$
$$E\big(\hat{V}(\bar{y}_n)\big) = V(\bar{y}_n).$$

Hence the theorem. □

Example 2.1 The experiment deals with the bamboo species *Melocanna bambusoides* with its fiber length. This experiment mainly deals with the variation of one factor dependency on other factors taken into account. Obtain a random sample (SRSWR) of size 20 of y = fiber length from a population of size 100 . Estimate the average fiber length (inches). Obtain the standard error of the estimate and 95% confidence interval for population mean.

82, 42, 29, 143, 85, 75, 116, 54, 147, 82, 168, 51, 73, 61, 115, 82, 106, 131, 63, 33, 94, 18, 148, 25, 66, 20, 115, 39, 151, 25, 65, 50, 15, 41, 56, 25, 56, 65, 100, 26, 133, 38, 88, 127, 89, 47, 110, 98, 99, 67, 29, 68, 73, 58, 35, 48, 38, 142, 70, 64, 67, 138, 65, 103, 98, 20, 61, 93, 101, 19, 72, 64, 49, 44, 46, 111, 75, 91, 72, 84, 18, 65, 60, 118, 21, 44, 77, 20, 37, 40, 149, 28, 23, 64, 113, 60, 27, 73, 87, 20.

A random sample of size 20 is drawn from a population of size 100 with replacement. The random sample is as follows: 65, 20, 48, 75, 110, 99, 65, 65, 35, 131, 65, 37, 151, 26, 75, 61, 94, 25, 25, 19.

Use the following R code to illustrate the procedure of selecting the sample and estimating the population parameters.

```
Z<-scan("clipboard")
#set.seed(100)
x<-sample(Z,20,replace=T)
mean1<-mean(x)
mean1;
se<-sqrt(var(x)/length(x));
se;
t<-qt((1-0.025),19)
t;
LL<-mean1-t*se;
UL<-mean1+t*se;
CL<-c(LL,UL);
CL;
```

The average fiber length is 64.55 with a standard error of 8.40, and its 95% of confidence interval is [46.97, 82.13] and population mean is 70.76. Population mean belongs to the confidence interval.

2.3   Simple Random Sampling Without Replacement

SRSWR has the drawback that one or more sampling units may occur more than once in a sample. Repeated units provide no extra information for the estimation of parameters. Simple random sampling without replacement (SRSWOR) is free from such loss of information and is thus preferred over SRSWR. The SRSWOR is a method of drawing n units from N units in the population such that at every draw, each unit in the population has the same chance of being selected, and draws are made without replacement of the selected units. The sample space S under SRSWOR consists of $\binom{N}{n}$ possible samples.

The selection probability of a unit at any given draw in the case of SRSWOR is $\frac{1}{N}$, i.e. $P_i^{(r)} = P_i^{(1)} = \frac{1}{N}$ $\forall i \in U; r = 1, 2, \ldots, n; s \in S$.

The probability of selecting a specific sample of size n under SRSWOR is $p(s) = 1/\binom{N}{n}$.

An unbiased estimator of population mean under SRSWOR is

$$\bar{y}_n = \frac{1}{n} \sum_{i=1}^{n} y_i. \tag{2.7}$$

The variance of the estimator (2.7) under SRSWOR is given by

$$V(\bar{y}_n) = \frac{(N-n)}{Nn(N-1)} \sum_{i=1}^{N} (y_i - \bar{Y})^2$$

$$= \frac{(N-n)}{Nn} S_y^2 \tag{2.8}$$

where $S_y^2 = \frac{1}{(N-1)} \sum_{i=1}^{N} (y_i - \bar{Y})^2$.

An unbiased estimator of (2.8) under SRSWOR is given by

$$\hat{V}(\bar{y}_n) = \frac{(N-n)}{Nn} s_y^2 \tag{2.9}$$

where $s_y^2 = \frac{1}{(n-1)} \sum_{i=1}^{n} (y_i - \bar{y}_n)^2$.

Theorem 2.4 *In SRSWOR, sample mean $\bar{y}_n = \frac{1}{n} \sum_{i=1}^{n} y_i$ is an unbiased estimator of population mean $\bar{Y} = \frac{1}{N} \sum_{i=1}^{N} y_i$.*

Proof The sample mean for SRSWOR can be written as

$$\bar{y}_n = \frac{1}{n} \sum_{i=1}^{N} \alpha_i y_i \tag{2.10}$$

where

$$\alpha_i = \begin{cases} 1 & ; \text{if the ith unit of population is included in the sample} \\ 0 & ; otherwise. \end{cases}$$

Taking expectation on both sides of (2.10),

$E(\bar{y}_n) = \frac{1}{n} \sum_{i=1}^{N} E(\alpha_i) y_i.$

The random variable α_i takes the value 1 with probability $\frac{n}{N}$ and 0 with probability $(1 - \frac{n}{N})$.

$E(\alpha_i) = P(\alpha_i) = \frac{n}{N}.$

$$E(\bar{y}_n) = \frac{1}{n}\frac{n}{N}\sum_{i=1}^{N} y_i$$

$$= \frac{1}{N}\sum_{i=1}^{N} y_i$$

$$= \bar{Y}.$$

Hence the theorem. □

Theorem 2.5 *In SRSWOR, the variance of the sample mean is given by*
$V(\bar{y}_n) = \frac{N-n}{Nn} S_y^2.$

Proof Using the definition of variance,
$$V(\bar{y}_n) = E\,[\bar{y}_n - E(\bar{y}_n)]^2$$
$$= E(\bar{y}_n^2) - (E(\bar{y}_n))^2.$$
By using Theorem 2.4, we have

$$E(\bar{y}_n) = \bar{Y}.$$
Then

$$V(\bar{y}_n) = E(\bar{y}_n^2) - \bar{Y}^2. \tag{2.11}$$

Let
$$E(\bar{y}_n^2) = E\left[\frac{1}{n}\sum_{i=1}^{n} y_i\right]^2$$
$$= \frac{1}{n^2}E\left[\sum_{i=1}^{n} y_i^2 + \sum_{i\neq j}^{n} y_i\,y_j\right]$$
$$= \frac{1}{n^2}E\left[\sum_{i=1}^{N} \alpha_i y_i^2 + \sum_{i\neq j}^{N} \alpha_i\alpha_j y_i\,y_j\right].$$
Hence

$$E(\bar{y}_n^2) = \frac{1}{n^2}\left[\sum_{i=1}^{N} E(\alpha_i) y_i^2 + \sum_{i\neq j}^{N} E(\alpha_i\alpha_j) y_i\,y_j\right] \tag{2.12}$$

where

$$\alpha_i\alpha_j = \begin{cases} 1 & \text{; if the ith and jth sampling units of population are included in the sample} \\ 0 & \text{; } otherwise. \end{cases}$$

$E(\alpha_i\alpha_j) = P(\alpha_i\alpha_j = 1) = P(\alpha_i = 1)\,P(\alpha_j = 1|\alpha_i = 1)$
where, $P(\alpha_j = 1|\alpha_i = 1)$ denotes the conditional probability of jth unit is in the sample given that ith unit is also selected in the sample.
We have
$P(\alpha_j = 1|\alpha_i = 1) = \frac{n-1}{N-1}.$
Hence
$E(\alpha_i\alpha_j) = \frac{n(n-1)}{N(N-1)}.$
So Eq. (2.12) becomes
$E(\bar{y}_n^2) = \frac{1}{n^2}\left[\frac{n}{N}\sum_{i=1}^{N} y_i^2 + \frac{n(n-1)}{N(N-1)}\sum_{i\neq j}^{N} y_i\,y_j\right].$

Hence

$$E(\bar{y}_n^2) = \frac{1}{nN}\left[\sum_{i=1}^{N} y_i^2 + \frac{(n-1)}{(N-1)}\sum_{i\neq j}^{N} y_i\, y_j\right]. \tag{2.13}$$

Also let

$$\sum_{i\neq j}^{N} y_i\, y_j = \left(\sum_{i=1}^{N} y_i\right)^2 - \sum_{i=1}^{N} y_i^2. \tag{2.14}$$

Using Eq. (2.14) in Eq. (2.13), we get

$$E(\bar{y}_n^2) = \frac{1}{nN}\left[\sum_{i=1}^{N} y_i^2\right] + \frac{n-1}{nN(N-1)}\left[\left(\sum_{i=1}^{N} y_i\right)^2 - \sum_{i=1}^{N} y_i^2\right]$$

$$= \frac{1}{nN}\left[\sum_{i=1}^{N} y_i^2\right] + \frac{n-1}{nN(N-1)}\left[N^2\bar{Y}^2 - N\bar{Y}^2 - \sum_{i=1}^{N} y_i^2 + N\bar{Y}^2\right]$$

$$= \frac{1}{nN}\left[\sum_{i=1}^{N} y_i^2\right] + \frac{n-1}{nN(N-1)}\left[N(N-1)\bar{Y}^2 - \sum_{i=1}^{N}(y_i - \bar{Y})^2\right]$$

$$= \frac{1}{nN}\left[\sum_{i=1}^{N} y_i^2\right] + \frac{n-1}{nN}\left[N\bar{Y}^2 - \frac{1}{N-1}\sum_{i=1}^{N}(y_i - \bar{Y})^2\right].$$

Hence

$$E(\bar{y}_n^2) = \frac{1}{nN}\left[\sum_{i=1}^{N} y_i^2\right] + \frac{n-1}{nN}\left[N\bar{Y}^2 - S_y^2\right]. \tag{2.15}$$

Using Eq. (2.15) in Eq. (2.11), we get

$$V(\bar{y}_n) = \frac{1}{nN}\left[\sum_{i=1}^{N} y_i^2\right] + \frac{n-1}{nN}\left[N\bar{Y}^2 - S_y^2\right] - \bar{Y}^2$$

$$= \frac{1}{nN}\left[\sum_{i=1}^{N} y_i^2\right] - \frac{\bar{Y}^2}{n} - \frac{n-1}{nN}S_y^2$$

$$= \frac{N-1}{nN(N-1)}\sum_{i=1}^{N}(y_i - \bar{Y})^2 - \frac{n-1}{nN}S_y^2$$

$$= \frac{N-n}{nN}S_y^2.$$

Hence the theorem.   □

Theorem 2.6 *In SRSWOR, $\hat{V}(\bar{y}_n)$ is an unbiased estimator of $V(\bar{y}_n)$.*

Proof We know

$$\hat{V}(\bar{y}_n) = \frac{(N-n)}{Nn}s_y^2 \tag{2.16}$$

where $s_y^2 = \frac{1}{(n-1)}\sum_{i=1}^{n}(y_i - \bar{y}_n)^2$.

Taking expectation on both sides of (2.16), we get

$$E\left(\hat{V}(\bar{y}_n)\right) = \frac{(N-n)}{Nn}E(s_y^2). \tag{2.17}$$

Now

$$E(s_y^2) = E\left[\frac{1}{(n-1)}\sum_{i=1}^{n}(y_i - \bar{y}_n)^2\right]$$

$$= \frac{1}{(n-1)}E\left[\sum_{i=1}^{n}(y_i - \bar{y}_n)^2\right]$$

$$= \frac{1}{(n-1)}E\left[\sum_{i=1}^{n}y_i^2 - n\bar{y}_n^2\right]$$

$$= \frac{1}{(n-1)}E\left[\sum_{i=1}^{N}\alpha_i y_i^2 - n\bar{y}_n^2\right]$$

$$= \frac{1}{(n-1)}\left[\sum_{i=1}^{N}E(\alpha_i)y_i^2 - nE(\bar{y}_n^2)\right]$$

$$= \frac{1}{(n-1)}\left[\frac{n}{N}\sum_{i=1}^{N}y_i^2 - nE(\bar{y}_n^2)\right].$$

$$E(s_y^2) = \frac{n}{(n-1)}\left[\frac{1}{N}\sum_{i=1}^{N}y_i^2 - E(\bar{y}_n^2)\right]. \qquad (2.18)$$

From Theorem 2.5,

$$E(\bar{y}_n^2) = \frac{1}{nN}\left[\sum_{i=1}^{N}y_i^2\right] + \frac{n-1}{nN}\left[N\bar{Y}^2 - S_y^2\right].$$

Using it in Eq. (2.18), we get

$$E(s_y^2) = \frac{n}{(n-1)}\left[\frac{1}{N}\sum_{i=1}^{N}y_i^2 - \frac{1}{nN}\sum_{i=1}^{N}y_i^2 - \frac{n-1}{nN}\left[N\bar{Y}^2 - S_y^2\right]\right]$$

$$= \frac{n}{(n-1)}\left[\frac{n-1}{nN}\sum_{i=1}^{N}y_i^2 - \frac{n-1}{nN}\left[N\bar{Y}^2 - S_y^2\right]\right]$$

$$= \frac{1}{N}\left[\frac{N-1}{N-1}\sum_{i=1}^{N}(y_i - \bar{Y})^2 + S_y^2\right]$$

$$= \frac{1}{N}\left[(N-1)S_y^2 + S_y^2\right].$$

Using this in Eq. (2.17), we get

$$E\left(\hat{V}(\bar{y}_n)\right) = \frac{N-n}{Nn}S_y^2$$

$$= V(\bar{y}_n).$$

Hence the theorem.   □

Example 2.2 Obtain a random sample (SRSWOR) of size 10 of y = male rural workers. Estimate the average male main rural workers. Obtain the standard error of the estimate and 95% confidence interval for the population mean (Table 2.1).

Use the following R code to illustrate the procedure:

```
Z<-scan("clipboard")
#set.seed(100)
x<-sample(Z,10,replace=F)
mean1<-mean(x)
mean1;
se<-sqrt(var(x)/length(x));
```

Table 2.1 The total number of rural male workers

Sr. No.	State	Male rural worker
1	Jammu and Kashmir	2005936
2	Himachal Pradesh	1990429
3	Punjab	5256654
4	Chandigarh*	40178
5	Uttaranchal	2148356
6	Haryana	5089381
7	Delhi	390512
8	Rajasthan	13117682
9	Uttar Pradesh	38005129
10	Bihar	17794140
11	Sikkim	164919
12	Arunachal Pradesh	214600
13	Nagaland	493665
14	Manipur	595828
15	Mizoram	163339
16	Tripura	922411
17	Meghalaya	444076
18	Assam	6873533
19	West Bengal	18480653
20	Jharkhand	5318986
21	Orissa	9845113
22	Chhattisgarh	5093104
23	Madhya Pradesh	13477218
24	Gujarat	9796390
25	Daman and Diu*	49190
26	Dadra and Nagar Haveli*	51085
27	Maharashtra	19897241
28	Telangana	4369204
29	Andhra Pradesh	11593150
30	Karnataka	10741327
31	Goa	266336
32	Lakshadweep*	13414
33	Kerala	9400183
34	Tamil Nadu	11954965
35	Pondicherry*	119908
36	Andaman and Nicobar	94740

*Union territories

```
se;
t<-qt((1-0.025),9)
t;
LL<-mean1-t*se;
UL<-mean1+t*se;
CL<-c(LL,UL);
CL;
```

The average number of rural male workers is 391432.4 with a standard error of 22172.4, and its 95% of confidence interval is [347619.4, 435245.5].

2.4 Estimation of Population Proportions

Consider a dichotomous population; the study variable y takes value 1 if it possesses the attribute A with respect to the population units and it takes value 0 otherwise. For example, a sample survey of HIV-infected individuals in India is conducted to estimate the proportion of HIV-infected individuals in India. The individuals are selected by using SRSWR or SRSWOR. Let the size of the population under study be denoted by N and the sample size by n.
Define

$$
y_i =
\begin{cases}
1 \; ; \text{if ith unit possesses the attribute A} \\
0 \; ; otherwise.
\end{cases}
$$

Suppose in the sample of size n drawn by SRSWOR, there are r number of units possessing the attribute A,

$$
\bar{y}_n = \frac{1}{n} \sum_{i=1}^{n} y_i = \frac{r}{n} = \hat{P}. \tag{2.19}
$$

Note that $\sum_{i=1}^{n} y_i = \sum_{i=1}^{n} y_i^2 = r$
and

$$
\hat{V}(\bar{y}_n) = \frac{N - n}{Nn} s_y^2 \tag{2.20}
$$

where
$$
\begin{aligned}
s_y^2 &= \frac{1}{n-1} \sum_{i=1}^{n} (y_i - \bar{y}_n)^2 \\
&= \frac{1}{n-1} \left(\sum_{i=1}^{n} y_i^2 - n \bar{y}_n^2 \right) \\
&= \frac{1}{n-1} \left(r - n\hat{P}^2 \right) \\
&= \frac{1}{n-1} \left(n\hat{P} - n\hat{P}^2 \right)
\end{aligned}
$$

$$= \frac{1}{n-1} n \hat{P}(1 - \hat{P})$$
$$= \frac{1}{n-1} n \hat{P} \hat{Q}$$

where $\hat{Q} = 1 - \hat{P}$.

Hence

$$\hat{V}(\bar{y}_n) = \frac{N-n}{Nn} \frac{n \hat{P} \hat{Q}}{n-1}$$
$$= \frac{N-n}{N(n-1)} \hat{P} \hat{Q}.$$

The population mean is given by

$$\bar{Y} = \frac{1}{N} \sum_{i=1}^{N} y_i = \frac{N_A}{N} = P \tag{2.21}$$

where N_A denotes the number of individuals in the population possessing attribute A.

The variance of the sample mean is given by

$$V(\bar{y}_n) = \frac{N-n}{Nn} S_y^2 \tag{2.22}$$

where $S_y^2 = \frac{1}{N-1} \left(\sum_{i=1}^{N} y_i^2 - N \bar{Y}^2 \right)$
$$= \frac{1}{N-1} \left(NP - NP^2 \right)$$
$$= \frac{1}{N-1} NP(1 - P)$$
$$= \frac{NPQ}{N-1}$$

where $Q = 1 - P$.

Hence

$$V(\bar{y}_n) = \frac{N-n}{Nn} \frac{NPQ}{N-1}$$

$$= \frac{(N-n)PQ}{n(N-1)}.$$

For SRSWR, the sample mean is

$$\bar{y}_n = \frac{1}{n} \sum_{i=1}^{n} y_i = \frac{r}{n} = \hat{P}$$

and

$$V(\bar{y}_n) = \frac{\sigma_y^2}{n} = V(\hat{P})$$

where

$$\sigma_y^2 = \frac{1}{N} \sum_{i=1}^{N} \left(y_i - \bar{Y} \right)^2$$
$$= \frac{1}{N} \left(\sum_{i=1}^{N} y_i^2 - N\bar{Y}^2 \right)$$
$$= \frac{1}{N} \left(NP - NP^2 \right)$$
$$= \frac{1}{N} NP(1 - P)$$
$$= PQ.$$

Hence

$$V(\bar{y}_n) = \frac{PQ}{n} = V(\hat{P}).$$

The unbiased variance estimator in case of SRSWR is

$$V(\bar{y}_n) = \frac{s_y^2}{n}$$

where

$$s_y^2 = \frac{1}{n-1} \sum_{i=1}^{n} (y_i - \bar{y}_n)^2$$
$$= \frac{1}{n-1} \left(\sum_{i=1}^{n} y_i^2 - n\, \bar{y}_n^2 \right)$$
$$= \frac{1}{n-1} \left(n\hat{P} - n\hat{P}^2 \right)$$
$$= \frac{1}{n-1} n\hat{P}\hat{Q}.$$

Hence
$$V(\hat{P}) = \frac{1}{n} \frac{n\hat{P}\hat{Q}}{n-1}$$
$$= \frac{\hat{P}\hat{Q}}{n-1}.$$

2.5   Determination of Sample Size

The precision of confidence interval depends on the width of the interval. Usually the interval with small width is preferred. The sample size under SRSWR is determined by assuming the selected sample which is drawn from a normal population with mean μ and variance σ_y^2. It is assumed that σ_y^2 is known.

In such a case, a $100(1 - \alpha)$ % confidence interval for μ is given as
$$\left(\bar{y}_n - z_{\frac{\alpha}{2}} \frac{\sigma_y}{\sqrt{n}} \quad \bar{y}_n + z_{\frac{\alpha}{2}} \frac{\sigma_y}{\sqrt{n}} \right).$$

The maximum error E in estimating the value of μ that is acceptable is defined as
$$E = |\bar{y}_n - \mu| = z_{\frac{\alpha}{2}} \frac{\sigma_y}{\sqrt{n}}.$$

So, the sample size (n) required can be obtained by solving the equation
$$\frac{|\bar{y}_n - \mu|}{\frac{\sigma_y}{\sqrt{n}}} = z_{\frac{\alpha}{2}}.$$

That is,
$$\frac{E}{\frac{\sigma_y}{\sqrt{n}}} = z_{\frac{\alpha}{2}}.$$

Squaring both sides and rearranging the terms, we get
$$n = \frac{z_{\frac{\alpha}{2}}^2 \sigma_y^2}{E^2}.$$

In some situations, we do not know the population distribution. In those cases, we use the non-parametric approach (Chebyshev's inequality) as follows to determine the sample size:
$$P\{|\bar{y}_n - \mu| \le E\} \ge 1 - \frac{\sigma_y^2}{nE^2}$$
$$\ge 1 - \alpha .$$

Hence
$$\alpha = \frac{\sigma_y^2}{nE^2}$$
where α is the level of significance.

Example 2.3 Determine the sample size for the normal data and to have 95% assurance that error in the estimated mean is less than 20 percent of the standard deviation.

Solution
We have normal population. So

$$n = \frac{z_{\frac{\alpha}{2}}^2 \; \sigma_y^2}{E^2}.$$

Here

$$z_{\frac{\alpha}{2}} = 1.96$$

and

$$E = 0.2 \; \sigma_y.$$

Thus

$$n = \frac{(1.96)^2 \; \sigma_y^2}{(0.2 \; \sigma_y)^2}$$
$$= 96.04$$
$$\approx 97.$$

Similarly, for SRSWOR a $100 \, (1 - \alpha) \, \%$ confidence interval for μ is given as

$$\left(\bar{y}_n - z_{\frac{\alpha}{2}} \sqrt{\frac{N-n}{Nn}} \; S_y \quad \bar{y}_n + z_{\frac{\alpha}{2}} \sqrt{\frac{N-n}{Nn}} \; S_y \right).$$

The maximum error E in estimating the value of μ is acceptable and is defined as

$$E = |\bar{y}_n - \mu| = z_{\frac{\alpha}{2}} \sqrt{\frac{N-n}{Nn}} \; S_y.$$

The required sample size can be obtained by solving the following equation:

$$\frac{|\bar{y}_n - \mu|}{\sqrt{\frac{N-n}{Nn}} \; S_y} = z_{\frac{\alpha}{2}}.$$

That is,

$$\frac{E}{\sqrt{\frac{N-n}{Nn}} \; S_y} = z_{\frac{\alpha}{2}}.$$

Squaring on both sides and rearranging the terms, we get

$$n = \frac{N \, z_{\frac{\alpha}{2}}^2 \; S_y^2}{N E^2 + z_{\frac{\alpha}{2}}^2 \; S_y^2}.$$

For the non-parametric setup under SRSWOR,

$$P\{|\bar{y}_n - \mu| \le E\} \ge 1 - \frac{S_y^2}{\frac{N-n}{Nn} E^2}$$
$$\ge 1 - \alpha.$$

Hence

$$\alpha = \frac{S_y^2}{\frac{N-n}{Nn} E^2}.$$

After rearranging the terms, we get

$$n = \frac{N \, \alpha^2 \; E^2 + N^2 \, \alpha \; S_y^2}{E^2}.$$

2.6 Use of Auxiliary Information

Many a time in the survey, additional information is available on another variable, related to the study variable. This variable is known as an auxiliary variable. Generally, the auxiliary variable is highly correlated with the study variable. It is also assumed that the information on the auxiliary variable is complete, i.e. the information on auxiliary variable is available for all the population units. Further, the information on the auxiliary variable is available before the sample is selected. For instance, the study variable in a survey may be holding-wise irrigated area, and it is proposed to estimate the total irrigated area in a district. The auxiliary variable in this case could be the holding-wise cultivated area in the district. A high correlation is expected between the variable irrigated area and cultivated area. This a priori information on holding-wise cultivated area can be used at the estimation stage to develop a more precise estimator of the total irrigated area in the district. Similarly, let the study variable be the number of tube wells in different villages, and the objective is to estimate the total number of tube wells in a district. Both the variable village-wise number of tube wells observed in the previous year and village-wise net irrigated area in the previous year in the district are expected to be correlated with the village-wise number of tube wells (study variable) observed in the current year. The two variables can be taken as auxiliary variables for the study variable to improve the precision of the estimator of the total number of tube wells in a district in the current year. Likewise, if the study variable is household-wise quantity of milk produced and the objective is to estimate the quantity of milk produced in a large area, the variable household-wise number of cattle is expected to be correlated with the study variable and thus may be used as an auxiliary variable for improving the precision of the estimator. Information on the auxiliary variable can be collected relatively inexpensively when not available and can be gainfully employed for improving the precision of the estimator. The auxiliary information, if available, can be utilized in many ways in the surveys. As described above in the section on unequal probability sampling, the auxiliary information can be used in the selection of the sample. Many a time, the sampling units are found to be of varying sizes. For example, in a sample survey conducted for the estimation of crop production, the larger villages may have a greater area under the crop and thus the contribution of larger villages to the total production is expected to be more. For such situations, the precision of the estimator can be increased by selecting sampling units (villages) with unequal probabilities. The sampling units can be selected with probabilities directly proportional to their sizes, i.e. $P_i \propto X_i$ or $P_i = \frac{X_i}{X_1 + X_2 + \cdots + X_i + \cdots + X_N}$. Thus the available auxiliary information on the size of the sampling units can be made use of in increasing the precision of the estimator. The auxiliary information is also frequently used at the estimation stage to improve the precision of the estimators. Some examples of estimators of the population mean (or population total) that make use of auxiliary information at the estimation stage are ratio and regression estimators. These estimators are very precise when there is high correlation between the study and auxiliary variables. In what follows, the relevant details on the ratio and regression estimators are provided in Sects. 2.6.1 and 2.6.2.

2.6.1 Ratio Method of Estimation

The ratio estimator is commonly used for the estimation of population total Y or population mean $\bar{Y}$.

Suppose that a SRSWOR of size n is drawn from a population U containing N distinct and identifiable units. Let $(y_1, x_1), (y_2, x_2), \ldots, (y_i, x_i), \ldots, (y_n, x_n)$ denote the observations on the study and the auxiliary variable, respectively. The ratio estimator of population mean is given by $\bar{y}_R = \frac{\bar{y}_n}{\bar{x}_n} \bar{X}$, where $\bar{y}_n$ and $\bar{x}_n$ are the sample means of y and x, respectively.

The following notations are used for the ratio method of estimation.

$r_i = \frac{y_i}{x_i}$ is the ratio of observation on y variable to the observation on x variable for the ith sampling unit:

$\bar{r}_N = \frac{1}{N} \sum_{i=1}^{N} r_i$ is the mean of the ratios of all units in the population.

$\bar{r}_n = \frac{1}{n} \sum_{i=1}^{n} r_i$ is the mean of the ratios of all units in the sample.

$R_N = \frac{\bar{Y}}{\bar{X}} = \frac{Y}{X}$ where Y is the population total of y variable, and X is the population total of x variable, respectively,

$R_n = \frac{\bar{y}_n}{\bar{x}_n} = \frac{\sum_{i=1}^{n} y_i}{\sum_{i=1}^{n} x_i}$.

The ratio estimator of the population mean $\bar{Y}$ is defined as

$\bar{y}_R = \frac{\bar{y}_n}{\bar{x}_n} \bar{X}$.

Theorem 2.7 *Bias in the ratio estimator $\bar{y}_R$ is given by*

$B(\bar{y}_R) = \frac{N-n}{Nn} \bar{Y} \left[C_x^2 - \rho\, C_y\, C_x \right]$.

Proof Let $y_i = \bar{Y} + \epsilon_i$ for $i = 1, 2 \ldots n$.

Hence

$\bar{y}_n = \bar{Y} + \bar{\epsilon}_n$.

Hence $E(\bar{\epsilon}_n) = 0$ and $E(\bar{\epsilon}_n^2) = \frac{N-n}{Nn} S_y^2$.

Also let $x_i = \bar{X} + \epsilon'_i$ for $i = 1, 2 \ldots n$.

Hence

$\bar{x}_n = \bar{X} + \bar{\epsilon}'_n$.

Therefore

$E(\bar{\epsilon}'_n) = 0$ and $E(\bar{\epsilon}'^2_n) = \frac{N-n}{Nn} S_x^2$.

Now $R_n = \frac{\bar{y}_n}{\bar{x}_n}$

$= \frac{\bar{Y}\left(1 + \frac{\bar{\epsilon}_n}{\bar{Y}}\right)}{\bar{X}\left(1 + \frac{\bar{\epsilon}'_n}{\bar{X}}\right)}$.

Assume that $\left|\frac{\bar{\epsilon}'_n}{\bar{X}}\right| < 1$.

Hence $\left(1 + \frac{\bar{\epsilon}'_n}{\bar{X}}\right)^{-1}$ can be expressed as a power series in $\bar{\epsilon}'_n$.

$R_n = R_N \left(1 + \frac{\bar{\epsilon}_n}{\bar{Y}}\right) \left(1 + \frac{\bar{\epsilon}'_n}{\bar{X}}\right)^{-1}$

$= R_N \left(1 + \frac{\bar{\epsilon}_n}{\bar{Y}}\right) \left[1 - \frac{\bar{\epsilon}'_n}{\bar{X}} + \frac{\bar{\epsilon}'^2_n}{\bar{X}^2} - \frac{\bar{\epsilon}'^3_n}{\bar{X}^3} + \cdots \right]$.

Neglecting the terms of order 3 or more in ϵ'_n, the above expression can be written as

$R_n = R_N \left(1 + \frac{\bar{\epsilon}_n}{\bar{Y}} - \frac{\bar{\epsilon}'_n}{\bar{X}} + \frac{\bar{\epsilon}'^2_n}{\bar{X}^2} - \frac{\bar{\epsilon}_n \bar{\epsilon}'_n}{\bar{Y}\bar{X}}\right)$.

Taking expectation on both sides, we get

$E(R_n) = R_N + R_N \left(\frac{E(\bar{\epsilon}'^2_n)}{\bar{X}^2} - \frac{E(\bar{\epsilon}_n \bar{\epsilon}'_n)}{\bar{Y}\bar{X}}\right)$.

Consider

$E(\bar{\epsilon}_n \bar{\epsilon}'_n) = E\left(\left[\frac{1}{n} \sum_{i=1}^{n} \epsilon_i\right]\left[\frac{1}{n} \sum_{i=1}^{n} \epsilon'_{i'}\right]\right)$

$$= \tfrac{1}{n^2} E\left(\sum_{i=1}^{n} \epsilon_i\, \epsilon_{i'} + \sum_{i\neq j'}^{n} \epsilon_i\, \epsilon_{j'}\right).$$

By using SRSWOR,

$$E(\bar\epsilon_n\, \bar\epsilon'_n) = \tfrac{1}{n^2}\left\{\tfrac{n}{N}\sum_{i=1}^{N} \epsilon_i\, \epsilon_{i'} + \tfrac{n\,(n-1)}{N\,(N-1)}\sum_{i\neq j'} \epsilon_i\, \epsilon_{j'}\right\}$$

$$= \tfrac{1}{nN}\sum_{i=1}^{N} \epsilon_i\, \epsilon_{i'} + \tfrac{(n-1)}{n\,N\,(N-1)}\left\{\left(\sum_{i=1}^{N}\epsilon_i\right)\left(\sum_{i=1}^{N}\epsilon_{i'}\right) - \sum_{i=1}^{N}\epsilon_i\, \epsilon_{i'}\right\}$$

$$= \tfrac{N-n}{Nn}\,\tfrac{1}{N-1}\left\{\sum_{i=1}^{N}\epsilon_i\, \epsilon_{i'}\right\}$$

$$= \tfrac{N-n}{Nn}\, S_{xy}.$$

But

$S_{xy} = \rho\, S_x\, S_y$, where ρ is the coefficient of correlation between x and y.

Hence

$$E(\bar\epsilon_n\, \bar\epsilon'_n) = \tfrac{N-n}{Nn}\,\rho\, S_x\, S_y.$$

Therefore,

$$E(R_n) = R_N + R_N\left(\tfrac{N-n}{Nn}\,\tfrac{S_x^2}{\bar{X}^2} - \tfrac{N-n}{Nn}\,\rho\,\tfrac{S_x\, S_y}{\bar{Y}\bar{X}}\right)$$

$$= R_N + R_N\,\tfrac{N-n}{Nn}\left(C_X^2 - \rho\, C_Y\, C_X\right).$$

Bias in $R_n = E(R_n) - R_N$

$$= R_N\,\tfrac{N-n}{Nn}\left(C_X^2 - \rho\, C_Y\, C_X\right).$$

Hence

$$B(\bar{y}_R) = E(\bar{y}_R) - \bar{Y}$$

$$= E(R_n)\,\bar{X} - \bar{Y}$$

$$= \tfrac{N-n}{Nn}\,\bar{Y}\left(C_X^2 - \rho\, C_Y\, C_X\right).$$

Hence the theorem.   □

Theorem 2.8 *The mean squared error of the ratio estimator $\bar{y}_R$ of population mean is given by*

$$MSE(\bar{y}_R) = \tfrac{N-n}{Nn}\,\bar{Y}^2\left(C_X^2 + C_Y^2 - 2\,\rho\, C_Y\, C_X\right).$$

Proof By using the definition of mean square error (MSE), we get

$$MSE(\bar{y}_R) = E(\bar{y}_R - \bar{Y})^2$$

$$\approx E(\tfrac{\bar{y}_n}{\bar{x}_n}\bar{X} - \bar{Y})^2$$

$$= E\left[\bar{Y}\,\frac{(1+\frac{\bar\epsilon_n}{\bar{Y}})}{(1+\frac{\bar\epsilon'_n}{\bar{X}})} - \bar{Y}\right]^2$$

$$= \bar{Y}^2\, E\left[(1 + \tfrac{\bar\epsilon_n}{\bar{Y}})(1 + \tfrac{\bar\epsilon'_n}{\bar{X}})^{-1} - 1\right]^2.$$

Expanding $(1 + \frac{\bar\epsilon'_n}{\bar{X}})^{-1}$ as a power series in $\bar\epsilon'_n$ and neglecting terms of power 3 or more, we get

$$MSE(\bar{y}_R) = \bar{Y}^2\, E\left[1 + \tfrac{\bar\epsilon_n}{\bar{Y}} - \tfrac{\bar\epsilon'_n}{\bar{X}} + \tfrac{\bar\epsilon'^2_n}{\bar{X}^2} - \tfrac{\bar\epsilon_n\,\bar\epsilon'_n}{\bar{X}\,\bar{Y}} + \cdots - 1\right]^2$$

$$= \bar{Y}^2\, E\left[\tfrac{\bar\epsilon_n}{\bar{Y}} - \tfrac{\bar\epsilon'_n}{\bar{X}} + \tfrac{\bar\epsilon'^2_n}{\bar{X}^2} - \tfrac{\bar\epsilon_n\,\bar\epsilon'_n}{\bar{X}\,\bar{Y}} + \cdots\right]^2$$

$$= \bar{Y}^2\, E\left[\tfrac{\bar\epsilon^2_n}{\bar{Y}^2} + \tfrac{\bar\epsilon'^2_n}{\bar{X}^2} - 2\tfrac{\bar\epsilon_n\,\bar\epsilon'_n}{\bar{X}\,\bar{Y}}\right]$$

$$= \bar{Y}^2\left[\tfrac{N-n}{Nn}\,\tfrac{S_y^2}{\bar{Y}^2} + \tfrac{N-n}{Nn}\,\tfrac{S_x^2}{\bar{X}^2} - 2\tfrac{N-n}{Nn}\,\tfrac{S_{xy}}{\bar{Y}\,\bar{X}}\right]$$

Table 2.2 Sugarcane production and cultivation

Sr. No.	State	Cultivated area (in lakh acres)	Production (in lakh tonnes)
1	AP	0.99	79.48
2	AS	0.3	11.15
3	BH	2.43	165.11
4	CH	0.3	12.47
5	GJ	1.84	122.34
6	HR	1.14	87.29
7	JH	0.07	5.23
8	KR	3.7	299.02
9	KER	0.01	1.22
10	MP	0.98	54.3
11	MH	9.02	726.37
12	OD	0.05	3.41
13	PU	0.93	75.33
14	RJ	0.05	4.04
15	TN	1.83	165.62
16	TL	0.35	22.17
17	UP	22.34	1623.38
18	UT	1.02	71.42
19	WB	0.17	12.94
20	Others	0.19	8.68

$$= \bar{Y}^2 \, \tfrac{N-n}{Nn} \left(C_X^2 + C_Y^2 - 2 \, \rho \, C_Y \, C_X \right).$$

Hence the theorem.   □

Corollary 2.1 *An estimator of the $MSE(\bar{y}_R)$ is given by*
$$M\hat{S}E(\bar{y}_R) = \tfrac{N-n}{Nn} \left(s_y^2 + r^2 \, s_x^2 - 2r \, s_{xy} \right)$$
where r is sample correlation coefficient between x and y.

Example 2.4 The sugarcane production and cultivation of Indian states are presented in Table 2.2. Draw a random sample using SRSWOR of size 10 and estimate the average amount of sugarcane production using the ratio method of estimation and obtain the 95% confidence interval for the population mean.

Use the following R code to illustrate the procedure:

```
x<-scan("clipboard")
y<-scan("clipboard")
set.seed(25)
s1<-sample(1:20,10,replace=F)
```

```
dx=x[s1];
dy=y[s1];
xmean=mean(x)
mdx=mean(dx);
mdy=mean(dy);
vdx=var(dx);
vdy=var(dy);
cdxdy=cov(dx,dy);
r=mdy/mdx;
r;
YR=(mdy/mdx)*xmean;
YR;
Mse=((1/10)-(1/20))*(vdy+r^2*vdx-2*r*cdxdy)
Mse;
t=qt(0.975,5);
t;
LL=YR-t*sqrt(Mse)
UL=YR+t*sqrt(Mse)
LL;UL;
```

The average sugarcane production is 170.43, and its 95% confidence interval is [162.6099, 178.2548].

2.6.1.1 Efficiency of Ratio Estimator

The efficiency of ratio estimator of population mean of study variable Y relative to that in SRSWOR is given by

$$Efficiency = \frac{V(\bar{y}_n)}{MSE(\bar{y}_R)}$$

$$= \frac{\frac{N-n}{Nn} S_y^2}{\bar{Y}^2 \frac{N-n}{Nn} \left(C_X^2 + C_Y^2 - 2 \rho \ C_Y \ C_X \right)}$$

$$= \frac{C_Y^2}{C_X^2 + C_Y^2 - 2 \rho \ C_Y \ C_X}$$

$$= \frac{1}{1 + \frac{C_X^2}{C_Y^2} - 2 \rho \ \frac{C_X}{C_Y}}.$$

If $\rho > \frac{C_X}{2C_Y}$, then the ration estimator is more precise than the corresponding estimator in SRSWOR, otherwise the estimator in SRSWOR is more precise.

2.6.2 Regression Method of Estimation

It has been just described that the ratio method of estimation performs well when there is a linear relationship between the study variable and the auxiliary variable and the correlation coefficient between the two variables is positive and large. However, no mention was made about the intercept of the linear relationship. If the intercept is large, then the precision of estimation can be further improved by making use of the regression estimator.

Suppose that a SRSWOR of size n is drawn from a population U containing N distinct and identifiable units. Let $(y_1, x_1), (y_2, x_2), \ldots, (y_i, x_i), \ldots, (y_n, x_n)$ denote the observations on the study variable and the auxiliary variable, respectively. The regression estimator of population mean, $\bar{y}_l$, is given by
$\bar{y}_l = \bar{y}_n + \hat{\beta}(\bar{X} - \bar{x}_n)$ where β is the estimator of the population regression coefficient β and is given by
$\hat{\beta} = \frac{s_{xy}}{s_x^2}$.

Theorem 2.9 *The bias in the regression estimator of the population mean $\bar{Y}$ is given by*
$$B(\bar{y}_l) = -\beta \left[\frac{cov(\bar{x}_n, s_{xy})}{S_{xy}} - \frac{cov(\bar{x}_n, s_x^2)}{S_x^2} \right].$$

Proof Let
$\bar{x}_n = \bar{X} + \epsilon_1, s_{xy} = S_{xy} + \epsilon_2, s_x^2 = S_x^2 + \epsilon_3$
where $E(\epsilon_1) = E(\epsilon_2) = E(\epsilon_3) = 0$.
Then
$\bar{y}_l = \bar{y}_n + \frac{S_{xy} + \epsilon_2}{S_x^2 + \epsilon_3} (-\epsilon_1)$

$\qquad = \bar{y}_n - \beta \, \epsilon_1 \left(1 + \frac{\epsilon_2}{S_{xy}}\right) \left(1 + \frac{\epsilon_3}{S_x^2}\right)^{-1}$.

Assuming that $|\frac{\epsilon_3}{S_x^2}| < 1$ and expanding $\left(1 + \frac{\epsilon_3}{S_x^2}\right)^{-1}$ by using the power series and neglecting the higher order terms, we can obtain
$E(\bar{y}_l) = E\left[\bar{y}_n - \beta \left(\epsilon_1 + \frac{\epsilon_1 \epsilon_2}{S_{xy}} - \frac{\epsilon_1 \epsilon_3}{S_x^2} + \cdots\right)\right]$

$\qquad \approx E(\bar{y}_n) - \beta \left[\frac{E(\epsilon_1 \epsilon_2)}{S_{xy}} - \frac{E(\epsilon_1 \epsilon_3)}{S_x^2}\right]$

$\qquad = \bar{Y} - \beta \left[\frac{cov(\bar{x}_n, s_{xy})}{S_{xy}} - \frac{cov(\bar{x}_n, s_x^2)}{S_x^2}\right]$.

Hence
$B(\bar{y}_l) = E(\bar{y}_l) - \bar{Y}$

$\qquad = -\beta \left[\frac{cov(\bar{x}_n, s_{xy})}{S_{xy}} - \frac{cov(\bar{x}_n, s_x^2)}{S_x^2}\right]$.

Hence the theorem.   □

Theorem 2.10 *The mean squared error of the regression estimator $\bar{y}_l$ is given by*
$MSE(\bar{y}_l) = \frac{N-n}{Nn} S_y^2 (1 - \rho^2)$.

Proof By using the definition of mean squared error, we get

$$MSE(\bar{y}_l) = E(\bar{y}_l - \bar{Y})^2$$

$$= E\left[\bar{y}_n - \beta\left(\epsilon_1 + \frac{\epsilon_1\,\epsilon_2}{S_{xy}} - \frac{\epsilon_1\,\epsilon_3}{S_x^2}\right) - \bar{Y}\right]^2$$

$$= E\left[(\bar{y}_n - \bar{Y}) - \beta\,\epsilon_1 - \beta\left(\frac{\epsilon_1\,\epsilon_2}{S_{xy}} - \frac{\epsilon_1\,\epsilon_3}{S_x^2}\right)\right]^2.$$

Neglecting the terms of order 3 and above in ϵ_1, we get

$$MSE(\bar{y}_l) = E\left[(\bar{y}_n - \bar{Y})^2 + \beta^2\,\epsilon_1^2 - 2\,\beta\,\epsilon_1\,(\bar{y}_n - \bar{Y})\right]$$

$$= E\,(\bar{y}_n - \bar{Y})^2 + \beta^2\,E(\epsilon_1^2) - 2\,\beta\,E(\epsilon_1\,(\bar{y}_n - \bar{Y}))$$

$$= V(\bar{y}_n) + \beta^2\,V(\bar{x}_n) - 2\,\beta\,cov(\bar{x}_n, \bar{y}_n)$$

$$= \frac{N-n}{Nn}\,(S_y^2 + \beta^2\,S_x^2 - 2\,\beta\,S_{xy})$$

$$= \frac{N-n}{Nn}\,S_y^2\,(1 - \rho^2)$$

where $\beta = \frac{S_{xy}}{S_x^2}$ and $\rho = \frac{S_{xy}}{S_x S_y}$.

Hence the theorem.   □

Corollary 2.2 *An estimator of the $MSE(\bar{y}_l)$ is given by*

$$M\hat{S}E(\bar{y}_l) = \frac{N-n}{Nn}\,s_y^2\,(1 - r^2)$$

where, r is the sample correlation coefficient between x and y.

Example 2.5 The sugarcane production and cultivation of Indian states are presented in Table 2.2. Draw a random sample using SRSWOR of size 10 and estimate the average amount of sugarcane production using the regression method of estimation and obtain the 95% confidence interval for the population mean.

Use the following R code to illustrate the procedure:

```
x<-scan("clipboard")
y<-scan("clipboard")
set.seed(25)
s1<-sample(1:20,10,replace=F)
dx=x[s1];
dy=y[s1];
xmean=mean(x)
mdx=mean(dx);
mdy=mean(dy);
vdx=var(dx);
vdy=var(dy);
cdxdy=cov(dx,dy);
beta=cdxdy/vdx;
beta;
yl=mdy+beta*(xmean-mdx);
yl;
cor=cdxdy/sqrt(vdx*vdy);
cor;
```

```
Mse=((1/10)-(1/20))*vdy*(1-cor^2)
Mse;
t=qt(0.975,5);
t;
LL=y1-t*sqrt(Mse)
UL=y1+t*sqrt(Mse)
LL;UL;
```

The average production of sugarcane is 176.4441, and its 95% of confidence interval is [168.8739, 184.0144].

2.6.2.1 Comparison Between Simple Regression Estimate with SRSWOR Estimate

The efficiency of simple regression estimate with SRSWOR estimate of the population mean is given by

$$Efficiency = \frac{V(\bar{y}_l)}{V(\bar{y}_n)}$$

$$= \frac{\frac{N-n}{Nn} S_y^2 (1-\rho^2)}{\frac{N-n}{Nn} S_y^2}$$

$$= 1 - \rho^2 < 1.$$

Thus, the regression estimator is always efficient than the SRSWOR estimate of the population mean.

2.6.2.2 Comparison of Regression Estimate with Ratio Estimate

The efficiency of regression estimate of the population mean with respect to that of the ratio estimate is given by

$$Efficiency = \frac{V(\bar{y}_l)}{V(\bar{y}_R)}$$

$$= \frac{\frac{N-n}{Nn} S_y^2 (1-\rho^2)}{\frac{N-n}{Nn} (S_y^2 - 2 R_N \rho S_x S_y + R_N^2 S_x^2)}.$$

The regression estimate is efficient than the ratio estimate if $V(\bar{y}_l) < V(\bar{y}_R)$.

That is, if $S_y^2 (1 - \rho^2) < S_y^2 - 2 R_N \rho S_x S_y + R_N^2 S_x^2$

That is if $(\rho S_y - R_N S_x)^2 > 0$

which is always true unless $R_N = \rho \frac{S_y}{S_x}$ when equality holds.

The regression estimate is always efficient as compared to the ratio estimate unless the regression line passes through the origin.

2.7   Exercises

1. Define Simple Random Sampling. Write a short note on the procedure for the selection of a simple random sample.
2. Explain the difference between SRSWOR and SRSWR.
3. In SRSWOR, obtain the confidence limits for population proportion.
4. Define ratio estimator for the population mean. Obtain its expectation and mean squared error.
5. Compare the ratio estimator with SRSWOR in terms of their precision.
6. In SRSWR for proportion, obtain the minimum sample size for the desired precision.
7. In SRSWOR, establish that sample mean is an unbiased estimator of the population mean. Obtain the sampling variance of the sample mean.
8. Define ratio estimator $\bar{y}_R$ of the population mean. Obtain its expectation and mean squared error.
9. In SRSWR, obtain the confidence limits for population mean, when S_y^2 is not known.
10. Define regression estimator of the population mean. Obtain its expectation and mean squared error.
11. Compare the ratio estimator with regression estimator in terms of their relative precision.
12. Prove that in SRSWOR, sample mean is an unbiased estimator of population mean and sample mean square is an unbiased estimate of population mean square.
13. Obtain the confidence limits for population proportion under SRSWOR.
14. Derive the formulae for sample mean and variance of sample mean under SRSWR.
15. Explain the ratio method of estimation for SRSWOR and obtain the estimate of variance.
16. Obtain the variance of regression estimate by using SRSWOR when population mean of the auxiliary variable is known.
17. Write a short note on the estimation of proportion for simple random sampling.
18. What is SRSWR for estimating population proportions? Estimate the sample size under desired precision.
19. Define ratio estimator $\bar{y}_R$ of the population mean. Obtain its expectation and mean squared error.
20. Can the ratio estimator $\bar{y}_R$ be unbiased? State conditions for $\bar{y}_R$ to be unbiased.
21. Compare simple regression estimate with the ratio estimate.
22. Describe the situation in which either the ratio or regression estimator is best.
23. Propose an estimator of the population mean using the regression method. Show that it is a biased estimator and mean squared error of the estimator you have proposed.

24. A sample of 10 villages was drawn from a block having 40 villages. The size of 2011 census population of this block is 250311. The relevant data are given below.

Village	Census population	Cultivated area (in acres)	Village	Census population	Cultivated area (in acres)
1	4000	1000	6	8453	4715
2	1785	589	7	1346	9267
3	3578	1575	8	6798	3506
4	8642	2562	9	2879	1101
5	9632	4512	10	1206	965

   Estimate the total cultivated area and its standard error by (1) regression method (2) ratio method. Compare their relative efficiencies.

25. Given that $N = 2000$, $\epsilon = 0.08$, $t(\alpha/2, \infty) = 1.96$, and $p = 0.6$, obtain the sample size and estimate the exact 95 % confidence interval for P.

26. Use Table 2.1, obtain a random sample (SRSWR) of size 20 on variable $y=$ male rural workers. Estimate the total male marginal rural workers. Obtain the variance of the estimate and 95% confidence interval for population total.

27. Given $N = 1200$, $\epsilon = 0.07$, $t(\alpha/2, \infty) = 1.96$, and $p = 0.5$, obtain the sample size and estimate the exact 95% confidence interval for P.

Chapter 3
Stratified Random Sampling

3.1 Introduction

In case of simple random sampling without replacement (SRSWOR), the sampling variance of the sample mean is

$$V(\bar{y}_n) = \left(\frac{1}{n} - \frac{1}{N}\right) S_y^2.$$

Clearly, the variance decreases as the sample size (n) increases or the population variability S_y^2 decreases. However, a good sampling strategy is one which helps in reducing the sampling variance to the lowest possible extent. As a matter of fact, S_y^2 is a population parameter (in fact, it is a measure of population variance) and is a constant quantity. The other way of increasing the precision of estimation is to increase the sample size. The sample size cannot be increased because it would result in an increase in the cost of the survey. An increase in sample size may lead to other types of errors, like non-sampling errors. Instead, a way to increase the precision of estimation is through better representatives of the population in the sample. The population may be divided into several homogeneous groups called strata, and then sampling may be carried out independently within each stratum. This way, within stratum variability is reduced and the between strata variability is taken care of by independent sampling within each stratum. This procedure of sample selection is known as stratified sampling.

Stratified sampling is a very popular procedure in sample surveys. The procedure enables one to draw a sample with any desired degree of representation of the different parts of the population by taking them as strata. In stratified sampling, the population consisting of N units is first divided into K disjoint sub-populations of $N_1, N_2, \ldots, N_K$ units, respectively. These sub-populations are non-overlapping and together they comprise the whole of the population, i.e. $\sum_{i=1}^{K} N_i = N$. These sub-populations are called strata. To obtain the full benefit from stratification, the values

© The Author(s), under exclusive license to Springer Nature Singapore Pte Ltd. 2021
R. Latpate et al., *Advanced Sampling Methods*,
https://doi.org/10.1007/978-981-16-0622-9_3

of N_i's must be known. When the strata have been determined, a sample is drawn from each stratum, the drawings being made independently in different strata. If a simple random sample without replacement is taken from each stratum, then the procedure is termed as stratified random sampling. Since the sampling is done independently from each stratum, the precision of the estimator of the population mean or the population total from each stratum would depend upon the variability within each stratum. Thus, the stratification of the population should be done in such a way that the within stratum variability is as small as possible and the between strata variability is as large as possible. Thus, the strata are homogeneous within themselves with respect to the variable under study. However, in many practical situations, it is usually difficult to stratify with respect to the variable under consideration especially because of geographical and cost considerations. Generally, the stratification is done according to administrative groupings, geographical regions, and on the basis of auxiliary characters correlated with the character under study.

Suppose that a SRSWOR sample of size n is drawn from a population containing N units divided into K strata of respective sizes $N_1, N_2, \ldots, N_K$, such that $\sum_{i=1}^{K} N_i = N$. Let y_{ij} denote the jth observation in the ith stratum. The population mean can be expressed as

$$\bar{Y} = \frac{1}{N} \sum_{i=1}^{K} \sum_{j=1}^{N_i} y_{ij}.$$

Again the population mean square is $S_y^2 = \frac{1}{N} \sum_{i=1}^{K} \sum_{j=1}^{N_i} (y_{ij} - \bar{Y})^2$.

We define population mean square for the ith stratum as $S_i^2 = \frac{1}{(N_i-1)} \sum_{j=1}^{N_i} (y_{ij} - \bar{Y}_{N_i})^2$

where $\bar{Y}_{N_i} = \frac{1}{N_i} \sum_{j=1}^{N_i} y_{ij}$ is the population mean for the ith stratum.

Draw SRSWOR of sizes $n_1, n_2, \ldots, n_i, \ldots n_k$, respectively, such that $n_1 + n_2 + \cdots + n_i + \cdots + n_k = n$ (sample size).

The population mean $\bar{Y}$ can be rewritten as

$$\bar{Y} = \frac{1}{N} \sum_{i=1}^{K} N_i \bar{Y}_{N_i} = \sum_{i=1}^{K} W_i \bar{Y}_{N_i}$$

where $W_i = \frac{N_i}{N}$ is the stratum weight. Since the sampling has been carried out independently within each stratum by SRSWOR, therefore sample mean for the ith stratum, $\bar{y}_{n_i} (= \frac{1}{n_i} \sum_{j=1}^{n_i} y_{ij})$, is an unbiased estimator of $\bar{Y}_{N_i}$ and obviously the stratified sample mean

$$\bar{y}_{st} = \frac{1}{N} \sum_{i=1}^{K} N_i \bar{y}_{n_i} = \sum_{i=1}^{K} W_i \bar{y}_{n_i}$$

which is the weighted mean of the strata sample means with strata size as the weights, will be an appropriate estimator of the population mean $\bar{Y}$.

Theorem 3.1 *If every sample stratum mean is an unbiased estimator for $\bar{Y}_{N_i}$, then $\bar{y}_{st}$ is an unbiased estimator of the population mean $\bar{Y}$.*

Proof Let

$$E(\bar{y}_{st}) = E(\sum_{i=1}^{K} W_i \bar{y}_{n_i}) = \sum_{i=1}^{K} W_i E(\bar{y}_{n_i})$$

$$= \sum_{i=1}^{K} W_i \bar{Y}_{N_i}$$

$$= \sum_{i=1}^{K} \frac{N_i}{N} \cdot \frac{1}{N_i} \sum_{j=1}^{N_i} y_{ij}$$

$$= \frac{1}{N} \sum_{i=1}^{K} \sum_{j=1}^{N_i} y_{ij}$$

$$= \bar{Y}.$$

Hence the theorem.   □

Theorem 3.2 *For stratified random sampling, the variance of estimate $\bar{y}_{st}$ is*

$$V(\bar{y}_{st}) = \sum_{i=1}^{K} W_i^2 \left(\frac{1}{n_i} - \frac{1}{N_i} \right) S_i^2 = \sum_{i=1}^{K} W_i^2 \frac{S_i^2}{n_i}(1 - f_i) \text{ where } f_i = \frac{n_i}{N_i} \text{ (sampling fraction).}$$

Proof Since $\bar{y}_{n_i}$ is an unbiased estimator for $\bar{Y}_{N_i}$, $\forall i$ and within stratum draws are SRSWOR, then

$V(\bar{y}_{n_i}) = \frac{N_i - n_i}{N_i n_i} S_i^2$.

Let

$$\bar{y}_{st} = \sum_{i=1}^{K} W_i \bar{y}_{n_i}$$

$$V(\bar{y}_{st}) = \sum_{i=1}^{K} W_i^2 V(\bar{y}_{n_i}).$$

Since the sampling has been carried out independently within each stratum by SRSWOR,

$$V(\bar{y}_{st}) = \sum_{i=1}^{K} W_i^2 \left(\frac{N_i - n_i}{N_i n_i} \right) S_i^2$$

$$= \sum_{i=1}^{K} W_i^2 \left(\frac{1}{n_i} - \frac{1}{N_i} \right) S_i^2$$

$$= \sum_{i=1}^{K} W_i^2 \frac{S_i^2}{n_i}(1 - f_i).$$

Hence the theorem.   □

Corollary 3.1 *If sampling fraction $\frac{n_i}{N_i}$ are negligible in all strata, then $V(\bar{y}_{st}) = \sum_{i=1}^{K} W_i^2 \frac{S_i^2}{n_i}$.*

Theorem 3.3 *If s_i^2 is an unbiased estimator for each stratum S_i^2, then the estimator of variance $\hat{V}(\bar{y}_{st}) = \sum\limits_{i=1}^{K} W_i^2 \frac{s_i^2}{n_i}(1 - f_i)$ is an unbiased estimator for $V(\bar{y}_{st}) = \sum\limits_{i=1}^{K} W_i^2 \frac{S_i^2}{n_i}(1 - f_i)$.*

Proof We know in SRSWOR, s_i^2 is an unbiased for each stratum S_i^2, i.e. $E(s_i^2) = S_i^2$.
Then,

$$
\begin{aligned}
E(\hat{V}(\bar{y}_{st})) &= E[\sum_{i=1}^{K} W_i^2 \frac{s_i^2}{n_i}(1 - f_i)] \\
&= \sum_{i=1}^{K} W_i^2 \frac{E(s_i^2)}{n_i}(1 - f_i) \\
&= \sum_{i=1}^{K} W_i^2 \frac{S_i^2}{n_i}(1 - f_i) \\
&= V(\bar{y}_{st}).
\end{aligned}
$$

Hence the theorem.   □

An unbiased estimate of the variance can be expressed as

$$
\hat{V}(\bar{y}_{st}) = \sum_{i=1}^{K} W_i^2 (\frac{1}{n_i} - \frac{1}{N_i}) s_i^2 = \sum_{i=1}^{K} W_i^2 \frac{s_i^2}{n_i} - \sum_{i=1}^{K} W_i \frac{s_i^2}{N}.
$$

From the above expression, it is clear that the sampling variance of stratified sample mean depends on S_i^2's, the variability within the strata. It, therefore, follows that homogeneous strata will lead to a greater precision of the stratified sample mean.

3.2 Stratified Random Sampling with Proportional Allocation

The estimate $\bar{y}_{st}$ is not equal to the sample mean $(\bar{y})$. The sample mean can be expressed as $\bar{y} = \sum\limits_{i=1}^{k} \frac{n_i}{n} \bar{y}_{n_i}$ where $\bar{y}_{n_i} = \frac{1}{n_i} \sum\limits_{j=1}^{n_i} y_{ij}$. The $\bar{y}_{st}$ is estimated based on their weights $\frac{N_i}{N}$. $\bar{y}_{st}$ coincides with $\bar{y}$ provided that every stratum satisfies the following conditions, $\frac{n_i}{n} = \frac{N_i}{N}$ or $\frac{n_i}{N_i} = \frac{n}{N}$ or $f_i = f$, i.e. sampling fraction is the same for all the strata. Such stratification is called stratification with proportional allocation of the n_i.

Corollary 3.2 *For stratification with proportional allocation, i.e. $n_i = \frac{nN_i}{N}$, $\forall i$ then*

$$
V(\bar{y}_{st})_{prop} = (\frac{1}{n} - \frac{1}{N}) \sum_{i=1}^{K} W_i S_i^2.
$$

Proof Since the variance of $\bar{y}_{st}$ is

$$
V(\bar{y}_{st})_{prop} = \sum_{i=1}^{K} W_i^2 (\frac{1}{n_i} - \frac{1}{N_i}) S_i^2
$$

$$= \sum_{i=1}^{K} W_i^2 \left(\frac{1}{\frac{nN_i}{N}} - \frac{1}{N_i} \right) S_i^2$$

$$= \left(\frac{1}{n} - \frac{1}{N} \right) \sum_{i=1}^{K} W_i S_i^2 \qquad \qquad \Box$$

Example 3.1 Table 3.1 gives India's census data of 2011, in which the number of females per thousand is measured district-wise and strata are regions. Draw a random sample by using SRSWOR of size 18 under the proportional allocation of stratified sampling. Estimate the average sex ratio and obtain the variance of the estimator.

Use the following R code to illustrate the procedure:

```
SR<-scan("clipboard")
N<-c(5,13,20,26,31,36)
dN<-c(0,N)
NS<-diff(dN)
n1<-(NS/sum(NS))*18;
n1<-round(n1)
s=list();
YS=list();
mean1=c();
var1=c();
set.seed(15)
for(i in 1:6)
{
    s[[i]]<-sample((dN[i]+1):dN[i+1],n1[i],replace=F)
    YS[[i]]<-SR[s[[i]]]
    mean1[i]<-mean(YS[[i]])
    var1[i]<-var(YS[[i]])
}
s;
YS;
mean1;#stratum mean
var1;#stratum variance
#mean of proportional allocation
Wh<-NS/sum(NS); #strata weights
mean_Ybarst=sum(Wh*mean1);
mean_Ybarst;
#variance of proportional allocation
sf<-(1-n1/NS)#sampling fraction
var_Ybarst=sum(Wh^2*var1*sf/n1);
var_Ybarst;
```

The average sex ratio is 946 and its variance is 60.81.

Table 3.1 Sex ratio of different districts of Maharashtra as per census 2011

Sr. No.	Region	Districts	No. of females per thousand males
1	Vidarbha (Varhad)	1. Akola	942
		2. Amravati	947
		3. Buldana	928
		4. Yavatmal	947
		5. Washim	926
2	Marathwada	6. Aurangabad	917
		7. Beed	912
		8. Jalna	929
		9. Osmanabad	920
		10. Nanded	937
		11. Latur	924
		12. Parbhani	940
		13. Hingoli	935
3	Konkan	14. Mumbai City	838
		15. Mumbai Suburban	857
		16. Thane	880
		17. Palghar	880
		18. Raigad	955
		19. Ratnagiri	1123
		20. Sindhudurg	1037
4	Vidarbha	21. Bhandara	984
		22. Chandrapur	959
		23. Gadchiroli	975
		24. Gondia	996
		25. Nagpur	948
		26. Wardha	946
5	Khandesh	27. Dhule	941
		28. Jalgaon	922
		29. Nandurbar	972
		30. Nashik	931
		31. Ahmednagar	934
6	Paschim Maharashtra	32. Kolhapur	953
		33. Pune	910
		34. Sangli	964
		35. Satara	986
		36. Solapur	932

3.3 Stratified Random Sampling with Optimal Allocation

In stratified sampling, having decided upon the number of strata, the strata bound-aries, and the total sample size n to be drawn, the next question which a survey statistician has to answer is regarding the allocation of the total sample of size n to the respective strata and also the method of selection of the allocated sample within each stratum. The expression for the variance of stratified sample mean shows that the precision of a stratified sample for given strata depends upon the n_i's, which can be fixed at will. The guiding principle in the determination of the n_i's is to choose them in such a manner so as to provide an estimate of the population mean with the desired degree of precision for a minimum cost or to provide an estimate with max-imum precision for a given cost, thus making the most effective use of the available resources. The allocation of the sample to different strata made according to this principle is called the principle of optimum allocation.

The cost of a survey is a function of strata sample sizes, n_i's, just as the variance is. However, the purpose of the survey and the nature of the study variable will dic-tate the manner in which the cost of the survey will vary with the total sample size. Similarly, the allocation of the sample size to different strata will also depend upon the purpose of the survey and the nature of the study variable. In yield estimation surveys, the major item in the survey cost consists of labor charges for the harvesting of produce and as such, survey cost is found to be approximately proportional to the number of crop cutting experiments (CCE). The cost per CCE may, however, vary in different strata depending upon labor availability. Under such situations, the total cost may be represented by

$$C = \sum_{i=1}^{K} c_i \, n_i,$$

where c_i is the total cost of a CCE in the ith stratum. When c_i is the same from stratum to stratum, i.e. $c_i = c \;\; \forall i = 1, 2, \ldots, K$, then the total cost of a survey is given by

$$C = c \, n.$$

The cost function will change in the form if the travel cost, salary of the field staff, the cost involved in statistical analysis of data, etc. are to be paid for.

The optimum values of n_i's can be obtained by minimizing the $V(\bar{y}_{st})$ for the fixed cost C as

$$\phi = V(\bar{y}_{st}) + \mu C$$

where μ is a Lagrangian multiplier and

$$V(\bar{y}_{st}) = \sum_{i=1}^{K} W_i^2 \left(\frac{1}{n_i} - \frac{1}{N_i} \right) S_i^2. \tag{3.1}$$

$$\phi = \sum_{i=1}^{K} W_i^2 \left(\frac{1}{n_i} - \frac{1}{N_i} \right) S_i^2 + \mu \sum_{i=1}^{K} c_i n_i. \tag{3.2}$$

Differentiate Eq. (3.2) with respect to n_i and equating to zero,

$$\frac{\partial \phi}{\partial n_i} = -\frac{W_i^2 S_i^2}{n_i^2} + \mu c_i = 0$$

$$\therefore \ n_i = \frac{W_i S_i}{\sqrt{\mu c_i}}$$

$$\frac{W_i S_i}{\sqrt{n_i}} = \sqrt{\mu \ c_i \ n_i}, \text{ or } n_i = \frac{W_i S_i}{\sqrt{\mu c_i}}, \ i = 1, 2, \dots, K.$$

From the above, one can easily infer that

1. the larger the stratum size, the larger should be the size of the sample to be selected from that stratum;
2. the larger the stratum variability, the larger should be the size of the sample from that stratum;
3. the cheaper the cost per sampling unit in a stratum, the larger should be the sample from that stratum.

The exact value of n_i, obtained after evaluating $\frac{1}{\sqrt{\mu}}$, the constant of proportionality for maximizing precision for a fixed cost C_0, is given by

$$C_0 = \sum_{i=1}^{K} c_i n_i = \sum_{i=1}^{K} c_i \cdot \frac{W_i S_i}{\sqrt{\mu c_i}}$$

$$C_0 = \sum_{i=1}^{K} \frac{W_i S_i \sqrt{c_i}}{\sqrt{\mu}}$$

$$\therefore \ \sqrt{\mu} = \frac{\sum_{i=1}^{K} W_i S_i \sqrt{c_i}}{C_0}.$$

Hence,

$$n_i = \frac{W_i S_i}{\sqrt{c_i}} \cdot \frac{C_0}{\sum_{i=1}^{K} W_i S_i \sqrt{c_i}}, i = 1, 2, \dots, K.$$

This is a sample for the ith strata.

The allocation of sample size n according to above equation is known as optimum allocation. When c_i is the same from stratum to stratum, i.e. $c_i = c \ \forall i = 1, 2, \dots, K$, the cost function takes the form $C = cn$, or in other words, the cost of survey is proportional to the size of the sample. The optimum values of n_i's are then given by

$$n_i = n \frac{W_i S_i}{\sum_{i=1}^{K} W_i S_i} = \frac{n N_i S_i}{\sum_{i=1}^{K} N_i S_i}, \quad i = 1, 2, \dots, K. \tag{3.3}$$

The allocation of the sample according to the above formula is known as Neyman Allocation. Using n_i from (3.3) in the $V(\bar{y}_{st})$ expression in (3.1) gives

$$V(\bar{y}_{st})_{opt} = \sum_{i=1}^{K} W_i^2 S_i^2 \left(\frac{1}{n_i} - \frac{1}{N_i}\right) = \sum_{i=1}^{K} W_i^2 S_i^2 \left(\frac{\sum_{i=1}^{K} W_i S_i}{n W_i S_i} - \frac{1}{N_i}\right)$$

$$V(\bar{y}_{st})_{opt} = \frac{1}{n} \left(\sum_{i=1}^{K} W_i S_i\right)^2 - \frac{1}{N} \sum_{i=1}^{K} W_i S_i^2.$$

Example 3.2 The census data of 2011 is given in Table 3.1 in which the number of females per thousand is measured district-wise and strata are regions. Draw a random sample by using SRSWOR of size 18 under optimal allocation of stratified sampling. Estimate the average sex ratio and obtain the variance of the estimator.

Use the following R code to illustrate the procedure:

```
SR<-scan("clipboard")
N<-c(5,13,20,26,31,36)
dN<-c(0,N)
NS<-diff(dN)
Y=list();
sd1=c();
for(i in 1:6)
{Y[[i]]<-SR[(dN[i]+1):dN[i+1]]
    sd1[i]<-sd(Y[[i]])
}
Y;
sd1;
nh<-18*NS*sd1/sum(NS*sd1)
nh<-round(nh);nh; #to calculate strata sizes and
reformulate strata


N<-c(20,26,36)
dN<-c(0,N)
NS<-diff(dN)
n1<-c(13,2,3);
n1<-round(n1)
s=list();
YS=list();
mean1=c();
var1=c();
set.seed(46)
for(i in 1:3)
{
    s[[i]]<-sample((dN[i]+1):dN[i+1],n1[i],replace=F)
    YS[[i]]<-SR[s[[i]]]
    mean1[i]<-mean(YS[[i]])
    var1[i]<-var(YS[[i]])
}
s;
YS;
mean1;#stratum mean
var1;#stratum variance
#mean of optimum allocation
Wh<-NS/sum(NS); #strata weights
```

```
mean_Ybarst=sum(Wh*mean1);
mean_Ybarst;
#variance of optimum allocation
sf<-(1-n1/NS)#sampling fraction
var_Ybarst=sum(Wh^2*var1*sf/n1);
var_Ybarst;
```

The average sex ratio is 947 and its variance is 40.78.

Theorem 3.4 *If stratum sizes are large, i.e. $\frac{1}{N_i} \to 0$, $\forall i$ then,*
$V(\bar{y}_{st})_{opt} \leq V(\bar{y}_{st})_{prop} \leq V(\bar{y}_n)$,
where optimum allocation is fixed for n, i.e. $n_i \propto N_i S_i$.

Proof The variance of simple random sampling without replacement is

$$V(\bar{y}_n) = \left(\frac{1}{n} - \frac{1}{N}\right) S_y^2. \tag{3.4}$$

The variance of stratified random sampling under proportional allocation for sample mean is

$$V(\bar{y}_{st})_{prop} = \left(\frac{1}{n} - \frac{1}{N}\right) \sum_{i=1}^{K} W_i S_i^2. \tag{3.5}$$

And the variance of stratified random sampling under optimum allocation for sample mean is

$$V(\bar{y}_{st})_{opt} = \frac{1}{n}\left(\sum_{i=1}^{K} W_i S_i\right)^2 - \frac{1}{N} \sum_{i=1}^{K} W_i S_i^2. \tag{3.6}$$

The population mean square is given by

$$(N-1)S_y^2 = \sum_{i=1}^{K} \sum_{j=1}^{N_i} (y_{ij} - \bar{Y})^2$$

$$= \sum_{i=1}^{K} \sum_{j=1}^{N_i} (y_{ij} - \bar{Y}_i)^2 + \sum_{i=1}^{K} N_i (\bar{Y}_i - \bar{Y})^2$$

$$= \sum_{i=1}^{K} (N_i - 1)S_i^2 + \sum_{i=1}^{K} N_i (\bar{Y}_i - \bar{Y})^2.$$

For large population, $N - 1 \approx N$ and $N_i - 1 \approx N_i$.

$$S_y^2 = \sum_{i=1}^{K} W_i S_i^2 + \sum_{i=1}^{K} W_i (\bar{Y}_i - \bar{Y})^2. \tag{3.7}$$

Hence,

$$V(\bar{y}_n) = \left(\frac{1}{n} - \frac{1}{N}\right)S_y^2 = \left(\frac{1}{n} - \frac{1}{N}\right) \sum_{i=1}^{K} W_i S_i^2 + \left(\frac{1}{n} - \frac{1}{N}\right) \sum_{i=1}^{K} W_i (\bar{Y}_i - \bar{Y})^2.$$

$$V(\bar{y}_n) = V(\bar{y}_{st})_{prop} + \left(\frac{1}{n} - \frac{1}{N}\right) \sum_{i=1}^{K} W_i(\bar{Y}_i - \bar{Y})^2. \tag{3.8}$$

From Eq. (3.8), the right-hand side is positive (≥ 0),

$$V(\bar{y}_n) \geq V(\bar{y}_{st})_{prop}. \tag{3.9}$$

Let

$$V(\bar{y}_{st})_{prop} - V(\bar{y}_{st})_{opt} = \left(\frac{1}{n} - \frac{1}{N}\right) \sum_{i=1}^{K} W_i S_i^2 - \left[\frac{1}{n}\left(\sum_{i=1}^{K} W_i S_i\right)^2 - \frac{1}{N} \sum_{i=1}^{K} W_i S_i^2\right]$$

$$= \frac{\sum_{i=1}^{K} W_i S_i^2}{n} - \frac{\sum_{i=1}^{K} W_i S_i^2}{N} - \frac{1}{n}\left(\sum_{i=1}^{K} W_i S_i\right)^2 + \frac{1}{N} \sum_{i=1}^{K} W_i S_i^2$$

$$= \frac{\sum_{i=1}^{K} W_i S_i^2}{n} - \frac{1}{n}\left(\sum_{i=1}^{K} W_i S_i\right)^2$$

$$= \frac{1}{n} \sum_{i=1}^{K} W_i (S_i - \bar{S})^2$$

where $\bar{S} = \sum_{i=1}^{K} W_i S_i$ is weighted mean of the S_i.

$$V(\bar{y}_{st})_{prop} = V(\bar{y}_{st})_{opt} + \frac{1}{n} \sum_{i=1}^{K} W_i (S_i - \bar{S})^2. \tag{3.10}$$

From Eq. (3.10), the remaining part of the R.H.S. is greater than or equal to zero. Hence,

$$V(\bar{y}_{st})_{prop} \geq V(\bar{y}_{st})_{opt}. \tag{3.11}$$

From Eqs. (3.9) and (3.11), we get
$V(\bar{y}_{st})_{opt} \leq V(\bar{y}_{st})_{prop} \leq V(\bar{y}_n)$.
Hence the theorem.   □

3.4   Post-Stratified Random Sampling

Holt and Smith (1979) presented the problem in social survey where the census information is available for large subgroups but there is no information on the individual
units, i.e. age, sex, occupation, and education. In this case, none of the information is available for stratification at the individual level prior to sampling. However,
censuses give information on all of these variables at an aggregate level. In this
sampling method, firstly units are selected by using SRSWOR and then units are
cross-classified according to these factors. Then the aggregate of these estimates

provides the overall population estimate which is known as post-stratified random sampling.

The population of size N is divided into K strata of sizes $N_1, N_2, \ldots, N_K, \sum_{i=1}^{K} N_i = N$. A study variable $y_{ij}, i = 1, 2, \ldots, K, j = 1, 2, \ldots, N_i$. The samples are drawn by using SRSWOR of size n and after drawing samples, they falls into the strata according to vector $\mathbf{n} = (n_1, n_2, \ldots, n_K), \sum_{i=1}^{K} n_i = n$. The components of $\mathbf{n}$ are not known until after the samples is selected.

Theorem 3.5 *The estimator of mean of post-stratified random sampling is an unbiased estimator for the population mean.*

Proof The estimate of mean of post-stratified random sampling is given by
$$\bar{y}_{pst} = \sum_{i=1}^{K} W_i \bar{y}_i.$$
And,
$$\begin{aligned}
E(\bar{y}_{pst}) &= E_1[E_2(\bar{y}_{pst})] \\
&= E_1[E_2(\sum_{i=1}^{K} W_i \bar{y}_i | n_i)] \\
&= E_1(\sum_{i=1}^{K} W_i \bar{Y}_i) \\
&= E_1(\bar{Y}) = \bar{Y}.
\end{aligned}$$
Hence the theorem.   □

Theorem 3.6 *The variance of the estimate of mean of post-stratified random sampling is*
$$V(\bar{y}_{pst}) = \sum_{i=1}^{K} W_i^2 \left(\frac{1}{n_i} - \frac{1}{N_i} \right) S_i^2.$$

Proof By using conditional mean and variance result, we have
$$\begin{aligned}
V(\bar{y}_{pst}) &= E_1[V_2(\bar{y}_{pst})] + V_1[E_2(\bar{y}_{pst})] \\
&= E_1[V_2(\sum_{i=1}^{K} W_i \bar{y}_i | n_i)] + V_1[E_2(\sum_{i=1}^{K} W_i \bar{y}_i | n_i)] \\
&= E_1\left[\sum_{i=1}^{K} W_i^2 \left(\frac{1}{n_i} - \frac{1}{N_i} \right) S_i^2 \right] + V_1(\sum_{i=1}^{K} W_i \bar{Y}_i) \\
&= E_1\left[\sum_{i=1}^{K} W_i^2 \left(\frac{1}{n_i} - \frac{1}{N_i} \right) S_i^2 \right] + V_1(\bar{Y}) \\
&= \sum_{i=1}^{K} W_i^2 \left(\frac{1}{n_i} - \frac{1}{N_i} \right) S_i^2.
\end{aligned}$$
Hence the theorem.   □

Corollary 3.3 *The estimate of variance of post-stratified random sampling is*

$$\hat{V}(\bar{y}_{pst}) = \frac{(N-1)}{(N-n)}\hat{V}(\bar{y}) - \sum_{i=1}^{K}\frac{W_i}{n_i}\sum_{j=1}^{n_i}y_{ij}^2 + \bar{y}_{pst}^2,$$

where $\hat{V}(\bar{y}) = \sum_{i=1}^{K}\frac{W_i}{n_i}\sum_{j=1}^{n_i}y_{ij}^2 - \bar{y}_{pst}^2 + \sum_{i=1}^{K}\frac{W_i^2 s_i^2}{n_i}.$

3.5 Stratified Random Sampling for Proportions

Whenever the population is dichotomous and we want to estimate the proportion of units in the population, it has certain attributes within the stratum.
Define

$$y_{ij} = \begin{cases} 1 & \text{; if the jth unit in ith stratum has attribute C} \\ 0 & \text{; } otherwise. \end{cases}$$

Then the population mean of ith stratum is

$$\bar{Y}_i = \frac{1}{N_i}\sum_{j=1}^{N_i}y_{ij} = \frac{N_i'}{N_i} = P_i$$

where N_i' denotes the number of individuals in ith stratum that possesses the attribute C, and P_i is the proportion of units in the ith stratum having the attribute C in the population. And sample mean is

$$\bar{y}_i = \frac{1}{n_i}\sum_{j=1}^{n_i}y_{ij} = \frac{n_i'}{n_i} = \hat{p}_i$$

where $\hat{p}_i$ is the proportion of units in the ith stratum having the attribute C in the sample.
The estimate under stratified random sampling is

$$p_{st} = \sum_{i=1}^{K}\frac{N_i}{N}\hat{p}_i.$$

Theorem 3.7 *For stratified random sampling, the variance of* p_{st} *is*

$$V(p_{st}) = \frac{1}{N^2}\sum_{i=1}^{K}N_i^2\frac{(N_i-n_i)}{N_i-1}\frac{P_iQ_i}{n_i},$$

where $Q_i = 1 - P_i.$

Proof Since the $\bar{y}_i$ is an unbiased estimator of each ith stratum mean $\bar{Y}_i$ and by using the definition of proportion, the $\hat{p}_i$ is an unbiased estimator of each ith stratum proportion P_i and samples are selected from each stratum independently. Then,

$$V(p_{st}) = \sum_{i=1}^{K}\left(\frac{N_i}{N}\right)^2 V(\hat{p}_i). \tag{3.12}$$

Samples are drawn with each stratum by using SRSWOR and using the result from Sect. 2.4, and we get

$V(\hat{p}_i) = \frac{N_i - n_i}{N_i n_i} \frac{N_i P_i Q_i}{N_i - 1}.$

Using this result in Eq. (3.12), it becomes

$V(p_{st}) = \sum_{i=1}^{K} \left(\frac{N_i}{N}\right)^2 \frac{N_i - n_i}{N_i n_i} \frac{N_i P_i Q_i}{N_i - 1}$

$= \frac{1}{N^2} \sum_{i=1}^{K} N_i^2 \frac{(N_i - n_i)}{N_i - 1} \frac{P_i Q_i}{n_i}.$

Hence the theorem.   □

Corollary 3.4 *An unbiased estimator of variance of stratified random sampling for proportions is*

$\hat{V}(p_{st}) = \frac{1}{N^2} \sum_{i=1}^{K} N_i \frac{(N_i - n_i)}{n_i - 1} \hat{p}_i \hat{q}_i$

where $\hat{q}_i = 1 - \hat{p}_i.$

Proof Since the $\bar{y}_i$ is an unbiased estimator of each ith stratum mean $\bar{Y}_i$ and by using the definition of proportion, the $\hat{p}_i$ is an unbiased estimator of each ith stratum proportion P_i and samples are selected from each stratum independently. Then,

$$\hat{V}(p_{st}) = \sum_{i=1}^{K} \left(\frac{N_i}{N}\right)^2 \hat{V}(\hat{p}_i). \qquad (3.13)$$

Samples are drawn with each stratum by using SRSWOR and using the result from Sect. 2.4, and we get

$\hat{V}(\hat{p}_i) = \frac{N_i - n_i}{N_i n_i} \frac{n_i \hat{p}_i \hat{q}_i}{n_i - 1}.$

Using this result in Eq. (3.13), it becomes

$\hat{V}(p_{st}) = \sum_{i=1}^{K} \left(\frac{N_i}{N}\right)^2 \frac{N_i - n_i}{N_i n_i} \frac{n_i \hat{p}_i \hat{q}_i}{n_i - 1}$

$= \frac{1}{N^2} \sum_{i=1}^{K} N_i \frac{(N_i - n_i)}{n_i - 1} \hat{p}_i \hat{q}_i.$ □

Corollary 3.5 *For stratified random sampling with proportional allocation for large stratum size N_i, i.e. $N_i - 1 \approx N_i$, the variance of p_{st} is*

$V(p_{st}) = \frac{(1-f)}{n} \sum_{i=1}^{K} \frac{N_i}{N} P_i Q_i$

where $Q_i = 1 - P_i.$

Proof The variance of p_{st} for stratified random sampling by using Theorem 3.7 is

$V(p_{st}) = \sum_{i=1}^{K} \left(\frac{N_i}{N}\right)^2 \frac{N_i - n_i}{N_i n_i} \frac{N_i P_i Q_i}{N_i - 1}.$

Under proportional allocation, we have $n_i = \frac{n N_i}{N}.$

Then,

$V(p_{st}) = \sum_{i=1}^{K} \left(\frac{N_i}{N}\right)^2 \left(\frac{1}{n_i} - \frac{1}{N_i}\right) \frac{N_i P_i Q_i}{N_i - 1}$

$= \sum_{i=1}^{K} \left(\frac{N_i}{N}\right)^2 \left(\frac{1}{\frac{n N_i}{N}} - \frac{1}{N_i}\right) \frac{N_i P_i Q_i}{N_i - 1}.$

For large stratum size N_i, i.e. $N_i - 1 \approx N_i$

$$V(p_{st}) = \sum_{i=1}^{K} \left(\frac{N_i}{N}\right)^2 \left(\frac{1}{\frac{nN_i}{N}} - \frac{1}{N_i}\right) \frac{N_i P_i Q_i}{N_i}$$

$$= \sum_{i=1}^{K} \frac{N_i}{N} \left(\frac{1}{n} - \frac{1}{N}\right) P_i Q_i$$

$$= \frac{(1-f)}{n} \sum_{i=1}^{K} \frac{N_i}{N} P_i Q_i. \qquad \square$$

3.6 Exercises

1. Define the term Stratified Random Sampling and give an illustration. What is the required sample size for each stratum for estimating the sampling variances?
2. What is post-stratification? Show that the sample mean of post-stratified sampling is an unbiased estimator for population mean. Also obtain its variance.
3. Describe proportional allocation in stratified random sampling. Obtain the variance of an unbiased estimator of population mean under proportional allocation.
4. Describe stratified sampling for estimating population proportions. Obtain its variance and 95% confidence interval for population mean.
5. Describe optimum allocation in stratified random sampling and obtain total sample size. Obtain the variance of an unbiased estimator of population mean under optimum allocation.
6. If terms in $\frac{1}{N_h}$ are ignored relative to unity, then prove that $V(\bar{y}_{st})_{opt} \leq V(\bar{y}_{st})_{prop} \leq V(\bar{y}_n)$ where the optimum allocation is fixed for n that is with $n_h \propto N_h S_h$.
7. Describe the stratified random sampling for proportions. Prove that $V(p_{st}) = \frac{1}{N^2} \sum_{h=1}^{L} \frac{N_h^2(N_h - n_h)P_h Q_h}{(N_h - 1)n_h}$.
8. What is stratification? Explain the optimum allocation criteria and obtain the variance under optimum allocation.
9. Estimate the mean fuelwood consumption of the households of forest villages in the Dehradun district of the Uttarakhand state and its standard error using the method of stratified random sampling. Three strata were constructed using the criterion of distance from the forest having 110, 60, and 80 households the chosen households 10, 5, and 8, respectively, from corresponding stratum and their fuelwood consumption are shown in Table 3.2. Find the estimated mean, total, and their variances.
10. ICFRE has implemented a tree improvement programme of *Gmelina arborea* and has planned to release high productive pest tolerance clones. These clones were tested in five different forest types of the country. In the second year, the survivals of the clones were noted in Table 3.3.

Table 3.2 Fuel wood consumption

Stratum No.	Selected unit	Fuelwood consumption (per HH/M)
I	3	5.40
	33	4.87
	6	4.61
	44	3.26
	58	4.96
	1	4.73
	92	4.39
	4	2.34
	45	4.74
	42	2.85
II	21	4.79
	51	4.57
	18	4.89
	5	4.42
	22	3.44
III	75	7.41
	69	3.70
	8	5.45
	42	7.01
	26	3.83
	51	5.25
	72	4.50
	68	6.51

Table 3.3 Survivals of clones

Stratum No.	No. of plots	Average no. of survived clones	S.D. of survived clones
I	200	5.40	2.5
II	300	16.5	15.5
III	250	25.3	20.5
IV	430	15.6	12.5
V	520	7.5	3.0

For a sample of 300 plots for the genotypic and phenotypic studies based upon the survival of the plants, compute the sample size in each stratum under proportional and Neyman allocation. Calculate the sampling variance of the survived trees from the sample, if the plots were selected under (i) proportional allocation and SRSWOR (ii) Neyman allocation and SRSWOR.

Reference

Holt, D., Smith, T.M.F.: Post stratification. J. R. Stat. Soc. **A142**(1), 33–46 (1979)

Chapter 4
Systematic Sampling

4.1 Introduction

In this sampling method, the first unit is selected by using a random number and the rest of the units are selected by a predetermined pattern. This method is called systematic sampling. Suppose a population consists of N units which are serially numbered from 1 to N. Population size N is expressed as the product of two integers k and n, where n is the sample size and k is a sampling interval such that $N = nk$. Also, $k = \frac{N}{n}$ and it is an integer. Firstly, we select one unit, r, at random from 1 to k, then every kth unit is selected automatically. The initial unit r is called random start. Thus, systematic sample $S = \{r, r + k, r + 2k, \ldots, r + (n - 1)k\}$ of size n is selected. The probability of selecting each of the k possible systematic samples is $\frac{1}{k}$. For example, $N = 300, n = 10$, so $k = \frac{300}{10} = 30$. Select a random number from 1 to 30; for instance, if it is 3, then the units selected are $3, 33, 63, 93, 123, 153, 183, 213, 243, 273$.

Let the characteristic of jth unit for random start r be $y_{jr} = y_{r+(j-1)k}; j = 1, 2, \ldots, n; r = 1, 2, \ldots, k$. In fact, systematic sampling stratifies N units into n strata, which contain first k units, second k units, and so forth. The resulting sampling scheme is represented in Tables 4.1 and 4.2.

The systematic sample mean for random start r is given by

$$\bar{y}_r = \frac{1}{n} \sum_{j=1}^{n} y_{jr}.$$

Theorem 4.1 *The systematic sample mean for $N = nk$ is an unbiased estimator for the population mean.*

Proof By using the definition of expectation, we get

$$E(\bar{y}_r) = \frac{1}{n} \sum_{j=1}^{n} E(y_{jr})$$

$$= \frac{1}{n} \sum_{j=1}^{n} \frac{1}{k} \sum_{r=1}^{k} y_{jr}$$

© The Author(s), under exclusive license to Springer Nature Singapore Pte Ltd. 2021
R. Latpate et al., *Advanced Sampling Methods*,
https://doi.org/10.1007/978-981-16-0622-9_4

Table 4.1 Schematic diagram of the serial number of the unit in the population for systematic sampling

Stratum number	Sample number 1	2	.	.	.	r	.	.	.	k
1	1	2	.	.	.	r	.	.	.	k
2	$1+k$	$2+k$	.	.	.	$r+k$	.	.	.	$2k$
3	$1+2k$	$2+2k$	.	.	.	$r+2k$	.	.	.	$3k$
.	.	.	.	.	.	.	.	.	.	.
.	.	.	.	.	.	.	.	.	.	.
.	.	.	.	.	.	.	.	.	.	.
j	$1+(j-1)k$	$2+(j-1)k$	.	.	.	$r+(j-1)k$	.	.	.	jk
.	.	.	.	.	.	.	.	.	.	.
.	.	.	.	.	.	.	.	.	.	.
.	.	.	.	.	.	.	.	.	.	.
n	$1+(n-1)k$	$2+(n-1)k$	.	.	.	$r+(n-1)k$	.	.	.	nk

Table 4.2 Schematic diagram of corresponding characteristics of the unit in the population for systematic sampling

Stratum number	y-values 1	2	.	.	.	r	.	.	.	k	Stratum mean
1	y_{11}	y_{12}	.	.	.	y_{1r}	.	.	.	y_{1k}	$\bar{y}_{1.}$
2	y_{21}	y_{22}	.	.	.	y_{2r}	.	.	.	y_{2k}	$\bar{y}_{2.}$
3	y_{31}	y_{32}	.	.	.	y_{3r}	.	.	.	y_{3k}	$\bar{y}_{3.}$
.	.	.	.	.	.	.	.	.	.	.	.
.	.	.	.	.	.	.	.	.	.	.	.
.	.	.	.	.	.	.	.	.	.	.	.
j	y_{j1}	y_{j2}	.	.	.	y_{jr}	.	.	.	y_{jk}	$\bar{y}_{j.}$
.	.	.	.	.	.	.	.	.	.	.	.
.	.	.	.	.	.	.	.	.	.	.	.
.	.	.	.	.	.	.	.	.	.	.	.
n	y_{n1}	y_{n2}	.	.	.	y_{nr}	.	.	.	y_{nk}	$\bar{y}_{n.}$
Systematic sample mean	$\bar{y}_1$	$\bar{y}_2$	.	.	.	$\bar{y}_r$	.	.	.	$\bar{y}_k$	

$$= \frac{1}{nk} \sum_{j=1}^{n} \sum_{r=1}^{k} y_{jr}$$

$$= \frac{1}{N} \sum_{j=1}^{n} \sum_{r=1}^{k} y_{jr}$$

$$= \bar{Y}.$$

Hence the theorem.

Theorem 4.2 *The variance between k systematic sample means is given by*

$$V(\bar{y}_r) = \frac{1}{k} \sum_{r=1}^{k} (\bar{y}_r - \bar{Y})^2.$$

Proof By using the definition of variance, we get

$$V(\bar{y}_r) = E(\bar{y}_r^2) - \left(E(\bar{y}_r)\right)^2$$

$$= \frac{1}{k} \sum_{r=1}^{k} \bar{y}_r^2 - \bar{Y}^2$$

$$= \frac{1}{k} \sum_{r=1}^{k} (\bar{y}_r - \bar{Y})^2.$$

Hence the theorem.

The systematic sampling method is easy to apply and provides more spread compared to other sampling methods. If the periodic trend is in the sampling frame, then systematic sampling could be biased.

Example 4.1 The experiment deals with the bamboo species *Melocanna bambusoides* with vessel lengths 27, 38, 53, 43, 32, 45, 25, 32, 43, 22, 38, 42, 39, 34, 25, 27, 33, 22, 34, 48, 41, 34, 23, 37, 32, 37, 44, 41, 23, 41, 20, 29, 28, 39, 32, 27, 22, 37, 23, 32, 27, 23, 31, 26, 35, 43, 26, 24, 34, 22, 27, 30, 19, 12, 11, 14, 24, 25, 27, and 20. Draw a systematic sample of size 6 and estimate the average vessel length.

Use the following R code to illustrate the procedure:

```
X<-scan("clipboard")
N=60;
n=6;
k=N/n;
set.seed(25)
ss<-sample(1:k,1,replace=F)
sample_index=c();
for(i in 1:6)
{
    sample_index[i]<-ss+(i-1)*k
}
sample_index
x<-X[sample_index]
mean(x);
```

The average vessel length is 27.83.

4.2   Comparison of Systematic Sampling with Simple Random Sampling

The variance of the means of simple random sampling without replacement (SRSWOR) is given by

$$V(\bar{y}_n) = \left(\frac{1}{n} - \frac{1}{N}\right)S_y^2$$
$$= \left(\frac{1}{n} - \frac{1}{nk}\right)S_y^2$$
$$= \frac{(k-1)}{nk}S_y^2$$

where $N = nk$.

The variance of the means of systematic sampling is given by

$$V(\bar{y}_r) = \frac{1}{k}\sum_{r=1}^{k}(\bar{y}_r - \bar{Y})^2$$

$$= \frac{1}{k}\sum_{r=1}^{k}\left(\frac{1}{n}\sum_{j=1}^{n}(y_{jr} - \bar{Y})\right)^2$$

$$= \frac{1}{n^2 k}\left[\sum_{r=1}^{k}\sum_{j=1}^{n}(y_{jr} - \bar{Y})^2 + \sum_{r=1}^{k}\sum_{j\neq j'=1}^{n}(y_{jr} - \bar{Y})(y_{j'r} - \bar{Y})\right]$$

$$= \frac{1}{n^2 k}\left[(nk - 1)S_y^2 + \sum_{r=1}^{k}\sum_{j\neq j'=1}^{n}(y_{jr} - \bar{Y})(y_{j'r} - \bar{Y})\right]$$

where

$$\sum_{r=1}^{k}\sum_{j\neq j'=1}^{n}(y_{jr} - \bar{Y})(y_{j'r} - \bar{Y}) = (n - 1)(kn - 1)\rho S_y^2$$

and, ρ is the intraclass correlation coefficient between units of a column.

$$V(\bar{y}_r) = \frac{1}{n^2 k}\left[(nk - 1)S_y^2 + (n - 1)(kn - 1)\rho S_y^2\right]$$
$$= \frac{(nk-1)S_y^2}{n^2 k}[1 + (n - 1)\rho].$$

The relative efficiency of systematic mean relative to the simple random sample mean without replacement is as follows:

$$\frac{V(\bar{y}_n)}{V(\bar{y}_r)} = \frac{(nk - 1)[1 + \rho(n - 1)]}{n(k - 1)}. \tag{4.1}$$

Equation (4.1) showed that relative efficiency depends upon the intraclass correlation ρ. If $\rho = \frac{-1}{kn-1}$, then two methods are equally efficient. When $\rho > \frac{-1}{kn-1}$, simple random sampling is more efficient and when $\rho < \frac{-1}{kn-1}$, systematic sampling is more efficient than simple random sampling.

4.3   Comparison of Systematic Sampling with Stratified Random Sampling

The variance of stratified random sample mean is obtained as follows:

$$V(\bar{y}_{st}) = V(\tfrac{1}{n} \sum_{j=1}^{n} y_{jr})$$

$$= \tfrac{1}{n^2} \sum_{j=1}^{n} V(y_{jr})$$

$$= \tfrac{1}{n^2 k} \sum_{j=1}^{n} \sum_{r=1}^{k} (y_{jr} - \bar{y}_{j.})^2$$

$$= \tfrac{\sigma_{wst}^2}{n}$$

where $\sigma_{wst}^2 = \tfrac{1}{nk} \sum_{j=1}^{n} \sum_{r=1}^{k} (y_{jr} - \bar{y}_{j.})^2.$

The variance of the means of systematic sampling is given by

$$V(\bar{y}_r) = \tfrac{1}{k} \sum_{r=1}^{k} (\bar{y}_r - \bar{Y})^2$$

$$= \tfrac{1}{k} \sum_{r=1}^{k} \left(\tfrac{1}{n} \sum_{j=1}^{n} (y_{jr} - \bar{y}_{j.}) \right)^2$$

$$= \tfrac{1}{n^2 k} \Big[\sum_{r=1}^{k} \sum_{j=1}^{n} (y_{jr} - \bar{y}_{j.})^2 + \sum_{r=1}^{k} \sum_{j \neq j'=1}^{n} (y_{jr} - \bar{y}_{j.})(y_{j'r} - \bar{y}_{j'.}) \Big]$$

$$= \tfrac{1}{n} \sigma_{wst}^2 + \tfrac{1}{n^2 k} \sum_{r=1}^{k} \sum_{j \neq j'=1}^{n} (y_{jr} - \bar{y}_{j.})(y_{j'r} - \bar{y}_{j'.})$$

$$= (1 + (n-1)\rho_{wst}) \tfrac{\sigma_{wst}^2}{n}$$

where

$$\rho_{st} = \frac{\sum_{r=1}^{k} \sum_{j \neq j'=1}^{n} (y_{jr} - \bar{y}_{j.})(y_{j'r} - \bar{y}_{j'.})}{nk(n-1)\sigma_{wst}^2}$$ and, ρ_{st} is the intraclass correlation coefficient between units of a systematic sample with deviation from the stratum mean.

The relative efficiency of stratified random sample relative to systematic mean is given by

$$\frac{V(\bar{y}_r)}{V(\bar{y}_{st})} = [1 + \rho_{st}(n-1)]. \tag{4.2}$$

If ρ_{st} is negative, then systematic sampling is more efficient than stratified random sampling and if ρ_{st} is positive, then stratified random sampling is more efficient.

When ρ_{st} is zero, both the methods are equally efficient. ρ_{st} can be evaluated by using overall population units.

$$V(\bar{y}_r) = \tfrac{1}{n^2 k} \big[(nk-1)S_y^2 + (n-1)(kn-1)\rho S_y^2 \big]$$

$$= \tfrac{(nk-1)S_y^2}{n^2 k} [1 + (n-1)\rho]$$

Random start	Systematic samples
1	1, 3, 5, 7, 9
2	2, 4, 6, 8, 10
3	3, 5, 7, 9, 1
4	4, 6, 8, 10, 2
5	5, 7, 9, 1, 3
6	6, 8, 10, 2, 4
7	7, 9, 1, 3, 5
8	8, 10, 2, 4, 6
9	9, 1, 3, 5, 7
10	10, 2, 4, 6, 8

Table 4.3 Schematic diagram of circular systematic sampling

4.4 Circular Systematic Sampling

When N is not a multiple of n, then circular systematic sampling is useful. The above difficulty can be removed if we take $k = \frac{N}{n}$ without rounding it off. In this sampling method, the sequential list of the population units is first prepared and k is obtained as an integer nearest to N/n and they are called *skip*. Select a random number from 1 to N and name it as random start. Select all units in the sample with serial numbers as follows:

$r + jk$ if $r + jk \leq N$

or

$r + jk - N$ if $r + jk > N$; $j = 0, 1, 2, \ldots, (n - 1)$.

Example 4.2 Consider a population of size $N = 10$, the target sample size $n = 5$, and $k = 10/5 = 2$. The schematic diagram of circular systematic sampling is given in Table 4.3.

4.5 Exercises

1. Define the term Systematic Sampling and give an illustration.
2. In systematic sampling, show that the sample mean is an unbiased estimator of the population mean. Hence obtain the sampling variance of the sample mean.
3. Write a short note on circular systematic sampling.
4. Write a short note on systematic sampling when N is not a multiple of n.
5. Derive the expression for the comparison between systematic sampling with SRSWOR.

Chapter 5
Cluster Sampling

5.1 Introduction

The smallest units into which the population can be divided are called the elements of the population, and groups of these elements are called clusters. A cluster may be a class of students or cultivator fields in a village. When a sampling unit is a cluster, the procedure of sampling is called cluster sampling. Cluster sampling is a sampling technique in which the entire population of interest is divided into clusters, and a sample of these clusters is selected by the simple random sampling (SRSWOR) technique.

The main reason for using cluster sampling is that it is usually much cheaper and more convenient to sample the clusters rather than individual units. Many a time, constructing a sampling frame that identifies every population element is too expensive or impossible to construct. For example, a list of all farms in a district is generally not available but the list of villages may be easily available. Considering villages as clusters, the selection of villages can be done by SRSWOR and then a complete enumeration of the selected villages can be done. Cluster sampling can also reduce the cost of a survey when the population elements are scattered over a wide area.

For a given number of sampling units, cluster sampling is more convenient and less costly than element sampling due to saving time in journeys, identification and contacts, etc. Cluster sampling, however, is generally less efficient than the SRSWOR sampling design with sample mean as the estimator of the population mean due to the tendency of the units in a cluster to be similar. In most practical situations, the loss in efficiency may be more than the offset by the reduction in the cost, and the efficiency per unit cost may be more in cluster sampling as compared to element sampling.

The basic task in cluster sampling is to specify appropriate clusters. Clusters are generally made up of neighboring units or of compact areas and, therefore, the units within the clusters tend to have similar characteristics. As a simple rule, the number of elements in a cluster should be small and the number of clusters should be large. The efficiency of cluster sampling decreases with the increase in the size of the cluster. Selection of the required number of clusters may be done by equal or unequal probabilities of selection, and all selected clusters are required to be completely enumerated. The selection of the clusters can also be made by first selecting randomly a unit, called the key unit, and then taking randomly the required number of neighboring clusters of the key clusters as the sample of clusters. For example, for estimating the milk production, a cluster of three villages may be formed by first selecting a key village at random and then taking two more villages among the villages in a circle of some specified radius with the key village as the center village.

5.2 Single-Stage Cluster Sampling

Let N denote the number of clusters in the population. Suppose that an SRSWOR of n clusters is to be selected in the sample. Let M_i denote the number of elements in the ith cluster, $i = 1, 2 \ldots, N$. We shall also denote the total number of units in the population, $\sum_{i=1}^{N} M_i$, by M_0 and the average number of elements per cluster in the population by $M = \frac{M_0}{N}$. Let y_{ij} denote the value of the character under study for the jth element in the ith cluster, $j = 1, 2, \ldots, M_i$. The population mean per element of the ith cluster is then given by $\bar{Y}_i = \frac{1}{M_i} \sum_{j=1}^{M_i} y_{ij}, i = 1, 2 \ldots, N$. Then the population mean of the cluster means is given by $\bar{Y}_N = \frac{1}{N} \sum_{i=1}^{N} \bar{Y}_i$. The population mean per element is $\bar{Y} = \frac{\sum_{i=1}^{N} \sum_{j=1}^{M_i} y_{ij}}{M_0}$.

5.2.1 Equal Cluster Sizes

For simplicity, the case of equal cluster sizes, that is, $M_i = M, \quad \forall i = 1, 2 \ldots, N$ may be considered first. When the selection is by SRSWOR, the unbiased estimator of the population mean $\bar{Y}$ for equal cluster sizes is given by
$\bar{y}_{cl} = \frac{1}{n} \sum_{i=1}^{n} \bar{Y}_i$.
The variance of $\bar{y}_{cl}$ is given by
$V(\bar{y}_{cl}) = \left(\frac{N-n}{Nn}\right) S_b^2$,
where $S_b^2 = \frac{\sum_{i=1}^{N}(\bar{Y}_i - \bar{Y}_N)^2}{N-1}$.
    An unbiased estimator of variance of $\bar{y}_{cl}$ is given by
$\hat{V}(\bar{y}_{cl}) = \left(\frac{N-n}{Nn}\right) s_b^2$,
where $s_b^2 = \frac{\sum_{i=1}^{n}(\bar{Y}_i - \bar{y}_{cl})^2}{n-1}$.
    For the n selected clusters, the estimator of population mean, $\bar{y}_{cl}$, is based on a sample of nM elements. Hence the relative efficiency of $\bar{y}_{cl}$ with respect to $\bar{y}_n$, the

sample mean based on a SRSWOR sample of size nM elements selected from a population of NM elements, is

$$RE(\bar{y}_{cl}) = \frac{V(\bar{y}_n)}{V(\bar{y}_{cl})}$$

$$= \frac{\left(\frac{M\ (N-n)}{NM^2 n}\right)S^2}{\frac{N-n}{Nn}S_b^2},$$

$$= \frac{S^2}{MS_b^2},$$

where $S^2 = \frac{\sum_{i=1}^{N}\sum_{j=1}^{M}(y_{ij}-\bar{Y})^2}{NM-1}$, the mean square between elements in the population.

Thus, it can be seen that the relative efficiency of the cluster sampling increases as the cluster size decreases. Similarly, the efficiency increases as the overall mean square between elements increases and the mean square between clusters decreases.

Theorem 5.1 *The estimator of the population mean is unbiased for the population mean per element.*

Proof The estimator of the population mean is
$$\bar{y}_{cl} = \frac{1}{n}\sum_{i=1}^{n}\bar{Y}_i.$$
It can also be written as
$$\bar{y}_{cl} = \frac{1}{n}\sum_{i=1}^{N}\alpha_i\bar{Y}_i$$
where

$$\alpha_i = \begin{cases} 1 \ ; \text{if ith cluster is selected in the sample} \\ 0 \ ; otherwise. \end{cases}$$

Then
$$E(\bar{y}_{cl}) = \frac{1}{n}E(\sum_{i=1}^{N}\alpha_i\bar{Y}_i)$$
$$= \frac{1}{n}\sum_{i=1}^{N}E(\alpha_i)\bar{Y}_i$$
$$= \frac{1}{n}\frac{n}{N}\sum_{i=1}^{N}\bar{Y}_i$$
$$= \frac{1}{N}\sum_{i=1}^{N}\bar{Y}_i$$
$$= \frac{1}{N}\sum_{i=1}^{N}\frac{1}{M}\sum_{j=1}^{M}y_{ij}$$
$$= \frac{1}{NM}\sum_{i=1}^{N}\sum_{j=1}^{M}y_{ij}$$
$$= \bar{Y}.$$
Hence the theorem. $\square$

Theorem 5.2 *The mean squares between and within clusters is an unbiased estimate of the mean squares between elements in the population.*

Proof The mean square between clusters is given by MS_b^2.
Hence,
$$E(Mean\ square\ between\ clusters) = E(MS_b^2)$$
$$= E\left[\frac{M}{N-1}\sum_{i=1}^{N}\left(\bar{Y}_i - \bar{Y}_N\right)^2\right]$$
$$= \frac{M}{N-1}E\left[\sum_{i=1}^{N}\bar{Y}_i^2 - N\bar{Y}_N^2\right]$$
$$= \frac{M}{N-1}\left[\sum_{i=1}^{N}E(\bar{Y}_i^2) - N\bar{Y}_N^2\right]$$

$$= \frac{M}{N-1} \left[\sum_{i=1}^{N} (\bar{Y}_N^2 + \frac{NM-M}{NM} \frac{S^2}{M}) - N\bar{Y}_N^2 \right]$$

$$= \frac{M}{N-1} \left[\frac{(N-1)S^2}{M} \right]$$

$$= S^2.$$

The mean square within clusters is given by $(= \sum_{i=1}^{N} \bar{S}_w^2)$

where $\bar{S}_w^2 = \frac{S_i^2}{N}$, and $S_i^2 = \frac{1}{M-1} \sum_{j=1}^{M} (y_{ij} - \bar{Y}_i)^2$

$$E(Mean\ square\ within\ clusters) = E \left[\frac{1}{N(M-1)} \sum_{i=1}^{N} \sum_{j=1}^{M} (y_{ij} - \bar{Y}_i)^2 \right]$$

$$= \frac{1}{N(M-1)} E \left[\sum_{i=1}^{N} \sum_{j=1}^{M} y_{ij}^2 - M \sum_{i=1}^{N} \bar{Y}_i^2 \right]$$

$$= \frac{1}{N(M-1)} \left[\sum_{i=1}^{N} \sum_{j=1}^{M} E(y_{ij}^2) - M \sum_{i=1}^{N} E(\bar{Y}_i^2) \right]$$

$$= \frac{1}{N(M-1)} \left[NM(\bar{Y}_N + \frac{NM-1}{NM} S^2) - MN(\bar{Y}_N + \frac{NM-M}{NM} \frac{S^2}{M}) \right]$$

$$= \frac{1}{N(M-1)} \left[(NM-1)S^2 - (NM-M)\frac{S^2}{M} \right]$$

$$= \frac{1}{N(M-1)} \left[(NM-N)S^2 \right]$$

$$= S^2.$$

Hence the theorem.   □

Theorem 5.3 *The cluster sampling is more efficient than simple random sampling if the intraclass correlation coefficient $\rho < 0$.*

Proof The intraclass correlation coefficient $= \rho = \frac{\sum_{i=1}^{N} \sum_{j\neq j'=1}^{M} (y_{ij} - \bar{Y}_N)(y_{ij'} - \bar{Y}_N)}{NM(M-1)S^2}$.

Now

$$\sum_{i=1}^{N} \left[\sum_{j=1}^{M} (y_{ij} - \bar{Y}_N) \right]^2 = \sum_{i=1}^{N} \left[\sum_{j=1}^{M} y_{ij} - M\bar{Y}_N \right]^2$$

$$= \sum_{i=1}^{N} \left[M\bar{Y}_i - M\bar{Y}_N \right]^2$$

$$= M^2 \sum_{i=1}^{N} (\bar{Y}_i - \bar{Y}_N)^2$$

$$= M^2 (N-1) S_b^2.$$

Thus,

$$\sum_{i=1}^{N} \left[\sum_{j=1}^{M} (y_{ij} - \bar{Y}_N) \right]^2 = M^2 (N-1) S_b^2. \tag{5.1}$$

Consider

$$\sum_{i=1}^{N} \left[\sum_{j=1}^{M} y_{ij} - \bar{Y}_N \right]^2 = \sum_{i=1}^{N} \left[\sum_{j=1}^{M} (y_{ij} - \bar{Y}_N)^2 + \sum_{j\neq j'=1}^{M} (y_{ij} - \bar{Y}_N)(y_{ij'} - \bar{Y}_N) \right]$$

$$= \sum_{i=1}^{N} \sum_{j=1}^{M} (y_{ij} - \bar{Y}_N)^2 + \sum_{i=1}^{N} \sum_{j\neq j'=1}^{M} (y_{ij} - \bar{Y}_N)(y_{ij'} - \bar{Y}_N).$$

Now

$$S^2 = \frac{1}{NM-1} \sum_{i=1}^{N} \sum_{j=1}^{M} (y_{ij} - \bar{Y}_N)^2.$$

Hence

$$\sum_{i=1}^{N} \left[\sum_{j=1}^{M} (y_{ij} - \bar{Y}_N) \right]^2 = (NM-1)S^2 + NM(M-1)\rho S^2. \tag{5.2}$$

Using Eqs. (5.1) and (5.2), we get
$$(NM - 1)S^2 + NM(M - 1)\rho S^2 = M^2 (N - 1) S_b^2.$$
Hence
$$M S_b^2 = \frac{(NM-1)S^2 + NM(M-1)\rho S^2}{M(N-1)}.$$
By using the definition of relative efficiency,
$$Relative\ efficiency = \frac{S^2}{M S_b^2}$$
$$= \frac{M(N-1)}{(NM-1)+NM(M-1)\rho}$$
$$= \frac{1-\frac{1}{N}}{(1-\frac{1}{NM})+(M-1)\rho}.$$
For large N and NM,
$$Relative\ efficiency = \frac{1}{1+(M-1)\rho}.$$
If $\rho < 0$, then relative efficiency is greater than 1. Hence in that case, cluster sampling is more efficient than SRSWOR. If $\rho = 0$, then both are equally efficient.

Hence the theorem. $\square$

Example 5.1 The water samples are collected from different locations of the Pune district of Maharashtra, India, and they are clustered accordingly as shown in Table 5.1. The aim of the study is to measure the quality of water. Based on 6 water quality parameters, draw a random sample of 3 clusters and obtain the average conductivity (ppm) of the water and obtain its variance.

Use the following R code to illustrate the procedure:

```
Y=scan("clipboard")
set.seed(5)
n=3
s1<-sample(1:6,n,replace=F)
N=c(0,6,6,6,6,6,6);
N=cumsum(N)
b=list();
yy=list();
for(i in 1:6)
{
b[[i]]=(N[i]+1):N[i+1]
yy[[i]]=Y[b[[i]]]
}
b;
yy;
yyy=list();
mean1=c();
for(i in 1:n)
{
for(j in 1:6)
{
if(j==s1[i])
```

```
{yyy[[i]]=yy[[j]]
mean1[i]=mean(yyy[[i]])
}}}
mean1; mean_clust=mean(mean1); mean_clust;
var_bet=(1/5)*sum((mean1-mean_clust)^2);
var_bet; Var_clust=((6-3)/6*3)*var_bet; Var_clust;
```

The average conductivity of the water is 506.67 and its variance is 5501.67.

5.2.2   Unequal Cluster Sizes

In the present situation of unequal cluster sizes, the estimator of population mean $\bar{Y}$ may be obtained by

$$\bar{y}_{cl}^{\star} = \frac{1}{n} \sum_{i=1}^{n} \frac{M_i}{M} \bar{Y}_i.$$

The variance and the estimator of the variance of $\bar{y}_{cl}^{\star}$ are obtained as
$$V(\bar{y}_{cl}^{\star}) = \frac{N-n}{Nn} S_b^{2'}$$
where $S_b^{2'} = \frac{\sum_{i=1}^{N} \left(\frac{M_i}{M} \bar{Y}_i - \bar{Y} \right)^2}{N-1}$
and the unbiased estimator $V(\bar{y}_{cl}^{\star})$ is given by
$$\hat{V}(\bar{y}_{cl}^{\star}) = \frac{N-n}{Nn} s_b^{2'}$$
where $s_b^{2'} = \frac{\sum_{i=1}^{n} \left(\frac{M_i}{M} \bar{Y}_i - \bar{y}_{cl}^{\star} \right)^2}{n-1}.$

Theorem 5.4 *Bias in* $\bar{y}_{cl}^{\star} = -\frac{N-1}{M_0} S_{M\bar{y}}.$

Proof $B(\bar{y}_{cl}^{\star}) = E(\bar{y}_{cl}^{\star}) - \bar{Y}$
$$= \frac{1}{N} \sum_{i=1}^{N} \bar{Y}_i - \sum_{i=1}^{N} \frac{M_i \bar{Y}_i}{M_0}$$
$$= -\frac{1}{M_0} \sum_{i=1}^{N} (M_i - \bar{M})(\bar{Y}_i - \bar{Y}_N)$$
$$= -\frac{N-1}{M_0} S_{M\bar{y}}.$$
Hence the theorem.   □

Corollary 5.1 The mean squared error of $\bar{y}_{cl}^{\star}$ is given by
$$MSE(\bar{y}_{cl}^{\star}) = V(\bar{y}_{cl}^{\star}) + \left[Bias(\bar{y}_{cl}^{\star}) \right]^2$$
$$= \frac{N-n}{Nn} S_b^{\star 2} + \frac{(N-1)^2}{M_0^2} S_{M\bar{y}}^2.$$

5.3   Multistage Cluster Sampling

Cluster sampling is a sampling procedure in which clusters are considered as sampling units, and all the elements of the selected clusters are enumerated. One of the main considerations of adopting cluster sampling is the reduction of travel cost because of the nearness of elements in the clusters. However, this method restricts the spread of the sample over the population which results generally in increasing the variance of the estimator. In order to increase the efficiency of the estimator with the given cost, it is natural to think of further sampling the clusters and selecting more number of clusters so as to increase the spread of the sample over the population. This type of sampling which consists of first selecting clusters and then selecting a specified number of elements from each selected cluster is known as subsampling or two-stage sampling, since the units are selected in two stages. In such sampling designs, clusters are generally termed as first-stage units (FSUs) or primary-stage units (PSUs) and the elements within clusters or ultimate observational units are termed as the second-stage units (SSUs) or ultimate-stage units (USUs). It may be noted that this procedure can be easily generalized to give rise to multistage sampling design, where the sampling units at each stage are clusters of units of the next stage, and the ultimate observational units are selected in stages, sampling at each stage being done from each of the sampling units or clusters selected in the previous stage. This procedure, being a compromise between uni-stage or direct sampling of units and cluster sampling, can be expected to be (i) more efficient than uni-stage sampling and less efficient than cluster sampling from considerations of operational convenience and cost, and (ii) less efficient than uni-stage sampling and more efficient than cluster sampling from the viewpoint of sampling variability, when the sample size in terms of the number of ultimate units is fixed.

It may be mentioned that multistage sampling may be the only feasible procedure in a number of practical situations, where a satisfactory sampling frame of ultimate observational units is not readily available, and the cost of obtaining such a frame is prohibitive or where the cost of locating and physically identifying the USUs is considerable. For instance, for conducting a socioeconomic survey in a region, where generally household is taken as the USU, a complete and up-to-date list of all the households in the region may not be available, whereas a list of villages and urban blocks, which are group of households, may be readily available. In such a case, a sample of villages or urban blocks may be selected first and then a sample of households may be drawn from each selected village and urban block after making a complete list of households. It may happen that even a list of villages is not available, but only a list of all tehsils (groups of villages) is available. In this case, a sample of households may be selected in three stages by selecting first a sample of tehsils, then a sample of villages from each selected tehsil after making a list of all the villages in the tehsil, and finally a sample of households from each selected village after listing all the households in it. Since the selection is done in three stages, this procedure is termed as three-stage sampling. Here, tehsils are taken as first-stage units (FSUs), villages as second-stage units (SSUs), and households as third or ultimate-stage units (TSUs).

One of the advantages of this type of sampling is that at the first stage, the frame of FSUs is required which is generally easily available and at the second stage the frame of SSUs is required for the selected FSUs only, and so on. Moreover, this method allows the use of different selection procedures in different stages. It is because of these considerations that multistage sampling is used in most of the large-scale surveys. It has been found to be very useful in practice. It is noteworthy that

Prof. P. C. Mahalanobis used this sampling procedure in crop surveys carried out in Bengal during the period 1937–1941, and he had termed this procedure as nested sampling. Cochran (1939), Hansen and Hurwitz (1943) have considered the use of this procedure in agricultural and population surveys, respectively. Lahiri (1954) discussed the use of multistage sampling in the Indian Sample Survey.

5.3.1 Two-stage Cluster Sampling

In this method, firstly the clusters are selected by using SRSWOR and then a sample of units are selected from the selected clusters. These two cases of two-stage sampling according to the size of clusters are as follows.

(i) Two-stage sampling with equal probabilities, with equal first-stage units
Let the population under study consist of NM units grouped into N first-stage units, with each first-stage unit containing M second-stage units.

Let us denote by y_{ij} the value of the characteristic under study for the jth second-stage unit (SSU) of the ith first-stage unit (FSU), $j = 1, 2, \ldots, M; i = 1, 2, \ldots, N$.

Let the mean of the ith FSU be denoted by

$\bar{Y}_{i.} = \frac{1}{M} \sum_{j=1}^{M} y_{ij}$ and the population mean by $\bar{Y}_{..} = \frac{\sum_{i=1}^{N} \sum_{j=1}^{M} y_{ij}}{NM} = \frac{1}{N} \sum_{i=1}^{N} \bar{Y}_{i.}$.

Further, let a sample of size nm be selected by first selecting n FSUs from N FSUs by SRSWOR and then selecting m ssu's by SRSWOR from each of the selected FSUs. Let us denote by $\bar{y}_{im} = \frac{1}{m} \sum_{j=1}^{m} y_{ij}$ the sample mean based on m selected ssu's from the ith FSU and by $\bar{y}_{nm} = \frac{1}{nm} \sum_{i=1}^{n} \sum_{j=1}^{m} y_{ij} = \frac{1}{n} \sum_{i=1}^{n} \bar{y}_{im}$, the sample mean based on all the nm units in the sample.

Clearly, $\bar{y}_{nm}$ is an unbiased estimator of $\bar{Y}$ with its variance given by

$V(\bar{y}_{nm}) = \frac{N-n}{Nn} S_b^2 + \frac{1}{n} \frac{M-m}{Mm} \bar{S}_w^2$

where $S_b^2 = \frac{\sum_{i=1}^{N} (\bar{Y}_{i.} - \bar{Y}_{..})^2}{N-1}$

and $\bar{S}_w^2 = \frac{\sum_{i=1}^{N} S_i^2}{N} = \frac{1}{N(M-1)} \sum_{i=1}^{N} \sum_{j=1}^{M} \left(y_{ij} - \bar{Y}_{i.}\right)^2$.

The estimator of $V(\bar{y}_{nm})$ is given by

$\hat{V}(\bar{y}_{nm}) = \frac{N-n}{Nn} s_b^2 + \frac{1}{n} \frac{M-m}{Mm} \bar{s}_w^2$

where $s_b^2 = \frac{\sum_{i=1}^{n} (\bar{y}_{im} - \bar{y}_{nm})^2}{n-1}$

and $\bar{s}_w^2 = \frac{\sum_{i=1}^{n} s_i^2}{n} = \frac{1}{n(m-1)} \sum_{i=1}^{n} \sum_{j=1}^{m} \left(y_{ij} - \bar{y}_{im}\right)^2$.

It is observed that the variance of sample mean $\bar{y}_{nm}$ in two-stage sampling consists of two components: the first representing the contribution arising from the sampling of first-stage units and the second arising from subsampling within the selected first-stage units. We note the following two cases:

Case (i) $n = N$ corresponds to stratified sampling with N first-stage units as strata and m units drawn from each stratum.

Case (ii) $m = M$ corresponds to cluster sampling.

Theorem 5.5 *The mean of the second-stage units is an unbiased estimator of the population mean.*

Proof We have
$$\bar{y}_{nm} = \frac{1}{nm} \sum_{i=1}^{n} \sum_{j=1}^{m} y_{ij} = \frac{1}{n} \sum_{i=1}^{n} \bar{y}_{im}.$$
At both the stages, samples are selected by using SRSWOR.
Hence
$$E(\bar{y}_{nm}) = E_1 E_2(\bar{y}_{nm}) = E_1 \left[\frac{1}{n} \sum_{i=1}^{n} \bar{Y}_{i.} \right]$$
$$= \frac{1}{N} \sum_{i=1}^{N} \bar{Y}_{i.} = \bar{Y}_{..}$$
Hence the theorem.   □

Theorem 5.6 *If the mean of the SSUs is an unbiased estimator of the population
mean, then*
$$V(\bar{y}_{nm}) = \frac{N-n}{Nn} S_b^2 + \frac{1}{n} \frac{M-m}{Mm} \bar{S}_w^2.$$

Proof By using the result on expectation and variance, we get

$$V(\bar{y}_{nm}) = V_1 \left[E_2(\bar{y}_{nm}) \right] + E_1 \left[V_2(\bar{y}_{nm}) \right]. \tag{5.3}$$

By using the theorem 5.5
$$E_2(\bar{y}_{nm}) = \frac{1}{n} \sum_{i=1}^{n} \bar{Y}_{i.}.$$
Then,
$$V_1 \left[E_2(\bar{y}_{nm}) \right] = V_1 \left[\frac{1}{n} \sum_{i=1}^{n} \bar{Y}_{i.} \right]$$
Here, sample size n is drawn from the population of size N by using SRSWOR.
Hence

$$V_1 \left[E_2(\bar{y}_{nm}) \right] = \frac{N-n}{Nn} S_b^2. \tag{5.4}$$

And in the second-stage m units are drawn from M units by using SRSWOR.

$$V_2(\bar{y}_{nm}) = \frac{M-m}{Mn} \frac{1}{nm} \sum_{i=1}^{n} S_i^2. \tag{5.5}$$

By taking an average on overall FSUs, we get

$$E_1 \left[V_2(\bar{y}_{nm}) \right] = E_1 \left[\frac{M-m}{Mn^2 m} \sum_{i=1}^{n} S_i^2 \right] = \frac{M-m}{M} \frac{\bar{S}_w^2}{mn}. \tag{5.6}$$

Using (5.4) and (5.6) in (5.3), we get
$$V(\bar{y}_{nm}) = \frac{N-n}{Nn} S_b^2 + \frac{1}{n} \frac{M-m}{Mm} \bar{S}_w^2.$$
Hence the theorem.   □

(ii) Two-stage sampling with unequal first-stage units

The situation of two-stage sampling where primary-stage units have varying numbers
of second-stage units is most likely to be encountered in practical situations. Thus,
villages in a district may have varying numbers of households/cultivator fields. Let
the population under consideration consist of N primary sampling units, and ith
primary sampling unit contains M_i secondary sampling units. Further, suppose that

a sample of n FSUs is selected from N FSUs by SRSWOR and from the ith selected FSU, a sample of m_i ssu's is selected from M_i ssu's by SRSWOR.

Let us denote by y_{ij} the value of the characteristic under study for the jth SSUs of the ith FSU, $j = 1, 2, \ldots, M_i; i = 1, 2, \ldots, N$.

Let the population mean of the ith FSU be denoted by $\bar{Y}_{i.} = \frac{1}{M_i} \sum_{j=1}^{M_i} y_{ij}$. The population mean of cluster means is given by $\bar{Y}_N = \frac{1}{N} \sum_{i=1}^{N} \bar{Y}_{i.}$ and the population mean per element by $\bar{Y}_{..} = \frac{\sum_{i=1}^{N} M_i \bar{Y}_{i.}}{\sum_{i=1}^{N} M_i}$. Further, let $\bar{M} = \frac{1}{N} \sum_{i=1}^{N} M_i$ denote the average number of ssu's per FSU and $\bar{y}_{im} = \frac{1}{m_i} \sum_{j=1}^{m_i} y_{ij}$, the sample mean for ith FSU.

An unbiased estimator of the population mean, $\bar{Y}_{..}$, in the present situation is given by
$$\bar{y}_{ts}^{\star} = \frac{1}{n\bar{M}} \sum_{i=1}^{n} \frac{M_i}{m_i} \sum_{j=1}^{m_i} y_{ij}.$$

The above estimator is useful in situations where information on the total number as SSUs within the selected FSUs is available, i.e. M_i is known for $i = 1, 2, \ldots, n$.

The variance of $\bar{y}_{ts}^{\star}$ is given by
$$V(\bar{y}_{ts}^{\star}) = \frac{N-n}{Nn} S_b^{2'} + \frac{1}{nN} \sum_{i=1}^{N} \frac{M_i^2}{\bar{M}^2} \left(\frac{M_i - m_i}{M_i m_i} \right) S_i^2$$
where $S_b^{2'} = \dfrac{\sum_{i=1}^{N} \left(\frac{M_i}{\bar{M}} \bar{Y}_{i.} - \bar{Y}_{..} \right)^2}{N-1}$
and $S_i^2 = \dfrac{\sum_{j=1}^{M_i} (y_{ij} - \bar{Y}_{i.})^2}{M_i - 1}$

The estimator of $V(\bar{y}_{ts}^{\star})$ is given by
$$\hat{V}(\bar{y}_{ts}^{\star}) = \frac{N-n}{Nn} s_b^{2'} + \frac{1}{nN} \sum_{i=1}^{n} \frac{M_i^2}{\bar{M}^2} \left(\frac{M_i - m_i}{M_i m_i} \right) s_i^2$$
where $s_b^{2'} = \dfrac{\sum_{i=1}^{n} (\bar{y}_{im} - \bar{y}_{ts}^{\star})^2}{n-1}$ and $s_i^2 = \dfrac{\sum_{j=1}^{m_i} (y_{ij} - \bar{Y}_{im})^2}{m_i - 1}$.

Another estimator of the population mean which is biased is given by

$$\bar{y}_{ts}^{\star\star} = \frac{1}{n} \sum_{i=1}^{n} \frac{1}{m_i} \sum_{j=1}^{m_i} y_{ij}.$$

It is important to note that the above estimator does not require the availability of information on the total number of ssu's in the selected FSUs. Since the above estimator is biased, relevant criterion for judging the suitability of the estimator is the mean squared error.

The mean squared error of $\bar{y}_{ts}^{\star\star}$ is given by

$$MSE(\bar{y}_{ts}^{\star\star}) = \frac{N-n}{Nn} S_b^2 + \frac{1}{nN} \sum_{i=1}^{N} \left(\frac{M_i - m_i}{M_i m_i} \right) S_i^2$$
where $S_b^2 = \dfrac{\sum_{i=1}^{N} (\bar{Y}_{i.} - \bar{Y}_{..})^2}{N-1}$.

The estimator of $MSE(\bar{y}_{ts}^{\star\star})$ is given by
$$\hat{M}SE(\bar{y}_{ts}^{\star\star}) = \frac{N-n}{Nn} s_b^2 + \frac{1}{nN} \sum_{i=1}^{n} \frac{M_i - m_i}{M_i m_i} s_i^2$$
where $s_b^2 = \dfrac{\sum_{i=1}^{n} (\bar{y}_{i.} - \bar{y}_{ts}^{\star\star})^2}{n-1}$.

When $M_i = M, \forall i$, both the estimators considered above are the same.

Theorem 5.7 *Mean of the SSUs is a biased estimator of the population mean per element when the size of each cluster is unknown. The bias is given by*

$$B(\bar{y}_{ts}^{**}) = -\frac{1}{NM} \sum_{i=1}^{N}(M_i - \bar{M})(\bar{Y}_{i.} - \bar{Y}_{N.})$$

Proof Let $E(\bar{y}_{ts}^{**}) = E_1(E_2(\frac{1}{n}\sum_{i=1}^{n} \bar{y}_{im}))$

$$= E_1(\frac{1}{n}\sum_{i=1}^{n} \bar{Y}_{i.})$$
$$= \frac{1}{n}\sum_{i=1}^{N} \bar{Y}_{i.}\frac{n}{N}$$
$$= \frac{1}{N}\sum_{i=1}^{N} \bar{Y}_{i.}$$
$$= \bar{Y}_{N.} \neq \bar{Y}_{..}$$

Thus the estimator $\bar{y}_{ts}^{**}$ is a biased estimator and its bias can be obtained as

$$Bias(\bar{y}_{ts}^{**}) = E(\bar{y}_{tS}^{**}) - \bar{Y}_{..}$$
$$= \bar{Y}_{N.} - \bar{Y}_{..}$$
$$= \frac{1}{N}\sum_{i=1}^{N} \bar{Y}_{i.} - \frac{1}{NM}\sum_{i=1}^{N} M_i\bar{Y}_{i.}$$
$$= -\frac{1}{NM}\sum_{i=1}^{N}(M_i - \bar{M})(\bar{Y}_{i.} - \bar{Y}_{N.})$$

Hence the theorem.   □

Theorem 5.8 *For two-stage sampling with unequal FSUs and unequal but known cluster sizes, the sample mean at the second stage is an unbiased estimator of the population mean per element.*

Proof The sample mean at the second stage when the cluster sizes M_i's are known is given by

$$\bar{y}_{ts}^{\star} = \frac{1}{n\bar{M}} \sum_{i=1}^{n} \frac{M_i}{m_i} \sum_{j=1}^{m_i} y_{ij}$$
$$= \frac{1}{n\bar{M}} \sum_{i=1}^{n} M_i \bar{y}_{im}.$$

Then

$$E(\bar{y}_{ts}^{\star}) = E_1\left[E_2\left[\frac{1}{n\bar{M}} \sum_{i=1}^{n} M_i \bar{y}_{im}|M_i \right] \right]$$
$$= E_1\left[\frac{1}{n\bar{M}} \sum_{i=1}^{n} M_i \bar{Y}_{i.} \right]$$
$$= \frac{1}{n\bar{M}} \left[\sum_{i=1}^{N} M_i \bar{Y}_{i.}\frac{n}{N} \right]$$
$$= \frac{1}{N\bar{M}} \sum_{i=1}^{N} M_i \bar{Y}_{i.}$$
$$= \frac{\sum_{i=1}^{N} M_i \bar{Y}_{i.}}{\sum_{i=1}^{N} M_i}$$
$$= \bar{Y}_{..}$$

Hence the theorem.   □

Theorem 5.9 *Consider two-stage cluster sampling with unequal FSUs and size of each cluster unknown. The sample mean at the second stage is an unbiased estimator of population mean. Then*

$$V(\bar{y}_{ts}^{\star}) = \frac{N-n}{Nn} S_b'^2 + \frac{1}{Nn} \sum_{i=1}^{N} \frac{M_i^2}{\bar{M}^2}\left(\frac{M_i-m_i}{M_i m_i}\right) S_i^2.$$

Proof By using the result of variance and expectation, we get

$$V(\bar{y}_{ts}^{\star}) = V_1\left[E_2(\bar{y}_{ts}^{\star}|M_i) \right] + E_1\left[V_2(\bar{y}_{ts}^{\star}|Mi) \right]$$
$$= V_1\left[E_2(\frac{1}{n\bar{M}} \sum_{i=1}^{n} M_i \bar{y}_{im}|M_i) \right] + E_1\left[V_2(\frac{1}{n\bar{M}} \sum_{i=1}^{n} M_i \bar{y}_{im}|M_i) \right].$$

By using the theorem 5.8, we get

$E_2(\frac{1}{nM} \sum_{i=1}^{n} M_i \bar{y}_{im}|M_i) = \frac{1}{nM} \sum_{i=1}^{n} M_i \bar{Y}_{i.}$

In the first stage, SRSWOR of size n is drawn from a population of size N.
Hence,

$$V_1 \left[E_2(\frac{1}{nM} \sum_{i=1}^{n} M_i \bar{y}_{im}|M_i) \right] = V_1 \left(\frac{1}{nM} \sum_{i=1}^{n} M_i \bar{Y}_{i.} \right)$$
$$= \frac{N-n}{Nn} S_b'^2$$

where

$$S_b'^2 = \frac{1}{N-1} \sum_{i=1}^{N} \left(\frac{M_i}{M} \bar{Y}_{i.} - \bar{Y}_{..} \right)^2$$

In this case, the sample size is m_i and the population size is M_i.
Hence

$$V_2(\frac{1}{nM} \sum_{i=1}^{n} M_i \bar{y}_{im}|M_i) = \frac{1}{n^2 M^2} \sum_{i=1}^{n} M_i^2 (\frac{M_i - m_i}{M_i m_i}) s_i^2$$

where

$$s_i^2 = \frac{1}{m_i - 1} \sum_{j=1}^{m_i} (y_{ij} - \bar{y}_{im})^2.$$

So

$$E_1 \left[V_2(\frac{1}{nM} \sum_{i=1}^{n} M_i \bar{y}_{im}|M_i) \right] = E_1 \left[\frac{1}{n^2 M^2} \sum_{i=1}^{n} M_i^2 (\frac{M_i - m_i}{M_i m_i}) s_i^2 \right]$$
$$= \frac{1}{n^2 M^2} \sum_{i=1}^{N} M_i^2 (\frac{M_i - m_i}{M_i m_i}) S_i^2 \frac{n}{N}$$
$$= \frac{1}{Nn} \sum_{i=1}^{N} \frac{M_i^2}{M^2} (\frac{M_i - m_i}{M_i m_i}) S_i^2$$

where

$$S_i^2 = \frac{1}{M_i - 1} \sum_{j=1}^{M_i} (y_{ij} - \bar{Y}_{i.})^2.$$

Hence, we get

$$V(\bar{y}_{ts}^*) = \frac{N-n}{Nn} S_b'^2 + \frac{1}{Nn} \sum_{i=1}^{N} \frac{M_i^2}{M^2} (\frac{M_i - m_i}{M_i m_i}) S_i^2.$$

Hence the theorem.   □

5.4 Exercises

1. Prove that an estimator of relative efficiency from the sample information for a large number of clusters is given by $Est.(R.E.) = \frac{S_b^2 + (1 - M^{-1})S_w^2}{M S_b^2}$.
2. Define two-stage samplings with equal first-stage units. Prove that the mean per second-stage unit in the sample is an unbiased estimate of population mean and obtain its variance.
3. Define the term Multistage Sampling.
4. Prove that in cluster sampling the relative efficiency with respect to SRSWOR in terms of intraclass correlation coefficient ρ is
   $RE = [1 + (M - 1)\rho]^{-1}$.
   Thus, cluster sampling is more efficient than SRS if $\rho < 0$. Obtain the optimum subsample size for two-stage sampling with equal cluster size.
5. Describe situations where two-stage sampling is appropriate. Describe similarities between stratified sampling, cluster sampling, and two-stage sampling.

6. Define two-stage sampling with unequal first-stage units. Obtain bias of the population mean for two-stage sampling with unequal first-stage units. and obtain its variance.

7. Prove that in cluster sampling the mean squares between (MS_b^2) and within clusters is an unbiased estimate of the mean square between elements in the population. Write a short note on the effect of cluster size on the relative efficiency.

8. Obtain the optimum cluster size of cluster sampling with fixed cost and fixed variance.

9. Write a short note on the Estimation of proportion for cluster sampling.

10. Prove that the estimate of variance of two-stage sampling is
$$\hat{V}(\bar{\bar{y}}_{nm}) = \frac{(N-n)S_b^2}{Nn} + \frac{(M-m)\bar{S}_w^2}{Mmn}.$$

11. Obtain the estimates of mean and variance of cluster sampling with unequal size clusters.

12. Prove that the mean per second-stage unit in the sample is an unbiased estimate of the population mean for two-stage sampling with equal first-stage units. Also, obtain the variance of the estimate of mean.

13. Define the following terms:
(i) First-stage unit (ii) Second-stage unit (iii) Cluster sampling (iv) Two-stage cluster sampling.

14. Write a short note on the effect of cluster size on the relative efficiency.

15. Prove that the simple arithmetic mean estimator for two-stage sampling with unequal first stage is a biased estimator of population mean. Obtain the bias and variance of the estimator you have proposed.

16. The water samples are collected from different locations of the Pune district and they are clustered accordingly. The aim of the study is to measure the quality of water as shown in Table 5.1. Draw a random sample of 3 clusters and obtain the average pH of the water and obtain its variance.

17. The water samples are collected from different locations of the Pune district and they are clustered accordingly. The aim of the study is to measure the quality of water as shown in Table 5.1. Draw a random sample of 3 clusters and obtain the average chloride of the water and obtain its variance.

18. The water samples are collected from different locations of the Pune district and they are clustered accordingly. The aim of the study is to measure the quality of water as shown in Table 5.1. Draw a random sample of 3 clusters and obtain the average sulfate of the water and obtain its variance.

19. The water samples are collected from different locations of the Pune district and they are clustered accordingly. The aim of the study is to measure the quality of water as shown in Table 5.1. Draw a random sample of 3 clusters and obtain the average phosphorus of the water and obtain its variance.

20. The water samples are collected from different locations of the Pune district and they are clustered accordingly. The aim of the study is to measure the quality

Table 5.1 Water quality at Pune district of Maharashtra

Sr. No.	Cluster	Sites	PH	Conductivity (ppm)	Chloride	Sulfate	Phosphorous	Nitrate
1	Baramati	1	7.4	730	332	157	0.167	15
		2	8.21	280	113	27	0.11	7
		3	7.48	460	119	838	0.043	30
		4	7.62	450	142	71	0.167	15
		5	7.97	400	151	192	0.05	34
		6	7.96	480	156	192	0.054	29
2	Daund	7	7.64	700	346	209	0.087	23
		8	7.83	680	326	196	0.07	34
		9	7.41	530	241	205	0.043	32
		10	7.26	910	381	265	0.07	38
		11	7.31	500	176	138	0.204	38
		12	7.42	370	131	94	0.17	40
3	Purandar	13	7.39	70	11	23	0.05	1
		14	7.31	550	176	126	0.157	19
		15	7.87	180	34	21	0.221	31
		16	8.24	150	28	15	0.094	3
		17	7.41	440	88	67	0.074	23
		18	7.84	440	85.2	60	0.127	18
4	Shirur	19	7.89	220	26	39	0.231	14
		20	7.72	420	159	138	0.334	30
		21	7.91	320	159	75	0.335	8
		22	7.44	460	173	166	0.388	36
		23	7.77	350	99	100	0.301	17
		24	7.78	830	488	182	0.04	26
5	Khed	25	8	390	62	49	0.033	34
		26	7.78	350	133	58	0.027	0
		27	8.06	400	156	67	0.047	35
		28	7.88	480	295	93	0.13	2
		29	7.42	780	329	178	0.16	4
		30	7.98	430	244	93	0.177	17
6	Indapur	31	7.17	1040	417	246	0.194	4
		32	7.73	510	190	210	0.181	40
		33	7.9	880	482	170	0.104	15
		34	7.31	620	301	112	0.14	10
		35	7.64	470	199	211	0.117	20
		36	7.5	580	227	185	0.161	4

of water as shown in Table 5.1. Draw a random sample of 3 clusters and obtain the average nitrate of the water and obtain its variance.

21. The water samples are collected from different locations of the Pune district and they are clustered accordingly. The aim of the study is to measure the quality of water as shown in Table 5.1. Draw a random sample of 3 clusters, again draw a random sample of size 3 from each of the selected clusters and obtain the average conductivity of the water and obtain its variance.

22. The water samples are collected from different locations of the Pune district and they are clustered accordingly. The aim of the study is to measure the quality of water as shown in Table 5.1. Draw a random sample of 3 clusters, again draw a random sample of size 3 from each of the selected clusters and obtain the average pH of the water and obtain its variance.

23. The water samples are collected from different locations of the Pune district and they are clustered accordingly. The aim of the study is to measure the quality of water as shown in Table 5.1. Draw a random sample of 3 clusters, again draw a random sample of size 3 from each of the selected clusters and obtain the average chloride of the water and obtain its variance.

24. The water samples are collected from different locations of the Pune district and they are clustered accordingly. The aim of the study is to measure the quality of water as shown in Table 5.1. Draw a random sample of 3 clusters, again draw a random sample of size 3 from each of the selected clusters and obtain the average sulfate of the water and obtain its variance.

25. The water samples are collected from different locations of the Pune district and they are clustered accordingly. The aim of the study is to measure the quality of water as shown in Table 5.1. Draw a random sample of 3 clusters, again draw a random sample of size 3 from each of the selected clusters and obtain the average phophorus of the water and obtain its variance.

26. The water samples are collected from different locations of the Pune district and they are clustered accordingly. The aim of the study is to measure the quality of water as shown in Table 5.1. Draw a random sample of 3 clusters, again draw a random sample of size 3 from each of the selected clusters and obtain the average nitrate of the water and obtain its variance.

References

Cochran, W.G.: The use of the analysis of variance in enumeration by sampling. J. Am. Stat. Assoc. **34**, 492–510 (1939)

Hansen, M.H., Hurwitz, W.N.: On the theory of sampling from finite populations. Ann. Math. Stat. **14**, 333–362 (1943)

Lahiri, D.B.: Technical paper on some aspects of the development of sample design. Sankhya **14**, 264–316 (1954)

Chapter 6
Double Sampling

6.1 Introduction

Many a time in the survey, additional information is available on another variable, related to the study variable. This variable is known as an auxiliary variable. Generally, the the auxiliary variable is highly correlated with the study variable. It is also assumed that the information on the auxiliary variable is not available for all the population units. The auxiliary information is collected with a relatively large sample by using SRSWR or SRSWOR from the population to get the population estimate and information of the study variable, and the auxiliary information is gathered from the selected units in the first phase or population with a smaller sample size compared to the first phase by using SRSWR or SRSWOR. Such a sampling procedure is called double sampling or two-phase sampling. If a sample is selected in more than two phases, it is called multiphase sampling. For example, the volume of a truck container and the weight of the truck container are correlated, and we want to estimate the volume of the truck container. By using SRSWR or SRSWOR, the large sample of the truck is selected and the weight of the truck container is measured, and from the selected truck container again by employing SRSWR or SRSWOR, the small sample of the truck container is selected and the volume of the truck container is measured. In this sampling scheme, the auxiliary information is utilized to improve the precision of the estimator, and the cost is reduced for the study variable by selecting a small sample.

6.2 Ratio Estimator in Double Sampling

We have defined the ratio estimator in Chap. 2, and the population mean of auxiliary variable X is unknown. We replace the population mean $\bar{X}$ by $\bar{x}^*$ from the first phase units of size m which is relatively larger than the sample size of the second phase,

R. Latpate et al., *Advanced Sampling Methods*,
https://doi.org/10.1007/978-981-16-0622-9_6

i.e. n. The following notations are used for ratio and regression estimators for double sampling.

Let

$\bar{y} = \frac{1}{n} \sum_{i=1}^{n} y_i$ be the sample mean of the study variable of the second phase units;

$\bar{x} = \frac{1}{n} \sum_{i=1}^{n} x_i$ be the sample mean of the auxiliary variable of the second phase units;

$\bar{x}^* = \frac{1}{m} \sum_{i=1}^{m} x_i^*$ be the sample mean of the auxiliary variable of the first phase units;

$s_y^2 = \frac{1}{n-1} \sum_{i=1}^{n} (y_i - \bar{y})^2$ be the sample mean square of the study variable of the second phase units;

$s_x^2 = \frac{1}{n-1} \sum_{i=1}^{n} (x_i - \bar{x})^2$ be the sample mean square of the auxiliary variable of the second phase units;

$s_x^{*2} = \frac{1}{m-1} \sum_{i=1}^{m} (x_i^* - \bar{x}^*)^2$ be the sample mean square of the auxiliary variable of the first phase units.

Estimates can be expressed in terms of errors as

$\bar{y} = \bar{Y}(1 + \epsilon_0)$

$\bar{x} = \bar{x}^*(1 + \epsilon_1)$

$\bar{x} = \bar{X}(1 + \epsilon_2)$

$\bar{x}^* = \bar{X}(1 + \epsilon_3)$

$s_y^2 = S_y^2(1 + \delta_0)$

$s_x^2 = S_x^{*2}(1 + \delta_1)$

where

$E(\epsilon_0) = E(\epsilon_1) = E(\epsilon_2) = E(\epsilon_3) = E(\delta_0) = E(\delta_1) = 0$

$E(\epsilon_0^2) = (\frac{1}{n} - \frac{1}{N})C_y^2$

$E(\epsilon_1^2) = (\frac{1}{n} - \frac{1}{m})C_x^2$

$E(\epsilon_2^2) = (\frac{1}{n} - \frac{1}{N})C_x^2$

$E(\epsilon_3^2) = (\frac{1}{m} - \frac{1}{N})C_x^2$

$E(\epsilon_0\epsilon_1) = (\frac{1}{n} - \frac{1}{m})\rho_{xy}C_yC_x$

$E(\epsilon_0\epsilon_2) = (\frac{1}{n} - \frac{1}{N})\rho_{xy}C_yC_x$

$E(\epsilon_0\epsilon_3) = (\frac{1}{m} - \frac{1}{N})\rho_{xy}C_yC_x$

$E(\epsilon_2\epsilon_3) = (\frac{1}{m} - \frac{1}{N})C_x^2.$

The resultant ratio estimator can be defined as follows:

$$\bar{y}_{RD} = \bar{y}\left(\frac{\bar{x}^*}{\bar{x}}\right). \tag{6.1}$$

The ratio estimator in terms of errors can be expressed as follows:

$\bar{y}_{RD} = \bar{Y}(1 + \epsilon_0)(1 + \epsilon_1)^{-1}.$

Assuming that $|\epsilon_i| < 1$, $i = 0, 1$ and by applying binomial expansion, we get

$$\hat{\bar{y}}_{RD} = \bar{Y}\,(1 + \epsilon_0)\,(1 - \epsilon_1 + \epsilon_1^2 - \cdots)$$

$$\bar{y}_{RD} = \bar{Y}\,(1 + \epsilon_0 - \epsilon_1 + \epsilon_1^2 - \epsilon_0\epsilon_1 + \cdots).\qquad(6.2)$$

Theorem 6.1 *In double sampling, the bias in ratio estimator $\bar{y}_{RD}$ to first-order approximation is given by*
$$B(\bar{y}_{RD}) = (\tfrac{1}{n} - \tfrac{1}{m})\bar{Y}(C_x^2 - \rho_{xy}C_yC_x).$$

Proof The ratio estimator is given by
$$\bar{y}_{RD} = \bar{y}\,(\tfrac{\bar{x}^*}{\bar{x}})$$
$$= \bar{Y}\,(1 + \epsilon_0 - \epsilon_1 + \epsilon_1^2 - \epsilon_0\epsilon_1 + \cdots).$$
Taking expectation on both sides, we get
$$E(\bar{y}_{RD}) = \bar{Y}\,[1 + E(\epsilon_0) - E(\epsilon_1) + E(\epsilon_1^2) - E(\epsilon_0\epsilon_1) + \cdots]$$
$$= \bar{Y}\,[1 + 0 - 0 + (\tfrac{1}{n} - \tfrac{1}{m})C_x^2 - (\tfrac{1}{n} - \tfrac{1}{m})\rho_{xy}C_yC_x]$$
$$= \bar{Y} + \bar{Y}(\tfrac{1}{n} - \tfrac{1}{m})(C_x^2 - \rho_{xy}C_yC_x).$$
The bias in ratio estimator is given by
$$B(\bar{y}_{RD}) = E(\bar{y}_{RD}) - \bar{Y}$$
$$= \bar{Y} + \bar{Y}(\tfrac{1}{n} - \tfrac{1}{m})(C_x^2 - \rho_{xy}C_yC_x) - \bar{Y}$$
$$= \bar{Y}(\tfrac{1}{n} - \tfrac{1}{m})(C_x^2 - \rho_{xy}C_yC_x).$$
Hence the theorem.   □

Theorem 6.2 *In double sampling, the mean squared error of the ratio estimator $\bar{y}_{RD}$ to the first order of approximation is given by*
$$MSE(\bar{y}_{RD}) = (\tfrac{1}{m} - \tfrac{1}{N})S_y^2 + (\tfrac{1}{n} - \tfrac{1}{m})[S_y^2 + R^2\,S_x^2 - 2\,R\,S_{xy}]$$
where $R = \tfrac{\bar{Y}}{\bar{X}}$ is the ratio of population means.

Proof By using the definition of mean squared error, we have
$$MSE(\bar{y}_{RD}) = E(\bar{y}_{RD} - \bar{Y})^2$$
$$= E[\bar{Y}(1 + \epsilon_0 - \epsilon_1 + \cdots) - \bar{Y}]^2$$
$$= \bar{Y}^2 E[\epsilon_0^2 + \epsilon_1^2 - 2\epsilon_0\epsilon_1]$$
$$= \bar{Y}^2[(\tfrac{1}{n} - \tfrac{1}{N})C_y^2 + (\tfrac{1}{n} - \tfrac{1}{m})(C_y^2 - 2\,\rho_{xy}\,C_xC_y)]$$
$$= (\tfrac{1}{m} - \tfrac{1}{N})S_y^2 + (\tfrac{1}{n} - \tfrac{1}{m})[S_y^2 + R^2\,S_x^2 - 2\,R\,S_{xy}].$$
Hence the theorem.   □

Corollary 6.1 *In double sampling, an estimator of the mean squared error of the ratio estimator $\bar{y}_{RD}$ to the first order of approximation is given by*
$$\hat{MSE}(\bar{y}_{RD}) = (\tfrac{1}{m} - \tfrac{1}{N})s_y^2 + (\tfrac{1}{n} - \tfrac{1}{m})[s_y^2 + r^2\,s_x^2 - 2\,r\,s_{xy}]$$
where $r = \tfrac{\bar{y}}{\bar{x}}$ is the ratio of sample means.

Example 6.1 Select a first phase sample of 20 units by SRSWOR sampling and note only the age of a person in a block from the selected units in the sample sample from Table 6.1. From the selected first phase sample of 20 units, select a subsample of 10 units and note the age of a person as well as sleeping hours. Estimate the average sleeping hours (yrs) in a block by using a ratio estimator in two-phase sampling. Deduce the 95% confidence interval.

Use the following R code to illustrate the procedure:

```
x<-scan("clipboard")
y<-scan("clipboard")
set.seed(3)
s1<-sample(1:50,20,replace=F)
X=x[s1];
mX=mean(X);
set.seed(3);
S1<-sample(s1,10,replace=F)
dx=x[S1];   dy=y[S1];
mdx=mean(dx); mdy=mean(dy);
vdx=var(dx);   vdy=var(dy);
cdxdy=cov(dx,dy);   r=mdy/mdx;  r;
ybrd=(mdy/mdx)*mX;  ybrd;
Mse=((1/20)-(1/50))*vdy+((1/10)-(1/20))*(vdy+r^2*vdx-2*r*cdxdy);  Mse;
t=qt(0.975,9);   t;
LL=ybrd-t*sqrt(Mse)
UL=ybrd+t*sqrt(Mse)
LL;UL;
```

The average sleeping hours in a year are 395.27 with mean squared error 369.89, and a 95% confidence interval is [351.76, 438.78].

6.3   Regression Estimator in Double Sampling

The simple linear regression estimate is obtained as
$$\bar{y}_{LD} = \bar{y} + \hat{\beta}(\bar{x}^* - \bar{x}).$$
$\hat{\beta}$ is the estimator of regression coefficient β of Y on X. Its mean squared error can be obtained by using Theorem 6.3.

Theorem 6.3 *In double sampling, the mean squared error of the linear regression estimator of population mean $\bar{Y}$ to the first order of approximation is given by*
$$MSE(\bar{y}_{LD}) = (\tfrac{1}{m} - \tfrac{1}{N})S_y^2 + (\tfrac{1}{n} - \tfrac{1}{m})S_y^2(1 - \rho_{xy}^2)$$
where ρ_{xy} is the correlation coefficient of X and Y.

Proof Define $\zeta = \frac{\hat{\beta}}{\beta} - 1$ and $E(\zeta) = 0$. Then the regression estimator can be expressed as
$$\bar{y}_{LD} = \bar{Y}(1 + \epsilon_0) + \beta\bar{X}(\epsilon_3 - \epsilon_2)(1 + \zeta).$$
Thus, the mean squared error of the estimator $\bar{y}_{LD}$ is given by
$$
\begin{aligned}
MSE(\bar{y}_{LD}) &= E[\bar{Y}(1 + \epsilon_0) + \beta\bar{X}(\epsilon_3 - \epsilon_2)(1 + \zeta) - \bar{Y}]^2 \\
&= E[\bar{Y}^2\epsilon_0^2 + \beta^2\bar{X}^2(\epsilon_3^2 + \epsilon_2^2 - 2\epsilon_3\epsilon_2) + 2\beta\bar{Y}\bar{X}(\epsilon_0\epsilon_3 - \epsilon_0\epsilon_2)] \\
&= (\tfrac{1}{n} - \tfrac{1}{N})\bar{Y}^2C_y^2 + \beta^2\bar{X}^2\{(\tfrac{1}{m} - \tfrac{1}{N})C_x^2 + (\tfrac{1}{n} - \tfrac{1}{N})C_x^2 - 2(\tfrac{1}{m} - \tfrac{1}{N})C_x^2\} + \\
& \quad 2\beta\bar{Y}\bar{X}\{(\tfrac{1}{m} - \tfrac{1}{N})\rho_{xy}C_xC_y - (\tfrac{1}{n} - \tfrac{1}{N})\rho_{xy}C_xC_y\} \\
&= (\tfrac{1}{n} - \tfrac{1}{N})S_y^2 + (\tfrac{1}{n} - \tfrac{1}{m})\{\beta^2\bar{X}^2C_x^2 - 2\beta\bar{Y}\bar{X}\rho_{xy}C_yC_x\} \\
&= (\tfrac{1}{n} - \tfrac{1}{m} + \tfrac{1}{m} - \tfrac{1}{N})S_y^2 + (\tfrac{1}{n} - \tfrac{1}{m})\{\beta^2S_x^2 - 2\beta S_{xy}\} \\
&= (\tfrac{1}{m} - \tfrac{1}{N})S_y^2 + (\tfrac{1}{n} - \tfrac{1}{m})\{S_y^2 + \beta^2S_x^2 - 2\beta S_{xy}\}
\end{aligned}
$$

$$= (\tfrac{1}{m} - \tfrac{1}{N})S_y^2 + (\tfrac{1}{n} - \tfrac{1}{m})\{S_y^2 + (\tfrac{S_{xy}}{S_x^2})^2 S_x^2 - 2(\tfrac{S_{xy}}{S_x^2})S_{xy}\}$$
$$= (\tfrac{1}{m} - \tfrac{1}{N})S_y^2 + (\tfrac{1}{n} - \tfrac{1}{m})S_y^2(1 - \rho_{xy}^2).$$

Hence the theorem. □

Corollary 6.2 *In double sampling, an estimator of the mean squared error of the regression estimator $\bar{y}_{LD}$ to the first order of approximation is given by*
$$\hat{MSE}(\bar{y}_{LD}) = (\tfrac{1}{m} - \tfrac{1}{N})s_y^2 + (\tfrac{1}{n} - \tfrac{1}{m})s_y^2(1 - r_{xy}^2)$$
where r_{xy} is the sample correlation coefficient between x and y.

Example 6.2 Select a first phase sample of 20 units by SRSWOR sampling and note only the age of a person in a block from the selected units in the sample (use Table 6.1). From the selected first phase sample of 20 units, select a subsample of 10 units and note the age of a person as well as sleeping hours. Estimate the average sleeping hours (yrs) in a block by using a regression estimator in two-phase sampling. Deduce the 95% confidence interval.

Use the following R code to illustrate the procedure:

```
x<-scan("clipboard")
y<-scan("clipboard")
set.seed(3)
s1<-sample(1:50,20,replace=F)
X=x[s1];
mX=mean(X);
set.seed(3);
S1<-sample(s1,10,replace=F)
dx=x[S1]; dy=y[S1];
mdx=mean(dx); mdy=mean(dy);
vdx=var(dx); vdy=var(dy);
cdxdy=cov(dx,dy);
beta=cdxdy/vdx; beta;
ybld=mdy+beta*(mX-mdx); ybld;
cor=cdxdy/sqrt(vdx*vdy); cor;
Mse=((1/20)-(1/50))*vdy+((1/10)-(1/20))*vdy*(1-cor^2); Mse;
```

Table 6.1 The total number of sleeping hours corresponding to different age of the patients

Sr. No.	Age	Number of hours of sleep
1	45.84	552.28
2	73.82	367.02
3	64.49	395.99
4	58.46	477.71
5	73.62	306.01
6	58.91	435.09
7	75.70	336.97
8	59.15	359.99
9	78.20	379.35
10	62.89	403.77

(continued)

Table 6.1 (continued)

Sr. No.	Age	Number of hours of sleep
11	65.65	366.45
12	64.94	401.95
13	71.50	403.39
14	62.92	412.02
15	75.96	309.47
16	74.63	394.52
17	79.12	352.69
18	65.80	347.07
19	63.25	388.45
20	70.06	347.54
21	67.10	383.84
22	62.63	400.17
23	72.67	369.27
24	70.85	361.34
25	84.86	279.81
26	64.06	421.00
27	70.33	332.56
28	62.61	397.37
29	76.42	327.21
30	57.91	425.10
31	68.26	436.99
32	58.82	488.72
33	80.57	280.05
34	77.63	276.77
35	63.06	381.98
36	71.14	326.33
37	68.55	383.53
38	84.53	240.07
39	51.64	434.50
40	76.62	343.46
41	64.89	419.98
42	76.66	310.65
43	63.50	419.73
44	56.01	454.06
45	70.84	354.54
46	58.23	395.14
47	72.68	314.81
48	60.06	425.12
49	68.19	367.02
50	58.66	412.70

```
t=qt(0.975,9); t;
LL=ybrd-t*sqrt(Mse)
UL=ybrd+t*sqrt(Mse)
LL;UL;
```
The average sleeping hours in a year are 380.16 with a mean squared error 70.05, and a 95% confidence interval is [376.34, 414.21].

6.4 Exercises

1. Define the term Double sampling and give an illustration.
2. Define ratio estimator in double sampling. Obtain the minimum cost of a ratio estimator in double sampling.
3. Define regression estimator in double sampling. Obtain the mean squared error of the ratio estimator in double sampling.
4. Define the term Multiphase sampling.
5. What is double sampling? Propose an estimator of the population mean (using the ratio method of estimation) in this case. Obtain the expected value and mean squared error of the estimator you have proposed.

Table 6.2 Income and expenditure of Tribals in Western Ghats of Maharashtra

Sr. No.	Income	Expenditure	Sr. No.	Income	Expenditure
1	24000	4700	21	60000	90000
2	25000	4000	22	19500	56400
3	31000	8500	23	11400	22700
4	21000	5000	24	20000	11050
5	27000	7000	25	59000	21400
6	35000	5500	26	25000	52500
7	14000	4000	27	55000	71100
8	41000	31400	28	60000	3800
9	19000	7400	29	10000	7400
10	40800	13400	30	55000	19000
11	70000	17500	31	30000	4000
12	14000	10000	32	30000	4000
13	14000	5500	33	25000	2000
14	13000	4300	34	18000	10500
15	10000	3000	35	15000	2000
16	154520	21500	36	30000	3000
17	88000	12700	37	100000	3000
18	203000	21200	38	6000	2000
19	372000	25600	39	20000	10000
20	21400	19000	40	20000	5000

6. Define ratio estimator in double sampling. Obtain the mean squared error of a ratio estimator in double sampling.
7. Find the asymptotic mean squared error of the linear regression estimator of a population mean in two-phase sampling.
8. Select a first phase sample of 20 units by SRSWOR from Table 6.2 and note only the income of tribals in Western Ghats of Maharashtra from the selected units in the sample. From the selected first phase sample of 20 units, select a subsample of 10 units and note the income as well as expenditure (in Rs.). Estimate the average expenditure by using ratio and regression estimators in two-phase sampling. Deduce their 95% confidence intervals.

Chapter 7
Probability Proportional to Size Sampling

7.1 Introduction

In sampling from a finite population, often the value of some auxiliary variable, closely related to the main characteristic of interest, is available for all the units of the population. This normalized value of the auxiliary variable may be taken as a measure of the size of a unit. For example, in agricultural surveys, the area under a crop may be taken as a size measure of farms for estimating the yield of crops. In such situations, sampling the units with probability proportional to size (PPS) measure with replacement or without replacement may be used in place of simple random sampling with replacement or simple random sampling without replacement (SRSWR or SRSWOR). Since the units with larger sizes are expected to have a bigger total of Y, it is expected that PPS measure sampling procedures will be more efficient than SRSWOR or SRSWR. Note that sampling units have an unequal probability of selection. In the sequel, we discuss some common unequal probability sampling procedures.

In this sampling method, different units in the population have different probabilities of being included in the sample. In this case, better estimates are obtained by including more important units with higher probability. Note that the probability of inclusion of units in the sample is not equal and depends on the size of units, and is called probability proportional to size (PPS).

7.2 Probability Proportional to Size Sampling with Replacement

In probability proportional to size with replacement (PPSWR) sampling, the units are selected with probabilities proportional to some measure of their size and with

© The Author(s), under exclusive license to Springer Nature Singapore Pte Ltd. 2021
R. Latpate et al., *Advanced Sampling Methods*,
https://doi.org/10.1007/978-981-16-0622-9_7

replacement of units. Let $P_i = \frac{X_i}{\sum\limits_{i=1}^{N} X_i}$ denote the normalized size measure of ith$(i = 1, 2, \ldots, N)$ unit, where X_i denotes the size measure of the ith unit. In PPSWR, ith unit is selected with probability P_i. Note that $\sum\limits_{i=1}^{N} P_i = 1$. The first- and second-order inclusion probabilities are given by $\pi_i = 1 - (1 - P_i)^n$; $i = 1, 2, \ldots, N$ and $\pi_{ij} = 1 - (1 - P_i)^n - (1 - P_j)^n + (1 - P_i - P_j)^n$; $i \neq j = 1, 2, \ldots, N$.

A PPSWR sample may be selected in two ways namely cumulative total method and Lahiri's method. In the cumulative total method, a cumulative total of sizes of the units is made. Let $T_i = \sum\limits_{j=1}^{i} X_j$. A random number between 1 and $T_N = \sum\limits_{j=1}^{N} X_j$ is drawn. Let the number drawn be r. If $T_{i-1} \leq r \leq T_i$, then the unit i is selected. This method is repeated until a sample of size n is selected. It may be seen that the method is cumbersome when the population size is large. Lahiri's method is used when the number of units is very large. The method involves drawing two random numbers i and r: the first random number i between 1 and N and the second random number r between 1 and M, where $M = max(X_i)$; $i \in U$. Select the unit i if $r \leq X_i$. Otherwise reject the pair of random numbers i and r and draw a fresh pair of random numbers. Repeat the trial till a unit is selected. This whole process is repeated until a sample of size n is selected.

Hansen and Hurwitz Estimator

Consider a population of N units is $U_1, U_2, \ldots, U_N$ and let y_i, $(i = 1, 2, \ldots, N)$ be the characteristic value of the study variable for ith population units. Here, the units are drawn by using PPS with probability P_i, where P_i is the probability of selecting the ith units of population at any draw.

The Hansen and Hurwitz estimator for population mean is defined as

$$\bar{y}_{HH} = \frac{1}{n} \sum_{i=1}^{n} \frac{y_i}{NP_i}.$$

Theorem 7.1 *The Hansen and Hurwitz estimate $\bar{y}_{HH}$ of mean under PPSWR is an unbiased estimator of the population mean.*

Proof The estimate of mean under PPSWR is given by

$$\bar{y}_{HH} = \frac{1}{n} \sum_{i=1}^{n} \frac{y_i}{NP_i}.$$

Then, $E(\bar{y}_{HH}) = \frac{1}{n} \sum\limits_{i=1}^{n} E\left(\frac{y_i}{NP_i}\right)$

$$= \frac{1}{n} n \sum_{i=1}^{N} P_i \left(\frac{y_i}{NP_i}\right)$$

$$= \sum_{i=1}^{N} \frac{y_i}{N} = \bar{Y}.$$

Hence the theorem.   □

Theorem 7.2 *The variance of the estimate of mean $\bar{y}_{HH}$ under PPSWR is given by*
$$V(\bar{y}_{HH}) = \frac{1}{n} \sum_{i=1}^{N} P_i \left(\frac{y_i}{NP_i} - \bar{Y} \right)^2.$$

Proof The estimate under PPSWR is given by
$$\bar{y}_{HH} = \frac{1}{n} \sum_{i=1}^{n} \frac{y_i}{NP_i}$$
and,
$$\begin{aligned}
V(\bar{y}_{HH}) &= E(\bar{y}_{HH}^2) - \left(E(\bar{y}_{HH}) \right)^2 \\
&= E\left(\left(\frac{1}{n} \sum_{i=1}^{n} \frac{y_i}{NP_i} \right)^2 \right) - \bar{Y}^2 \\
&= \frac{1}{n^2} E\left[\sum_{i=1}^{n} \left(\frac{y_i}{NP_i} \right)^2 + \sum_{i \neq j=1}^{n} \left(\frac{y_i}{NP_i} \right)\left(\frac{y_j}{NP_j} \right) \right] - \bar{Y}^2 \\
&= \frac{1}{n^2} \left[\sum_{i=1}^{n} E\left(\frac{y_i}{NP_i} \right)^2 + \sum_{i \neq j=1}^{n} E\left(\left(\frac{y_i}{NP_i} \right)\left(\frac{y_j}{NP_j} \right) \right) \right] - \bar{Y}^2.
\end{aligned}$$

Now,
$$E\left(\frac{y_i}{NP_i} \right)^2 = \sum_{i=1}^{N} P_i \left(\frac{y_i}{NP_i} \right)^2$$
and,
$$E\left(\left(\frac{y_i}{NP_i} \right)\left(\frac{y_j}{NP_j} \right) \right) = E\left(\frac{y_i}{NP_i} \right) E\left(\frac{y_j}{NP_j} \right).$$
Since draws are made with replacement, hence,
$$\begin{aligned}
E\left(\left(\frac{y_i}{NP_i} \right)\left(\frac{y_j}{NP_j} \right) \right) &= \sum_{i=1}^{N} P_i \left(\frac{y_i}{NP_i} \right) \cdot \sum_{j=1}^{N} P_j \left(\frac{y_j}{NP_j} \right) \\
&= \sum_{i=1}^{N} \left(\frac{y_i}{N} \right) \cdot \sum_{j=1}^{N} \left(\frac{y_j}{N} \right) \\
&= \bar{Y} \cdot \bar{Y} = \bar{Y}^2.
\end{aligned}$$

Then,
$$\begin{aligned}
V(\bar{y}_{HH}) &= \frac{1}{n^2} \left[n \sum_{i=1}^{N} P_i \left(\frac{y_i}{NP_i} \right)^2 + n(n-1)\bar{Y}^2 \right] - \bar{Y}^2 \\
&= \frac{n}{n^2} \left[\sum_{i=1}^{N} P_i \left(\frac{y_i}{NP_i} \right)^2 + (n-1)\bar{Y}^2 \right] - \bar{Y}^2 \\
&= \frac{1}{n} \left[\sum_{i=1}^{N} P_i \left(\frac{y_i}{NP_i} \right)^2 + (n-1)\bar{Y}^2 - n\bar{Y}^2 \right] \\
&= \frac{1}{n} \left[\sum_{i=1}^{N} P_i \left(\frac{y_i}{NP_i} \right)^2 - \bar{Y}^2 \right] \\
&= \frac{1}{n} \sum_{i=1}^{N} P_i \left(\frac{y_i}{NP_i} - \bar{Y} \right)^2.
\end{aligned}$$
Hence the theorem. $\qquad\qquad\square$

Theorem 7.3 *An estimator of variance of PPSWR* $\hat{V}(\bar{y}_{HH}) = \frac{1}{n(n-1)} \sum_{i=1}^{n} \left(\frac{y_i}{NP_i} - \bar{y}_{HH} \right)^2$ *is unbiased for* $V(\bar{y}_{HH})$.

Proof The estimate of variance $\hat{V}(\bar{y}_{HH})$ under PPSWR is given by

$$\hat{V}(\bar{y}_{HH}) = \frac{1}{n(n-1)} \sum_{i=1}^{n} \left(\frac{y_i}{NP_i} - \bar{y}_{HH}\right)^2.$$

Expanding and taking expectation on both sides, we get

$$E(\hat{V}(\bar{y}_{HH})) = \frac{1}{n(n-1)} E\left(\sum_{i=1}^{n}(\frac{y_i}{NP_i})^2 - n\bar{y}_{HH}^2\right)$$

$$= \frac{1}{n(n-1)}\left(\sum_{i=1}^{n} E(\frac{y_i}{NP_i})^2 - nE(\bar{y}_{HH}^2)\right).$$

By definition of variance,

$$V(\bar{y}_{HH}) = E(\bar{y}_{HH}^2) - (E(\bar{y}_{HH}))^2.$$

We get

$$E(\bar{y}_{HH}^2) = V(\bar{y}_{HH}) + \bar{Y}^2.$$

And,

$$E(\frac{y_i}{NP_i})^2 = \sum_{i=1}^{N} P_i(\frac{y_i}{NP_i})^2.$$

Then by using these results, we have

$$E(\hat{V}(\bar{y}_{HH})) = \frac{1}{n(n-1)}\left(n \sum_{i=1}^{N} P_i(\frac{y_i}{NP_i})^2 - n(V(\bar{y}_{HH}) + \bar{Y}^2)\right)$$

$$= \frac{n}{n(n-1)}\left(\sum_{i=1}^{N} P_i(\frac{y_i}{NP_i})^2 - \bar{Y}^2 - V(\bar{y}_{HH})\right)$$

$$= \frac{1}{(n-1)}\left(\sum_{i=1}^{N} P_i(\frac{y_i}{NP_i} - \bar{Y})^2 - V(\bar{y}_{HH})\right)$$

$$= \frac{1}{(n-1)}\left(nV(\bar{y}_{HH}) - V(\bar{y}_{HH})\right)$$

$$= \frac{(n-1)}{(n-1)} V(\bar{y}_{HH})$$

$$= V(\bar{y}_{HH}).$$

Hence the theorem.   □

Example 7.1 Estimate the total number of computer help requests for the last year in a large firm as shown in Table 7.1. The director of the computer support department plans to sample 3 divisions of a large firm that has 10 divisions, with varying numbers of employees per division. Since the number of computer support requests within each division should be highly correlated with the number of employees in that division, the director decides to use unequal probability sampling with replacement proportional to the number of employees in that division. Also, obtain the 95% of confidence interval (Table 7.1).

Use the following R code to illustrate the procedure:

```
y<-c(750,420,1550,500,1785,1225,175,2198,1145,975)
x<-c(1000,650,2100,860,2840,1910,390,3200,1500,1200)
N=sum(x);
p<-round(x/N,4);
Cum<-cumsum(x); Cum<-c(0,Cum)
n=3;
samp<-c(); Z=c();
for(i in 1:n)
{r=round(runif(1,0,N))
```

Table 7.1 Division-wise the number of help requests and of employees during the year 2017

Division	No. of Computer help requests	Number of employees
1	750	1000
2	420	650
3	1550	2100
4	500	860
5	1785	2840
6	1225	1910
7	175	390
8	2198	3200
9	1145	1500
10	975	1200

```
for(j in 1:10)
{if((r>Cum[j])&(r<=Cum[j+1]))
{samp[i]<-j
Z[i]<-y[samp[i]]/(p[samp[i]])}}}
samp; Z; mean(Z);
est_var<-(1/(n*(n-1)))*sum((Z-mean(Z))^2); est_var;
est_se<-sqrt(est_var); est_se;
t<-qt((1-0.025),(n-1)); t;
LL<-mean(Z)-t*est_se;
UL<-mean(Z)+t*est_se;
CI<-c(LL,UL);
CI;
```

The total number of requests is 10779.69 with a standard error 435.87 and a 95% confidence interval is [8904.25, 12655.12].

7.3 Probability Proportional to Size Sampling Without Replacement

In probability proportional to size without replacement (PPSWOR) sampling, the units are selected with probabilities proportional to some measure of their size and without replacement of units. PPSWOR can provide a more efficient estimator than PPSWR as the former has a larger effective sample size than the latter.

(i) Horvitz–Thompson Estimator

Let the population consist of N units $U_1, U_2, \ldots, U_N$ and $y_i, i = 1, 2, \ldots, N$ be the value of the characteristic under study for the ith unit in the population. A sample of size n is drawn without replacement with varying probability of selection.

While selecting a unit in every draw, there is defined a new probability distribution for the units available at that draw. The probability distribution at each draw may or may not depend upon the initial probabilities at the first draw.

Let $P_i^{(r)}$ denote the probability of selecting the ith unit at the rth draw, $i = 1, 2, \ldots, N;\ r = 1, 2, \ldots, n$.

$P_i^{(r)}$ satisfies the following conditions:

(i) $\sum_{i=1}^{N} P_i^{(r)} = 1$;

(ii) $0 < P_i^{(r)} < 1\ \ for\ any\ \ 1 \le r \le n$.

The first-order selection probability for unit i is the probability that unit i is selected in a sample of size n and it is given by

$\pi_i = \sum_{i \in s} P(S)$.

The second-order inclusion probability for units i and j is defined as the probability that the unit i and unit j are selected in the sample at the first and second draws, respectively, and it is given by

$\pi_{ij} = \sum_{i, j \in s} P(S)$.

For any sampling design d, the inclusion probabilities satisfy the following conditions:

(i) $\sum_{i=1}^{N} \pi_i = n$;

(ii) $\sum_{j=1}^{N} \pi_{ij} = (n - 1)\pi_i,\ \ for\ every\ \ i = 1, 2, \ldots, N$;

(iii) $\sum_{i=1}^{N} \sum_{i \ne j=1}^{N} \pi_{ij} = n(n - 1)$.

For an observed sample s, obtained using a sampling design $\{s, P(s); s \in S\}$, the Horvitz–Thompson (HT) estimator for the population mean is defined as

$\bar{y}_{HT} = \frac{1}{n} \sum_{i=1}^{n} n \frac{y_i}{\pi_i N} = \sum_{i=1}^{n} \frac{y_i}{\pi_i N}$.

Theorem 7.4 *The HT estimator of population mean under PPSWOR is unbiased for the population mean.*

Proof Define random variable $\alpha_i;\ i = 1, 2, \ldots, N$ as follows:

$$\alpha_i = \begin{cases} 1 & ; \text{if ith unit is included in the sample of size n} \\ 0 & ; otherwise. \end{cases} \qquad (7.1)$$

Hence

$E(\alpha_i) = 1\ P(i \in s) + 0\ P(i \notin s)$
$\qquad = 1\ \pi_i + 0\ (1 - \pi_i)$.

Now

$\bar{y}_{HT} = \sum_{i=1}^{n} \frac{y_i}{\pi_i N} = \sum_{i=1}^{N} \alpha_i \frac{y_i}{\pi_i N}$.

Hence

$E(\bar{y}_{HT}) = \sum_{i=1}^{N} E(\alpha_i) \frac{y_i}{\pi_i N}$
$\qquad = \sum_{i=1}^{N} \pi_i \frac{y_i}{\pi_i N}$
$\qquad = \bar{Y}$.

Hence the theorem. $\square$

Theorem 7.5 *In PPSWOR, the variance of the HT estimator $\bar{y}_{HT}$ is given by*

$$V(\bar{y}_{HT}) = \frac{1}{N^2} \left\{ \sum_{i=1}^{N} \frac{1-\pi_i}{\pi_i} y_i^2 + \sum_{i \neq j=1}^{N} \frac{(\pi_{ij} - \pi_i \pi_j)}{\pi_i \pi_j} y_i y_j \right\}.$$

Proof Using the definition of α_i in (7.1) above, it can be easily shown that
$E(\alpha_i^2) = E(\alpha_i) = \pi_i$
and
$E(\alpha_i \alpha_j) = \pi_{ij}.$
By using the definition of variance, we get

$$V(\bar{y}_{HT}) = E(\bar{y}_{HT}^2) - (E(\bar{y}_{HT}))^2. \tag{7.2}$$

We have
$$E(\bar{y}_{HT}^2) = E(\sum_{i=1}^{N} \alpha_i \frac{y_i}{\pi_i N})^2$$
$$= E\left(\sum_{i=1}^{N} \alpha_i^2 (\frac{y_i}{\pi_i N})^2 + \sum_{i \neq j=1}^{N} \alpha_i \alpha_j \frac{y_i}{\pi_i N} \frac{y_j}{\pi_j N} \right)$$
$$= \left(\sum_{i=1}^{N} E(\alpha_i^2)(\frac{y_i}{\pi_i N})^2 + \sum_{i \neq j=1}^{N} E(\alpha_i \alpha_j) \frac{y_i}{\pi_i N} \frac{y_j}{\pi_j N} \right)$$
$$= \frac{1}{N^2} \left(\sum_{i=1}^{N} E(\alpha_i)(\frac{y_i}{\pi_i})^2 + \sum_{i \neq j=1}^{N} E(\alpha_i \alpha_j) \frac{y_i}{\pi_i} \frac{y_j}{\pi_j} \right)$$
$$= \frac{1}{N^2} \left(\sum_{i=1}^{N} \frac{y_i^2}{\pi_i} + \sum_{i \neq j=1}^{N} \pi_{ij} \frac{y_i}{\pi_i} \frac{y_j}{\pi_j} \right)$$
and
$$(E(\bar{y}_{HT}))^2 = \left[\sum_{i=1}^{N} E(\alpha_i) \frac{y_i}{\pi_i N} \right]^2$$
$$= \sum_{i=1}^{N} \pi_i^2 (\frac{y_i}{\pi_i N})^2 + \sum_{i \neq j=1}^{N} E(\alpha_i) \frac{y_i}{\pi_i N} E(\alpha_j) \frac{y_j}{\pi_j N}$$
$$= \frac{1}{N^2} \left\{ \sum_{i=1}^{N} y_i^2 + \sum_{i \neq j=1}^{N} y_i y_j \right\}.$$
Substituting the values of these two terms in Eq. (7.2), we get

$$V(\bar{y}_{HT}) = \frac{1}{N^2} \left\{ \sum_{i=1}^{N} (\frac{y_i^2}{\pi_i} - y_i^2) + \sum_{i \neq j=1}^{N} (\pi_{ij} \frac{y_i}{\pi_i} \frac{y_j}{\pi_j} - y_i y_j) \right\}$$
$$= \frac{1}{N^2} \left\{ \sum_{i=1}^{N} \frac{(1-\pi_i)}{\pi_i} y_i^2 + \sum_{i \neq j=1}^{N} \frac{(\pi_{ij} - \pi_i \pi_j)}{\pi_i \pi_j} y_i y_j \right\}.$$
Hence the theorem. □

Corollary 7.1 *An unbiased estimate of the variance of the HT estimator is given by*
$$\hat{V}(\bar{y}_{HT}) = \frac{1}{N^2} \left\{ \sum_{i=1}^{n} \frac{(1-\pi_i)}{\pi_i^2} y_i^2 + \sum_{i \neq j=1}^{n} \frac{(\pi_{ij} - \pi_i \pi_j)}{\pi_{ij} \pi_i \pi_j} y_i y_j \right\}.$$

Example 7.2 Estimate the total number of computer help requests for the last year in a large firm. The director of the computer support department plans to sample 3 divisions of a large firm that has 10 divisions, with varying numbers of employees per division. Since the number of computer support requests within each division should be highly correlated with the number of employees in that division, the director decides to use unequal probability sampling without replacement proportional to the number of employees in that division (Table 7.1).

Use the following R code to illustrate the procedure:

```
y<-c(750,420,1550,500,1785,1225,175,2198,1145,975)
sum(y)
x<-c(0,1000,650,2100,860,2840,1910,390,3200,1500,1200)
N=sum(x); p<-round(x/N,4); Cum<-cumsum(x);
n=3;
samp<-c();Z=c(); rand=c();
for(i in 1:n)
{ r=round(runif(1,0,N))
    rand[i]=r;
    for(j in 1:10)
    { if((r>Cum[j])&(r<=Cum[j+1]))
        { samp[i]<-j
            x[j+1]=0 } }
    Cum<-cumsum(x)
    N=sum(x) }
samp; y<-y[samp]; y;
x<-c(1000,650,2100,860,2840,1910,390,3200,1500,1200)
N=sum(x); p<-round(x/N,4); pi=c();
for(i in 1:n)
{ pi[i]<-1-(1-p[samp[i]])^n }
pi;
tau<-sum(y/pi); tau;
pij=matrix(c(rep(0,9)),byrow=T,ncol=3);
for(i in 1:n)
{
for(j in 1:n)
{ if(i<j)
{pij[i,j]=pi[i]+pi[j]-(1-(1-p[samp[i]]-p[samp[j]])^2)
}}}
pij;#est_variance
S=matrix(c(rep(0,9)),byrow=T,ncol=3);
pij=t(pij)+pij;
for(i in 1:n)
{ for(j in 1:n)
{ if(i!=j)
{S[i,j]=((pij[i,j]-pi[i]*pi[j])/pij[i,j])*(y[i]/pi[i])*(y[j]/pi[j])
}}}
S; xx<-rowSums(S);
est_var<-sum(((1-pi)/pi^2)*y^2)+sum(xx); est_var;
est_se<-sqrt(est_var); est_se;
```

The total number of requests is 11895.78 with standard error 8974.269.

(ii) Yates and Grundy form of Variance

It should be noted that the variance of the HT estimator is not zero even when all $\frac{y_i}{\pi_i}$ are the same. This become a drawback of this estimator. Another drawback is that for some of the samples, the variance term may come negative.

Therefore, Yates and Grundy (1953) proposed a more elegant expression for the variance of the estimate.

If units that are included in the sample of size n take value 1 and units that are not included in the sample take value 0, then

$$\sum_{i=1}^{N} \alpha_i = n.$$

Taking the expectation on both sides, we get

$\sum_{i=1}^{N} E(\alpha_i) = \sum_{i=1}^{N} \pi_i = n$.

Also

$\sum_{i=1}^{N} E(\alpha_i^2) = \sum_{i=1}^{N} \pi_i = n$.

We get

$E(\sum_{i=1}^{N} \alpha_i)^2 = \sum_{i=1}^{N} E(\alpha_i^2) + \sum_{i\neq j=1}^{N} E(\alpha_i \alpha_j)$.

That is,

$E(n^2) = \sum_{i=1}^{N} E(\alpha_i) + \sum_{i\neq j=1}^{N} E(\alpha_i \alpha_j)$.

That is

$n^2 = n + \sum_{i\neq j=1}^{N} E(\alpha_i \alpha_j)$.

Hence

$\sum_{i\neq j=1}^{N} E(\alpha_i \alpha_j) = n(n-1)$.

Also

$$E(\alpha_i \alpha_j) = P(\alpha_i = 1, \alpha_j = 1)$$
$$= P(\alpha_j = 1).P(\alpha_i = 1|\alpha_j = 1)$$
$$= E(\alpha_j)E(\alpha_i|\alpha_j = 1).$$

Therefore,

$$\sum_{i\neq j=1}^{N} \left(E(\alpha_i \alpha_j) - E(\alpha_i)E(\alpha_j)\right) = \sum_{i\neq j=1}^{N} \left(E(\alpha_j)E(\alpha_i|\alpha_j = 1) - E(\alpha_i)E(\alpha_j)\right)$$
$$= E(\alpha_j) \sum_{i\neq j=1}^{N} \left[E(\alpha_i|\alpha_j = 1) - E(\alpha_i)\right]$$
$$= E(\alpha_j) \left\{ \sum_{i=1}^{N} \left[E(\alpha_i|\alpha_j = 1) - E(\alpha_i)\right] + E(\alpha_j) \right\}$$
$$= E(\alpha_j) \left\{ (n-1) - n + E(\alpha_j) \right\}$$
$$= \pi_j(\pi_j - 1)$$
$$= -\pi_j(1 - \pi_j).$$

Similarly,

$\sum_{j\neq i=1}^{N} \left(E(\alpha_i \alpha_j) - E(\alpha_i)E(\alpha_j)\right) = -\pi_i(1 - \pi_i)$.

The variance of the HT estimator is given by

$$V(\bar{y}_{HT}) = \frac{1}{N^2} \left\{ \sum_{i=1}^{N} \frac{1-\pi_i}{\pi_i} y_i^2 + \sum_{i\neq j=1}^{N} \frac{(\pi_{ij} - \pi_i \pi_j)}{\pi_i \pi_j} y_i y_j \right\}.$$

By using the definition of variance and the above results,

$$V(\bar{y}_{HT}) = \frac{1}{2N^2} \left\{ \sum_{i=1}^{N} \frac{1-\pi_i}{\pi_i} y_i^2 + \sum_{j=1}^{N} \frac{1-\pi_j}{\pi_j} y_j^2 - 2 \sum_{i\neq j=1}^{N} \sum_{j=1}^{N} \frac{(\pi_i \pi_j - \pi_{ij})}{\pi_i \pi_j} y_i y_j \right\}$$
$$= \frac{1}{2N^2} \left\{ \sum_{i=1}^{N} \pi_i(1 - \pi_i) \frac{y_i^2}{\pi_i^2} + \sum_{j=1}^{N} \pi_j(1 - \pi_j) \frac{y_j^2}{\pi_j^2} - 2 \sum_{i\neq j=1}^{N} \sum_{j=1}^{N} \frac{(\pi_i \pi_j - \pi_{ij})}{\pi_i \pi_j} y_i y_j \right\}$$
$$= \frac{1}{2N^2} \left\{ \sum_{j\neq i=1}^{N} \sum_{j=1}^{N} (-\pi_{ij} + \pi_i \pi_j) \frac{y_i^2}{\pi_i^2} + \sum_{i\neq j=1}^{N} \sum_{i=1}^{N} (-\pi_{ij} + \pi_i \pi_j) \frac{y_j^2}{\pi_j^2} \right\} +$$
$$\frac{1}{2N^2} \left\{ 2 \sum_{i\neq j=1}^{N} \sum_{j=1}^{N} (\pi_{ij} - \pi_i \pi_j) \frac{y_i}{\pi_i} \frac{y_j}{\pi_j} \right\}$$
$$= \frac{1}{2N^2} \left\{ \sum_{i\neq j=1}^{N} \sum_{j=1}^{N} (\pi_i \pi_j - \pi_{ij}) \left(\frac{y_i}{\pi_i} - \frac{y_j}{\pi_j} \right)^2 \right\}.$$

The above form of the HT estimator is known as the Yates and Grundy type of estimator.

An estimate of the Yates and Grundy variance is given by

$$\hat{V}(\bar{y}_{HT}) = \frac{1}{2n^2} \left\{ \sum_{i\neq j=1}^{n} \sum_{j=1}^{n} \frac{(\pi_i \pi_j - \pi_{ij})}{\pi_{ij}} \left(\frac{y_i}{\pi_i} - \frac{y_j}{\pi_j} \right)^2 \right\}.$$

7.4   Inclusion Probability Proportional to Size Sampling Designs

Inclusion probability proportional to size (IPPS) sampling is a sampling design in which units are selected without replacement and for which π_i, the probability of including the ith unit in the sample of size n, is nP_i, where P_i is the initial probability of selecting the ith unit in the population. An estimator which is commonly used to estimate the population total with IPPS sampling procedures is the well-known HT estimator (1952). Narain (1951) showed that a necessary condition for the HT estimator of the population total under IPPS to be better than the corresponding estimator under PPSWR is that $\phi_{ij} = \frac{\pi_{ij}}{\pi_i \pi_j} \leq \frac{2(n-1)}{n}, \forall i \neq j \in U.$

A sufficient condition for the HT estimator of the population total under IPPS to be better than the corresponding estimator under PPSWR is that $\phi_{ij} = \frac{\pi_{ij}}{\pi_i \pi_j} > \frac{(n-1)}{n}, \forall i \neq j \in U.$

There are a number of IPPS sampling designs available in the literature. Initially, the focus was on obtaining IPPS sampling schemes for sample size $n = 2$. For a general sample size, $n > 2$, not many IPPS schemes are available in the literature. Some important IPPS schemes for $n = 2$ are modified Midzuno–Sen (1952) strategy, Yates and Grundy (1953) sampling design, Brewer (1963) sampling design, Durbin (1967) ungrouped procedure and Rao (1965) rejective procedure, and IPPS sampling plans for $n \geq 2$ are Fellegi (1963) scheme, Sampford (1967) scheme, Tille's method, Midzuno–Sen (1952), Rao et al. (1962), and Gupta et al. (1982, 1984).

7.4.1   Sampford's (1967) IPPS Scheme

The most important sampling scheme available in the literature for obtaining an IPPS sample of size $n \geq 2$ is due to Sampford. This sampling scheme provides a nonnegative variance estimator which is better than the PPSWR scheme and provides a stable variance estimator. We describe the scheme as follows.

Let $\gamma_i = \frac{P_i}{1-P_i}$, $S(m) = \{i_1, i_2, \ldots, i_m : i_j's \ are \ all \ distinct, j = 1, 2, \ldots, m\}$, $L_m = \sum_{S(m)} \gamma_{i_1} \cdots \gamma_{i_m}; (1 \leq m \leq N)$, and $L_0 = 1$.

A sample $s(i_1, i_2, \ldots, i_n)$ is selected with probability

$$p(s) = K_n \sum_{u=1}^{n} P_{i_u} \sum_{v \neq u=1}^{n} \gamma_{i_v} = nK_n\gamma_{i_1} \cdots \gamma_{i_n} \left(1 - \sum_{u=1}^{n} P_{i_u}\right)$$

where $K_n = \left(\sum_{t=1}^{n} \frac{tL_{n-t}}{n^t}\right)^{-1}.$

The above probability of selection can be realized by drawing units one by one.

The above sampling scheme ensures $\pi_i = nP_i, \ for \ i = 1, 2, \ldots, N.$ For proofs, a reference may be made to Sampford (1967).

Sampford's IPPS sampling plan ensures that $\pi_{ij} > 0, \forall i \neq j = 1, 2, \ldots N$ and $\pi_{ij} < \pi_i\pi_j, \forall i \neq j = 1, 2, \ldots N$ (sufficient condition for nonnegativity of variance estimator).

7.4.2 Midzuno–Sen (1952) Sampling Strategy

It is one of the simplest sampling designs. Under this sampling design, the first unit is chosen with probability P_i and the remaining $(n-1)$ units are selected from the remaining $(N-1)$ population units by SRSWOR.

Under this scheme,

$$\pi_i = \frac{N-n}{N-1} P_i + \frac{n-1}{N-1}, \forall i \in U$$
$$\pi_{ij} = \frac{(n-1)(N-n)}{(N-1)(N-2)}(P_i + P_j) + \frac{(n-1)(n-2)}{(N-1)(N-2)}, \forall i \neq j \in U.$$

It is easy to see that if the first unit is selected with revised probability of selection

$\pi'_I = \frac{N-1}{N-n} n P_i - \frac{n-1}{N-n}$ and the remaining $(n-1)$ units are selected by SRSWOR from the remaining $(N-1)$ population units, then the plan reduces to an IPPS sampling plan. Since the probability of selection of every population unit should be different from zero, achieving an IPPS sampling plan with such revised probabilities requires the condition $\frac{n-1}{n(N-1)} \leq P_i \leq \frac{1}{n}$ to be satisfied.

7.4.3 Rao–Hartley–Cochran (RHC) Sampling Strategy

It is a very simple sampling strategy. In this scheme, first the population is randomly divided into n groups $G_1, G_2, \ldots, G_n$ of sizes $N_1, N_2, \ldots, N_n$ units such that $N = N_1 + N_2 + \cdots + N_n$. Let P_{tg} denote the initial selection probability of unit t from the gth group, $t = 1, 2, \ldots, N_g$ and $g = 1, 2, \ldots, n$. Further, $\sum_{t=1}^{N_g} P_{tg} = \pi_g$ and $\sum_{g=1}^{n} \pi_g = 1$. From the gth group, nth unit is selected with probability $\frac{P_{tg}}{\pi_g}, g = 1, 2, \ldots, n$.

An unbiased estimator of population total Y under RHC sampling is
$\hat{Y}_{RHC} = \sum_{g=1}^{n} \frac{y_{tg}}{P_{tg}} \pi_g.$

The variance of $\hat{Y}_{RHC}$ is given by

$$V(\hat{Y}_{RHC}) = \frac{n(\sum_{g=1}^{n} N_g^2 - N)}{N(N-1)}\left[\sum_{t=1}^{N} \frac{Y_t^2}{nP_t} - \frac{Y^2}{n}\right].$$

If $N_1 = N_2 = \cdots = N_n = \frac{N}{n}$,

$$V(\hat{Y}_{RHC}) = \left(1 - \frac{n-1}{N-1}\right)\left[\sum_{t=1}^{N} \frac{Y_t^2}{nP_t} - \frac{Y^2}{n}\right].$$

An unbiased estimator of $V(\hat{Y}_{RHC})$ is given by

$$\hat{V}(\hat{Y}_{RHC}) = \frac{\sum_{g=1}^{n} N_g^2 - N}{N^2 - \sum_{g=1}^{n} N_g^2} \sum_{g=1}^{n} \pi_g \left[\frac{y_{tg}}{P_{tg}} - \hat{Y}_{RHC}\right]^2.$$

7.5 Classes of Linear Estimators

Horvitz–Thompson (1952) described various sub-classes of estimators. Some sub-classes of estimators, for example, are the following:

(a) Considering the order of appearance of the elements, the first class of estimators is of the form $e_s = \sum_{r=1}^{n} \alpha_r y_r$, where the weight $\alpha_r (r = 1, 2, \ldots, n)$ is a function of the order of the drawing and is the weight to be attached to the element selected at the rth draw, $\alpha_r (r = 1, 2, \ldots, n)$ being defined in advance. Here the subscripts identifying the elements have been dropped from the $y's$, meaning thereby that y_r is the value of the characteristic (or the study variable) corresponding to the element selected at the rth draw. This estimator is called the T_1 class of estimators.

(b) Considering the presence or absence of an element in the sample, the second class of estimators of the form $e_s = \sum_{i=1}^{n} \beta_i y_i$, where β_i is the weight to be attached to the ith element whenever it appears in the sample and is defined in advance for all $i = 1, 2, \ldots, n$. This estimator is called the T_2 class of estimators.

(c) Considering the sample obtained as one of the sets of all possible distinct samples, the third class of estimators is of the form $e_s = \gamma_{s_n} \sum_{r=1}^{n} y_r$, where γ_{s_n} is the weight to be attached to the sth sample whenever it is selected and is defined for all $s \in S$. This estimator is called the T_3 class of estimators.

It is noteworthy that this classification of the estimators is not exhaustive. Koop (1957) described seven classes of estimators which included the three estimators of Horvitz and Horvitz and Thompson (1952). Godambe (1955) described the most general class of estimators and referred as to the generalized linear estimators.

Suppose that the parameter of interest is the population total, i.e. $\theta = \mathbf{Y}'\mathbf{1} = \sum_{i \in U} Y_i$. As mentioned above, the population size is N and the sample size is n.

In case of SRSWOR, the estimator of θ is $\frac{N}{n} \sum_{r \in s} y_r$. This estimator belongs to T_2 class of estimators. Similarly, in case of SRSWR, the usual estimator of θ is $e_s = \frac{N}{n} \sum_{i=1}^{n} y_i$. This estimator belongs to T_1 class of estimators. On the other hand, for SRSWR, an unbiased estimator of θ is $e_s = \sum_{r \in s} \frac{y_r}{[1-(1-\frac{1}{N})^n]}$.

This estimator belongs to T_2 class of estimators. Similarly, in case of stratified sampling with SRSWOR in each stratum, the usual unbiased estimator of θ is $e_s = \sum_{h=1}^{L} \frac{N_h}{n_h} \sum_{r \in s_h} y_r$. This estimator belongs to T_2 class of estimators. Here L denotes the number of strata in which the population is divided. N_h is the size of the hth$(h = 1, 2, \ldots, L)$ stratum, $\sum_{h=1}^{L} N_h = N$, and n_h is the size of the sample s_h drawn from the hth stratum, $\sum_{h=1}^{L} n_h = n$. Further, let x be the auxiliary variable, correlated with the study variable y. Then the usual ratio estimator of θ is $e_s = \frac{\sum_{r \in s} y_r}{\sum_{r \in s} x_r} \sum_{i \in U} x_i$. This estimator belongs to T_3 class of estimators. The famous HT estimator of the population total, $e_s = \sum_{i=1}^{n} \frac{y_i}{\pi_i}$, also belongs to T_2 class of estimators.

7.6 Exercises

1. Define the term Probability Proportional to Size Sampling and give an illustration.
2. Show that for the probability proportional to size with replacement (PPSWR) method, the Hansen and Hurwitz estimator is unbiased for the population mean and obtain its variance.
3. Write a short note on the Cumulative total method for the selection of a sample under PPSWR.
4. Show that for the probability proportional to size without replacement (PPSWOR) design, the Horvitz–Thomson estimator is unbiased for the population mean and obtain its variance.
5. Describe Lahiri's method for the selection of samples for PPSWR.
6. Define probability proportional to size sampling. Obtain the PPSWR estimator along with the variance of the estimator. Show that the sample mean square is unbiased for the population mean square under the same sampling scheme.
7. Describe the IPPS scheme in detail. Discuss Sampford's IPPS scheme.

References

Brewer K.R.W.: A model of systematic sampling with unequal probabilities. Aust. J. Stat. **5**, 5–13 (1963)

Durbin, J.: Estimation of sampling error in multi-stage survey. Appl. Stat. **16**, 152–164 (1967)

Fellegi, I.P.: Sampling with varying probabilities without replacement: rotating and non-rotating samples. J. Am. Stat. Assoc. **58**, 183–201 (1963)

Godambe, V.P.: A unified theory of sampling from finite populations. J. R. Stat. Soc.: Ser. B **17**, 267–278 (1955)

Gupta, V.K.: Use of combinatorial properties of block designs in unequal probability sampling. Unpublished Ph.D. thesis, IARI, New Delhi (1984)

Gupta, V.K., Nigam, A.K., Kumar, P.: On a family of sampling schemes with inclusion probability proportional to size. Biometrika **69**, 191–196 (1982)

Horvitz, D.G., Thompson, D.J.: A generalization of sampling without replacement from a finite universe. J. Am. Stat. Assoc. **47**, 663–685 (1952)

Koop, J.C.: Contributions to the general theory of sampling finite populations without replacement and with unequal probabilities. Unpublished Ph.D. thesis, North Carolina State College, Raleigh (University Microfilms, Ann Arbor) (1957)

Midzuno, H.: On the sampling system with probability proportional to sum of sizes. Ann. Inst. Stat. Math. **3**, 99–107 (1952)

Narain, R.D.: On sampling without replacement with varying probabilities. J. Indian Soc. Agric. Stat. **3**(2), 169–175 (1951)

Rao, J.N.K., Hartley, H.O., Cochran, W.G.: On a simple procedure of unequal probability sampling without replacement. J. R. Stat. Soc. **B24**, 482–491 (1962)

Rao, J.N.K.: On two simple schemes of unequal probability sampling without replacement. J. Indian Stat. Assoc. **3**, 173–18 (1965)

Sampford, M.R.: On sampling without replacement with unequal probabilities of selection. Biometrika **54**, 499–513 (1967)

Yates, F., Grundy, P.M.: Selection without replacement from within strata with probability proportional to size. J. R. Stat. Soc. **B15**(2), 53–61 (1953)

Chapter 8
Randomized Response Techniques

8.1 Introduction

Generally, in surveys, the randomization is done for the selection of the sample from a population for the estimation of some population parameter like population mean or total or proportion, ratios like birth or death rate, population correlation coefficient or regression coefficient, etc. The sampled units obtained through the randomization are approached, interviewed and the required information is obtained. It is assumed that the reporting is truthful, without bias. The estimate of the population parameter of interest is obtained along with the sampling variance.

However, if the nature of the characteristic to be observed is sensitive in nature, as would be in the studies to estimate the population proportion of persons in a city having HIV aids, or proportion of students in a University consuming marijuana or proportion of individuals in a district having extramarital relations, and so on, then in that case there would be a tendency on the part of the respondent (the interviewee) to either escape giving an answer by refusing to participate or giving untruthful answer. So there would be refusal bias and/or response bias in the estimate produced by direct questioning of the interviewee.

How to handle such a situation? What is the best survey methodology for such a situation? The answer lies in having two randomizations—one for selecting the sample from the population (to be interviewed or observed) and the second to get a response from the interviewee. The second randomization may further be split up into more randomizations depending upon the methodology adopted for eliciting the truthful reporting. The purpose is to reduce the potential bias due to nonresponse and social desirability when being questioned about sensitive behaviors and beliefs. The respondents use a randomization device like tossing a coin, or flipping a six-faced dice, whose outcome is unobserved by the interviewer. The interviewee represented with two questions, one being sensitive and on which the answer is desired and the other a non-sensitive question at times totally unrelated with the study variable. For example, there may be two groups as A and B. Group A would be I consume mari-

R. Latpate et al., *Advanced Sampling Methods*,
https://doi.org/10.1007/978-981-16-0622-9_8

juana? or I had abortion/ or I have extramarital relations? Group B could be I do not consume marijuana or I did not have abortion or I do not have extramarital relations. Depending upon the design, the randomization device decides which question the respondent answers or in other words he answers to Group A or Group B as chosen by the randomization device. Without telling the interviewer (hiding it completely) the interviewee answers True or False or Yes or No to the question selected by the randomization device thus providing his true status and at the same time protecting her/his identity.

The Group B may even have a totally unrelated question, unrelated with the attribute under study. For instance, the Group B could be having a question like I own a four-wheel drive. There could be several randomization devices and several designs for eliciting truthful reporting protecting the identity of the interviewee.

Since the inception of randomized response technique (RRT) by Warner (1965), many theoretical developments have been made. First important development in terms of modification has been made by Greenberg et al. (1969) using an unrelated question design (UQD). The statistical efficiency of this technique was compared with the Warner's technique under the truthful and untruthful responses. Horvitz et al. (1967) modified Warner's method based on the two samples with a different set of selection probabilities for the two questions for each sample. Moors (1971) suggested an optimized method over Warner's method based on the choice of the parameters of the model. Raghavarao (1978) proposed a new estimator and compared it with Warner's estimator. Mangat and Singh (1990) proposed a new randomized response method which requires the use of two randomization devices. They recommended this method for the cases when the respondents are truthful and when they are not completely truthful in their answers. Thereafter, Mangat (1994) proposed simpler method for the different suggested conditions. In order to judge the effect on the accuracy in estimation, an unbiased estimator of mean square error for direct response survey was obtained by Haung (2004). Kim and Warde (2004) extended their proposed model to stratified sampling. The results pertaining to investigations concerning randomized response sampling were presented by Chang et al. (2004). Blair et al. (2015) reviewed standard designs available to applied researchers along with developing various multivariate regression techniques for substantive analyses.

For more theories on the randomized response method, one may go through Boruch (1971); Chi et al. (1972); Reinmuth and Geurts (1975); Locander et al. (1976); Fidler and Kleinknecht (1977); Lamb and Stem (1978); Tezcan and Omran (1981); Tracy and Fox (1981); Edgell et al. (1982); Volicer and Volicer (1982); Fox and Tracy (1986); Chaudhuri and Mukerjee (1988); Kuk (1990); Umesh and Peterson, R.A. (1991); Van der Heijden and van Gils (1996); Van der Heijden et al. (2000); Elffers et al. (2003); Lara et al. (2006); Gjestvang and Singh (2006); Cruyff et al. (2007); Himmelfarb (2008); De Jong et al. (2010).

In comparison, there is less development that has been seen in the real application of RRT (Blair et al. 2015), however, some of the real applications are included in Sect. 8.4.

8.2 Warner's Randomized Response Design

In most of the sample surveys related to the perception of the individuals, the respondents do not provide the correct answers or do not reply to certain questions. The resulting evasive answer bias was potentially removed through the revolutionary article " Randomized Response: A survey Technique for Eliminating Evasive Answer Bias" of Warner (1965). Although this was the innovative approach but limited to the appeal of RRT, subsequently, many more theoretical improvements have been made by various researchers to make the significance enhance in the potential of this method. Warner's binary approach may not reduce all evasive answer bias; however, it clearly indicates an approach based on a certain probability to sensitive surveys useful for minimizing response biases. This classical design of Warner is popularly known as *Mirrored Question Design* which means that the design randomize whether or not a respondent answers the sensitive question or its inverse.

To understand Warner's randomized response method, one example of illegal forest destruction is taken. We take to examine whether people of *forest villages* of India cuts trees illegally from the reserved forest for domestic or commercial purpose. Reserve forests are those forests accorded a certain degree of protection and forest villages are those which are allowed to be located inside the reserve forest and the inhabitants had to work for the forest department free of cost. These people can be categorized into two categories A and B as:

Category A: The people of forest villages cut trees illegally from the reserve forest.
Category B: The people of forest villages do not cut trees illegally from the reserve forest.

Every person in a population belongs to either group B or group B and it is required to estimate by survey the proportion belonging to group A. A SRSWR of size n is drawn from the adult population of forest villages. Before the interview, each interviewee is furnished with an identical spinner with a face marked so that the spinner points to letter A with probability p and to the letter B with probability $(1 - p)$. Then in each interview, the interviewee is asked to spin the spinner unobserved by the interviewer and reporting only whether or not the spinner points to the letter representing the group to which the interviewee belongs. In other words, the interviewee is required only to say YES or NO according to whether or not the spinner points to the correct group; the interviewee does not report the group to which the spinner points and also does not reveal whether he/she belongs to the stigmatizing category. Assuming that the YES or NO reports are made truthfully, maximum likelihood estimates of the true population proportion are straight forward.

It is straightforward to see that the response probability for the sensitive question is identified and A and B are mutually exclusive and exhaustive events. The value of p may not be $1/2$ and is known to the surveyor based on the random device (spinner).

Let π denotes the true probability of A in the population, i.e. the population proportion of respondents who belong to the sensitive category A. Again suppose λ denotes the number of population proportion of YES responses. Our aim is to

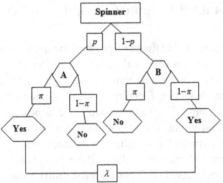

Note: **A** could be the question "*I am HIV positive*" presented with probability *p* and **B** could
be the question "*I am HIV negative*" presented with probability 1–*p*

Fig. 8.1 Tree diagram for probability used in Warner's approach

estimate π and find the variance of estimate and its estimate. Under the Warner's
approach, the probability diagram is suitable to understand (Fig. 8.1).

The respondents have the option YES or NO presented to them. The interviewer
is unaware of which question the respondent has answered but knows the probability
p and $1 - p$ with which the two statements A and B are presented. Here, with a
random sample (SRSWOR) of n respondents, the interviewer records a binomial
estimate $\hat{\lambda} = n_1/n$ of the proportion λ of YES answers, where n_1 is the number of
YES answers out of n.

If the questions are answered truthfully then, then under Warner's approach the
relation between λ and π, given that p is known, is given by the relation.

$$P(YES\ response) = \lambda = p\pi + (1 - p)(1 - \pi) = (2p - 1)\pi + (1 - p)$$

and

$$P(NO\ response) = 1 - \lambda = p(1 - \pi) + (1 - p)\pi$$

or $\pi = \frac{\lambda - (1 - p)}{(2p - 1)}$, $p \neq 1/2$.

The unbiased estimator $\hat{\pi}$ of sensitive proportion π is obtained as

$\hat{\pi} = \frac{\hat{\lambda} - (1 - p)}{(2p - 1)}$

Since the second term, $(1 - p)/(2p - 1)$, is a constant and so its variance is zero.
Therefore the sample variance $V(\hat{\pi})$ is obtained as

$V(\hat{\pi}) = V(\hat{\lambda})/(2p - 1)^2 = \frac{V(n_1)}{n^2(2p - 1)^2}$,

Thus, $V(\hat{\pi}) = \lambda(1 - \lambda)/n(2p - 1)^2$, using $V(n_1) = n\lambda(1 - \lambda)$.
The simplified form of $V(\hat{\pi})$ in terms of π can be written as,

$V(\hat{\pi}) = \frac{\pi(1 - \pi)}{n} + \frac{p(1 - p)}{n(2p - 1)^2}$

It may be noted that $n\hat{\lambda} = n_1$ follows a binomial distribution with probability of
success (YES)$= \lambda$ and probability of failure as $(1 - \lambda)$.
The unbiased estimator of variance of $\hat{\pi}$ is given by

$$\hat{V}(\hat{\pi}) = \frac{\hat{\lambda}(1-\hat{\lambda})}{(n-1)(2p-1)^2} = \frac{1}{n-1}[\hat{\pi}(1-\hat{\pi}) + \frac{p(1-p)}{(2p-1)^2}]$$

In some situations, it may possible to take the two independent samples of same sizes n for the same purpose and same questions. For example, villagers from two different states for the same purpose to estimate the proportion of individuals who cuts trees from reserve forest illegally. In this case, the p may be taken same in both the samples and we have

$$\lambda_i = p\pi_i + (1-p)(1-\pi_i) = (2p-1)\pi_i + (1-p), \text{ for } i = 1, 2 \text{ and}$$

$$\hat{\pi}_i = \frac{\hat{\lambda}_i(1-p)}{(2p-1)}, \text{ with}$$

$$V(\hat{\pi}_i) = \frac{\lambda_i(1-\lambda_i)}{n(2p-1)^2}, \quad i = 1, 2.$$

The pooled estimator (t) based on these two samples is given by simply taking their averages (Rao 2017) as

$$t = \frac{\hat{\pi}_1 + \hat{\pi}_2}{2}, \text{ with}$$

$$V(t) = \frac{1}{4}[V(\hat{\pi}_1) + V(\hat{\pi}_2)] = \frac{1}{2n(2p-1)^2}[\frac{\lambda_1(1-\lambda_1)+\lambda_2(1-\lambda_2)}{2}]$$

As long as $\hat{\pi}_1$ and $\hat{\pi}_2$ are not too different.

$$V(t) = \frac{\lambda(1-\lambda)}{2n(2p-1)^2}$$

The estimator t may help under different sampling technique which may not be possible for the single sample. The detailed discussion about independent samples, which are different from the above, may be seen in Horvitz et al. (1967).

8.3 Three Versions of Warner's Design

This section deals with the three important versions of the Warner's randomized response design with known probability (Blair et al. 2015) and are explained as follows.

8.3.1 Unrelated Question Design (Greenberg et al. 1969, 1971)

This design was a successful improvement upon. In Warner's approach, both the categories A and B are related to the sensitive question. An alternative randomized response device for estimating π was proposed by Greenberg et al. (1969) as follows. Let us first take two questions to report the response from the respondents.

A. Have you beaten your spouse within 6 months after the marriage?

B. Do you live in your own house?

The second question is not related with the first one but is innocuous by nature. In this design, the respondent to be instructed that they should answer either a sensitive question A or an unrelated, non-sensitive question B. This method has an advantage of reducing the extra variance as compared with Warner's method. Since the category B is the non-sensitive category, it is assumed that π_B, the proportion of persons

having YES response for B is known. A Sample of size n using SRSWR is drawn and report YES or NO truthfully about A and B with probability p and $(1-p)$, respectively. The sample proportion $\hat{\lambda}$ of YES response is an unbiased estimator of
$$P(YES\ response) = \lambda = p\pi_A + (1-p)\pi_B.$$
Since it is assumed that π_B is known, therefore using the above expression, an unbiased estimator of π_A is given by
$$\hat{\pi}_A = \frac{1}{p}[\hat{\lambda} - (1-p)\pi_B].$$
When the prevalence π_B is unknown, we need to take two independent non-overlapping random samples using SRSWR to compute an unbiased estimate of the sensitive characteristic $\hat{\pi}_A$ (Fox and Tracy 1986).

Let us take two samples sample 1 and sample 2 of size n_1 and n_2 with category A and B with probability p and $(1-p)$, respectively, in each sample as:

Sample 1:

A: Have you ever had an induced abortion?

B: Were you born in the lunar year of the horse?

Sample 2:

A: Were you a leader during the most recent campus election?

B: Did you ever arrange a data for another student during your last year of 10th standard?

Let π_A and π_B denote the true proportion of YES responses of the sensitive question and unrelated question, respectively, and suppose p_1 and $p_2(p_1 \neq p_2)$ are the probability that the respondent to answer YES for the sensitive question from sample 1 and sample 2, respectively. Obviously, the probability that the respondent to answer non-sensitive question from sample 1 and sample 2 will be $1 - p_1$ and $1 - p_2$, respectively. Therefore, the respective probabilities of affirmative responses (YES answers) in both the samples will be
$$P(Yes\ response) = \lambda_j = p_j\pi_A + (1-p_j)\pi_B, j = 1, 2.$$
$$P(No\ response) = 1 - \lambda_j = p_j(1-\pi_A) + (1-p_j)(1-\pi_B).$$
The unbiased estimator of π_A the probability of the sensitive characteristic in the population is given by
$$\hat{\pi}_A = \frac{\hat{\lambda}_1(1-p_2) - \hat{\lambda}_2(1-p_1)}{p_1 - p_2}, \text{ with variance}$$
$$V(\hat{\pi}_A) = \frac{1}{(p_1-p_2)^2}$$
$$\left[\frac{\lambda_1(1-\lambda_1)(1-p_2)^2}{n_1} + \frac{\lambda_2(1-\lambda_2)(1-p_1)^2}{n_2}\right].$$
The unbiased estimator of $V(\hat{\pi}_A)$ is given by
$$\hat{V}(\hat{\pi}_A) = \frac{1}{(p_1-p_2)^2}\left[\frac{\hat{\lambda}_1(1-\hat{\lambda}_1)(1-p_2)^2}{n_1-1} + \frac{\hat{\lambda}_2(1-\hat{\lambda}_2)(1-p_1)^2}{n_2-1}\right].$$
Moors (1971) proposes to make $p_2 = 0$, i.e. in the second sample the respondent's chance to answer the sensitive question is zero. He showed that this design is more efficient than the mirror question, and it allows for quantitative responses.

8.3.2   Forced Response Design (Boruch 1971)

Under the forced response method, randomization determines whether a respondent truthfully answers the sensitive question or simply replies with a forced answer, YES or NO. In this case, the randomization dice generates (i) the probability (i.e. p) of a respondent being directed to answer the sensitive question, and (ii) the probability (say p_1) of forced YES response or the probability ($p_0 = 1 - p - p_1$) of forced NO response independent of the true answer. Total of 2457 civilians affected by militant violence were interviewed and the 6 faced dice was used for the randomization process (Boruch 1971). The common instructions for answering the question " Now during the height of the conflict in 2007 and 2008, did you know militants like a family member, a friend or someone you talked to on a regular basis" was as
(i) Answer either YES or NO.
(ii) Consider the number of your dice throw. If number shows 1, tell NO.
(iii) If number 6 shows, tell YES.
(iv) If either of the numbers 2, 3, 4 or 5 shows, tell your opinion about the question that would be asked after you throw the dice.
(v) Please before you answer, take note of the number you rolled on the device.

From the above, let us take
$R_i = -1$ indicate that the respondent is forced to say NO.
$R_i = 1$ indicate that the respondent is forced to say YES.
$R_i = 0$ indicates that the respondent is answering Z_i (a binary response YES or NO to the question asked
Let $p_0 = P(R_i = -1)$, $p_1 = P(R_i = 1)$, then
$P(YES\ response) = p_1 + (1 - p_1 - p_0)\pi$, where $P(Z_i = 1) = \pi$
$P(NO\ response) = p_0 + (1 - p_1 - p_0)(1 - \pi)$

The unbiased estimator $\hat{\pi}$ is given by
$$\hat{\pi} = \frac{\left(\hat{\lambda} - (1-p)\right)p_1}{p}, \text{ with variance}$$
$$V(\hat{\pi}) = \frac{\lambda(1-\lambda)}{np^2}$$
The estimator of $V(\hat{\pi})$ is given by,
$$\hat{V}(\hat{\pi}) = \frac{\hat{\lambda}(1-\hat{\lambda})}{(n-1)p^2}$$
The procedure is easier for respondents to comprehend and follow along with to avoid any confusing them with two questions. Besides, the non-sensitive parameter n_1 is fixed by the design which implies that only $\hat{\pi}$ is to be estimated. Despite its simplicity, the forced response model may not relieve the privacy concerns of respondents to a large extent. Comparatively, this design is used by the majority of the researchers.

8.3.3 Disguised Response Design (Kuk 1990)

This design is basically the modification of Forced Response Design which does not require direct answers from the respondents since some respondents may feel uncomfortable in providing a particular response even though the surveyor does not know whether they are answering the sensitive question. This means that the disguised response design substitutes True/False or Yes/No responses against a sensitive question with some other neutral binary responses even though it is linked probabilistically to the sensitive characteristic under study. Further, this design may utilize repeated draws from a randomizing device. Both qualitative and quantitative characteristics are applicable under this method. Van der Heijden and van Gils (1996) suggested that a disguised response design would have been better suited given respondents had difficulties even giving a false YES response. This design is best understood with the following process.

The surveyor has the two decks of cards both have two color cards, red and black. Suppose the proportion of black cards in deck 1 and deck 2 are p_1 and $p_2(p_1 \neq p_2)$, respectively. Obviously, those for red cards will be $1 - p_1$ and $1 - p_2$. The respondent is advised to draw one card from each deck randomly and then indicate the color of the card selected from deck 1 if the sensitive characteristic applies and the color of the card drawn from deck 2 if that characteristic does not apply. In this process, the expected proportion of respondents indicating black card from the deck associated with their status on the sensitive question is given by

$\lambda = p_1 p_2 + p_1(1 - p_2)\pi + (1 - p_1)p_2(1 - \pi)$, or

$\lambda = p_1\pi + p_2(1 - \pi)$, where π is the prevalence of the sensitive characteristic. In this case, the unbiased estimator $\hat{\pi}$ is given by

$\hat{\pi} = \frac{\hat{\lambda}-p_2}{p_1-p_2}$, with variance

$V(\hat{\pi}) = \frac{\lambda(1-\lambda)}{n(p_1-p_2)^2}$

The estimator of $V(\hat{\pi})$ is given by,

$\hat{V}(\hat{\pi}) = \frac{\hat{\lambda}(1-\hat{\lambda})}{(n-1)(p_1-p_2)^2}$.

The process was generalized by Kuk (1990) by repeating the number of draws from the decks and respondents were instructed to make k $(k > 1)$ number of selections from each deck and report the exact number of black cards obtained from the deck corresponding to their status on the sensitive characteristic. Suppose, $\hat{\lambda}_k$ denotes the proportion of black cards drawn at the kth draw, the estimator and its sample variance is generalized to

$\hat{\pi} = \frac{\hat{\lambda}_k-p_2}{p_1-p_2}$, with the variance

$V_k(\hat{\pi}) = \frac{\lambda_k(1-\lambda_k)}{kn(p_1-p_2)^2} + \frac{\pi(1-\pi)(k-1)}{kn}$

In practical, it is quite difficult for the respondents to make large number of draws and keeping track of the number of cards matching a particular color, however, as k increases the variance of the estimator decreases.

8.4 Some Case Studies

Substantial theoretical developments have been found in the area of randomized response techniques. Due to its feasibility in reducing the potential bias caused by sensitive questions, the RRT has been applied on a wide range of sensitive topics in diversified areas of research. Most of the studies are found in the prevalence of drug used and stigmatizing behaviors. The direct questioning and randomized response method for estimating stigmatizing behaviors of sorority women were compared by Fidler and Kleinknecht (1977) using two non-overlapping random samples of 200. The RRT reduced respondents' inclination to refuse to respond or to lie and it was observed that more sensitive questions have a greater proportion of reporting stigmatizing behaviors when respondent assured about maintaining the privacy of information. Similar study on the prevalence of alcohol-related problems was taken by Duffy and Waterton (1988) based on the survey of 1500 adult from the City of Edinburgh. The interviewing strategies were adopted for two questions relating to stigmatizing attributes. There was significant increase in estimates of randomized response method in comparison with the other conventional methods they used. Stubbe et al. (2014) conducted a study on the use of performance enhancing drugs in Dutch fitness centers by taking a total of 718 athletes from 92 fitness centers. The conventional method showed that the prevalence is varying between 0% and 0.4% for the different types of drugs with an overall prevalence of 0.4%. The estimates of RRT were quite higher varying between 0.8% and 4.8% with an overall prevalence of 8.2%. The successful use of RRT based on the 1146 university students was carried out by Cobo et al. (2016) for the use of cannabis, the most widely illicit drug in developed countries, in Spanish universities. The result supported and validated the significance G by of RRT as it provides significantly fewer response refusals and significantly higher drug use estimates. Case study on substance use was conducted by Schröter et al. (2016) in which the prevalence of doping in recreational triathletes of Frankfurt and Wiesbaden of Germany was assessed. The survey based on 2,017 athletes concluded that twelve-month prevalence for physical and cognitive doping ranged from 4–12% and 1–9%, respectively. Kirtadze et al. (2018) used RRT to check response validity for prevalence of drug use based on the household survey (111 urban and 49 rural areas) on addictive substances in Georgia. As per the standard self-report estimates, an estimated 17.3% of Georgian household residents have tried cannabis which was significantly low in comparison with the RRT estimates of 29.9%. They agreed that standard non-RRT approaches produce under-estimates of cannabis users.

For the studies related to the employees of some organizations or society, the empirical estimates for employee-theft base rates using the methods including RRT have been provided by Wimbush and Dalton (1997). The RRT admission frequencies were greater than the conventional with the statistically different differences. Van der Heijden et al. (2000) used the disguised RRT to study fraud and malingering by employees regarding social welfare provisions in the Netherlands. The base rate of entry-level job applicant faking during the application process has been estimated by Donovan et al. (2003). It was shown that a substantial number of recent job applicants reported for engaging in varying degrees of misrepresentation. The base rate for faking was also strongly related to severity and verifiability of the deceptive behavior. Gingerich (2010) used RRT to measure corruption among public bureaucrats in three countries of South America, i.e. Bolivia, Brazil, and Chile. The survey was based on the 2859 bureaucrats from 30 different institutions. The superior performance of the RRT method relative to direct survey questioning for samples of moderate to large size was shown.

In order to determine the extent of purposive concealment of death in Misamis Oriental Province in the southern Philippines, the RRT was used by Madigan et al. (1976) by taking 4000 household (equal allocation of two zones rural and urban) for the survey. The result shows that approximately 75% of urban and 47% of rural deaths were not registered with municipal authorities. The proportion of farmers in north-eastern South Africa killing carnivores was estimated by St John et al. (2012). They showed that farmers' attitudes toward carnivores, question sensitivity and estimates of peers' behavior, in which the likelihood of farmers is killing the carnivores. It was recommended that attitude and estimates of peer-behavior were useful indicators of involvement in illicit behaviors which might be used while identifying the people to engage in interventions aimed at changing behavior.

Chaloupka (1985) estimated the permit noncompliance for the Capricornia Section of the Great Barrier Reef Marine Park. Due to the estimated high rate of noncompliance with specific conditions, it was a useful recommendation that there should not be unreservedly bias management decisions on usage data derived simply from permit issuance.

Krumpal (2012) estimated the prevalence of xenophobia and anti-semitism in Germany based on the sample size of 2041. He suggested that the RRT yielding more valid prevalence estimates than direct questioning and the benefits of using the RRT increases with the increasing topic sensitivity. Chen et al. (2014) used the RRT to investigate commercial sexual behavior in Beijing and provided the estimates and their variances for commercial sex proportion. Rosenfeld et al. (2016) estimated whether people voted for an anti-abortion referendum held during the 2011 Mississippi General Election. While validating these indirect estimates against the official election outcome, it was observed that direct questioning method got significant underestimation of sensitive votes against the referendum and the RRT yield estimates much closer to the actual vote count.

8.5   Exercises

1. Write a note on Randomized response sampling.
2. Describe the randomized response technique. Obtain Warner's estimator and variance.
3. Describe the Warner's randomized response technique. Suppose π denotes the true population proportion of respondents who belong to the sensitive category, then find the estimator and variance of π. Also obtain the estimate of the variance.

4. Under the question 3 above, if p denotes the probability that the respondent answers the sensitive question under study, then find the probability of YES response by a respondent for the sensitive question.
5. Describe Forced Response Design and Unrelated Question Design in randomized response techniques. Describe the situations where Forced Response Design is preferred over Unrelated Question Design.
6. Define Disguised Response Design in detail. Obtain the unbiased estimator of π along with its variance under this design.

References

Warner, S.L.: Randomized response: a survey technique for eliminating evasive answer bias. J. Am. Stat. Assoc. **60**(309), 63–69 (1965)

Greenberg, B.G., Abul-Ela, A., Simmons, W., Horvitz, D.G.: The unrelated question randomized response model: theoretical framework. J. Am. Stat. Assoc. **64**(326), 520–539 (1969)

Horvitz, D.G., Shah, B.V., Simmmons, W.R.: The unrelated question randomized response model. Proceedings of the Social Statistics Section American Statistical Association, pp. 65–72 (1967)

Moors, J.J.A.: Optimization of the unrelated question randomized response model. J. Am. Stat. Assoc. **66**, 627–629 (1971)

Raghavarao, D.: On an estimation problem in Warner's randomized response technique. Biometrics **34**, 87–90 (1978)

Mangat, N.S., Singh, R.: An alternative randomized response procedure. Biometrika **77**, 439–442 (1990)

Mangat, N.S.: An improved randomized response strategy. J. R. Stat. Soc. Ser. B **56**, 93–95 (1994)

Haung, K.: A survey technique for estimating the proportion and sensitivity in a dichotomous finite population. Stat. Neerl. **58**, 75–82 (2004)

Kim, J., Warde, W.D.: A mixed randomized response model. J. Stat. Plan. Inference **110**, 1–11 (2004)

Blair, G., Imai, K., Zhou, Y.Y.: Design and analysis of the randomized response technique. J. Am. Stat. Assoc. **110**(511), 1304–1319 (2015)

Boruch, R.F.: Assuring confidentiality of responses in social research: a note on strategies. Am. Sociol. **6**, 308–311 (1971)

Reinmuth, J.E., Geurts, M.D.: The collection of sensitive information using a two-stage, randomized response model. J. Mark. Res. **12**, 402–407 (1975)

Locander, W., Sudman, S., Bradburn, N.: An investigation of interview method, threat and response distortion. J. Am. Stat. Assoc. **71**, 269–275 (1976)

Lamb, C.W., Stem, D.E.: An empirical validation of the randomized response technique. J. Mark. Res. **15**, 616–621 (1978)

Tezcan, S., Omran, A.R.: Prevalence and reporting of induced abortion in Turkey: two survey techniques. Stud. Fam. Plan. **12**, 262–271 (1981)

Tracy, P.E., Fox, J.A.: The validity of randomized response for sensitive measurements. Am. Sociol. Rev. **46**, 187–200 (1981)

Edgell, S.E., Himmelfarb, S., Duchan, K.L.: Validity of forced responses in a randomized response model. Sociol. Methods Res. **11**, 89–100 (1982)

Chaudhuri, A., Mukerjee, R.: Randomized Response: Theory and Techniques. Marcel Dekker, New York (1988)

Kuk, A.: Asking sensitive questions directly. Biometrika **77**, 436–438 (1990)

Lara, D., García, S.G., Ellertson, C., Camlin, C., Suarez, J.: The measure of induced abortion levels in Mexico using random response technique. Sociol. Methods Res. **35**, 279–301 (2006)

Gjestvang, C.R., Singh, S.: A new randomized response model. J. R. Stat. Soc. Ser. B **68**, 523–530 (2006)

Cruyff, M.J., van den Hout, A., van der Heijden, P.G., Bockenholt, U.: Log-linear randomized-response models taking self-protective response behavior into account. Sociol. Methods Res. **36**, 266–282 (2007)

Himmelfarb, S.: The multi-item randomized response technique. Sociol. Methods Res. **36**, 495–514 (2008)

De Jong, M.G., Pieters, R., Fox, J.P.: Reducing social desirability bias through item randomized response: an application to measure under reported desires. J. Mark. Res. **47**, 14–27 (2010)

Rao, T.J.: Some questions on randomized response technique. Stat. Appl. **15**(1–2), 93–99 (2017)

Van der Heijden, P.G., van Gils, G.: Some logistic regression models for randomized response data. In: Proceedings of the 11th International Workshop on Statistical Modeling, pp. 15–19 (1996)

Duffy, J.C., Waterton, J.J.: Randomized response vs. direct questioning: estimating the prevalence of alcohol related problems in a field survey. Aust. J. Stat. **30**(1), 1–14 (1988)

Stubbe, J.H., Chorus, A.M., Frank, L.E., Hon, O., Heijden, P.G.: Prevalence of use of performance enhancing drugs by fitness centre members. Drug Test. Anal. **6**, 434–438 (2014)

Cobo, B., Rueda, M.M., López-Torrecillas, F.: Application of randomized response techniques for investigation cannabis use by Spanish university students. Int. J. Methods Psychiatr. Res. **26**, e1517 (2016)

Schröter, H., Barkley, J., Studzinski, B., Dietz, P., Ulrich, R., Striegel, H., Simon, P.A.: Comparison of the cheater detection and the unrelated question models: a randomized response survey on physical and cognitive doping in recreational triathletes. PLoS One **11**, e0155765 (2016)

Kirtadze, I., Otiashvili, D., Tabatadze, M., Vardanashvili, I., Sturua, L., Zabransky, T., Anthony, J.C.: Republic of Georgia estimates for prevalence of drug use: randomized response techniques suggest under-estimation. Drug Alcohol Depend. **187**, 300–304 (2018)

Wimbush, J.C., Dalton, D.R.: Base rate for employee theft: convergence of multiple methods. J. Appl. Psychol. **82**, 756–763 (1997)

Donovan, J.J., Dwight, S.A., Hurtz, G.M.: An Assessment of the prevalence, severity, and verifiability of entry-level applicant faking using the randomized response technique. Hum. Perform. **16**, 81–106 (2003)

Gingerich, D.W.: Understanding off-the-books politics: conducting inference on the determinants of sensitive behavior with randomized response surveys. Polit. Anal. **18**, 349–380 (2010)

Madigan, F.C., Abernathy, J.R., Herrin, A.N., Tan, C.: Purposive concealment of death in household surveys in Misamis Oriental Province. Popul. Stud. **30**, 295–303 (1976)

St John, F.A., Keane, A.M., Edwards-Jones, G., Jones, L., Yarnell, R.W., Jones, J.P.: Identifying indicators of illegal behaviour: carnivore killing in human-managed landscapes. Proc. R. Soc. B: Biol. Sci. **279**, 804–812 (2012)

Chaloupka, M.Y.: Application of the randomized response technique to marine park management: an assessment of permit compliance. Environ. Manag. **9**, 393–398 (1985)

Krumpal, I.: Estimating the prevalence of xenophobia and anti-semitism in Germany: a comparison of randomized response and direct questioning. Soc. Sci. Res. **41**, 1387–1403 (2012)

Chen, X., Du, Q., Jin, Z., Xu, T., Shi, J., Gao, G.: The randomized response technique application in the survey of homosexual commercial sex among men in Beijing. Iran. J. Public Health **43**, 416–422 (2014)

Rosenfeld, B., Imai, K., Shapiro, J.N.: An empirical validation study of popular survey methodologies for sensitive questions. Am. J. Polit. Sci. **60**(3), 783–802 (2016)

Adeleke, I.A., Esan, E.O., Okafor, R.O.: Horvitz-Thompson theorem as a tool for generalization of probability sampling techniques. Ghana J. Dev. Stud. **5**(1), 80–94 (2008)

Fidler, D.S., Kleinknecht, R.E.: Randomized response versus direct questioning: two data-collection methods for sensitive information. Psychol. Bull. **84**, 1045–1049 (1977)

Van der Heijden, P.G., van Gils, G., Bouts, J., Hox, J.J.: A comparison of randomized response, computer-assisted self-interview and face-to-face direct questioning eliciting sensitive information in the context of welfare and unemployment benefit. Sociol. Methods Res. **28**, 505–537 (2000)

Chang, H., Wang, C., Haung, K.: On estimating the proportion of a qualitative sensitive character using randomized response sampling. Qual. Quant. **38**, 675–680 (2004)

Chi, I.C., Chow, L., Rider, R.V.: The randomized response technique as used in the taiwan outcome of pregnancy study. Stud. Fam. Plan. **3**, 265–269 (1972)

Volicer, B.J., Volicer, L.: Randomized response technique for estimating alcohol use and noncompliance in hypertensives. J. Stud. Alcohol Drugs **43**, 739–750 (1982)

Fox, J.A., Tracy, P.E.: Randomized Response: A Method for Sensitive Surveys. Sage, Beverly Hills, CA (1986)

Elffers, H., Van Der Heijden, P., Hezemans, M.: Explaining regulatory non-compliance: a survey study of rule transgression for two Dutch instrumental laws, applying the randomized response method. J. Quant. Criminol. **19**, 409–439 (2003)

Umesh, U., Peterson, R.A.: A critical evaluation of the randomized response method: applications, validation, and research agenda. Sociol. Methods Res. **20**, 104–138 (1991)

Greenberg, B.G., Kuebler Jr., R.R., Abernathy, J.R., Horvitz, D.G.: Application of the randomized response technique in obtaining quantitative data. J. Am. Stat. Assoc. **66**, 243–250 (1971)

Efron, B., Tibshirani, R.J.: An Introduction to the Bootstrap. CRC Press, New York (1994)

Chapter 9
Resampling Techniques

9.1 Introduction

The classical theory of statistical inference involves estimating a parameter (or a parametric function) and then determining precision of the estimate using a random sample. An estimator can be obtained that have the specified desirable properties (like unbiasedness, minimum variance, etc.), but there is a conceptual problem with the method of determining precision of the estimate. Conceptually, precision is defined as a measure of variability of the estimator in terms of the variation in the value of the estimator from sample to sample. However, in practice, it is determined using observations in a single sample only. Ideally, precision of an estimator should be determined using values of the estimator from several random samples. This apparent discrepancy in the concept and practice is overcome by the method of resampling. Resampling methods are based on the principle of treating the observed sample as the target population and then sampling it repeatedly with replacement.

Resampling involves procedures that provide an economic way of using available data for estimating population parameters as well as precision of those estimates. Resampling methods are now well established in applied statistics. Two features of resampling techniques are instrumental in its popularity. First, it does not require high-level mathematics. Elementary mathematics is adequate to understand and implement resampling methods. Second, resampling methods are computationally intensive. As such these methods are gaining more acceptance because computational tools and technologies are becoming easily available and affordable, The emphasis, therefore, has shifted to strong concepts rather than rigorous mathematical treatment. Resampling methods have revolutionized statistics through its specialized branches like Monte Carlo, Bootstrap, and Jackknife.

R. Latpate et al., *Advanced Sampling Methods*,
https://doi.org/10.1007/978-981-16-0622-9_9

9.2 Monte Carlo Methods

The term Monte Carlo refers to all methods that use statistical sampling processes to approximate solutions to quantitative problems. It can be used for a wide variety of probabilistic problems ranging from numerical integration to optimization. These methods are used in many application domains such as economics, robotics, and nuclear engineering. Monte Carlo simulation is the use of experiments with random numbers to evaluate mathematical expressions. The experimental units are the random numbers. The expressions may be definite integrals, systems of equations, or more complicated mathematical models. In most cases, when mathematical expressions are to be evaluated, the standard approximations from numerical analysis are to be preferred, but Monte Carlo methods provide an alternative that is sometimes the only tractable approach. Monte Carlo is often the preferred method for evaluating integrals over high-dimensional domains. Very large and sparse systems of equations can sometimes be solved effectively by Monte Carlo methods. Random variables are defined and then simulated in order to solve a problem that is strictly deterministic.

A Monte Carlo method begins with the identification of a random variable such that the expected value of some function of the random variable is a parameter in the problem to be solved. To use the Monte Carlo method, we must be able to simulate samples from the random variable. The problem being addressed may be strictly deterministic. The evaluation of a definite integral, which is a deterministic quantity, provides a good example of a Monte Carlo method.

9.2.1 Evaluating an Integral

In its simplest form, Monte Carlo simulation is the evaluation of a definite integral (Gentle 2003)

$$\theta = \int_D f(x)dx$$

by identifying a random variable Y with support on D and density $p(y)$ and a function g such that the expected value of $g(Y)$ is θ:

$$E(g(Y)) = \int_D g(y)p(y)dy$$
$$= \int_D f(y)dy$$
$$= \theta.$$

Let us first consider the case in which D is the interval $[a, b]$, Y is taken to be a random variable with a uniform density over $[a, b]$, and g is taken to be f. In this case,

$$\theta = (b - a)E(f(Y)).$$

The problem of evaluating the integral becomes the familiar statistical problem of estimating a mean, $E(f(Y))$. The statistician quite naturally takes a random sample and uses the sample mean. For a sample of size m, an estimate of θ is

$$\hat{\theta} = (b - a)\frac{\sum_{i=1}^{m} f(y_i)}{m}, \tag{9.1}$$

where the y_i are values of a random sample from a uniform distribution over (a, b). The estimate is unbiased and is given by

$$E(\hat{\theta}) = (b - a)\frac{\sum E(f(Y_i))}{m}$$
$$= (b - a)E(f(Y))$$
$$= \int_a^b f(x)dx.$$

The variance is

$$V(\hat{\theta}) = (b - a)^2 \frac{\sum V(f(Y_i))}{m^2}$$
$$= \frac{(b - a)^2}{m} V(f(Y))$$
$$= \frac{(b - a)}{m} \int_a^b \left(f(x) - \int_a^b f(t)dt \right)^2 dx. \tag{9.2}$$

The above integral is a measure of the roughness of the function. There are various ways of defining roughness. Most definitions involve derivatives. The more derivatives that exist, the less rough the function. Suppose that the original integral can be written as

$$\theta = \int_D f(x)dx$$
$$= \int_D g(x)p(x)dx,$$

where $p(x)$ is a probability density over D. As with the uniform example considered earlier, it may require some scaling to get the density to be over the interval D. (In the uniform case, $D = (a, b)$, both a and b must be finite, and $p(x) = 1/(b - a)$.) Now, suppose that we can generate m random variates y_i from the distribution with density p. Then, our estimate of θ is just

$$\hat{\theta} = \frac{\sum g(y_i)}{m}. \tag{9.3}$$

Compare this estimator with the estimator in (9.1). The use of a probability density as a weighting function allows us to apply the Monte Carlo method to improper integrals. The first thing to note, therefore, is that the estimator (9.3) applies to integrals over general domains, while the estimator (9.1) applies only to integrals over finite intervals. Another important difference is that the variance of the estimator in Eq. (9.3) is likely to be smaller than that of the estimator in Eq. (9.1).

9.2.2 Variance of Monte Carlo Estimators

The variance of a Monte Carlo estimator has important uses in assessing the quality of the estimate of the integral. The expression for the variance, as in Eq. (9.2), is likely to be very complicated and to contain terms that are unknown. We therefore need methods for estimating the variance of the Monte Carlo estimator. Estimating the variance of a Monte Carlo estimate usually has the form of the estimator of θ in Eq. (9.1):

$$\hat{\theta} = c \frac{\sum f_i}{m}.$$

The variance of the estimator has the form of Eq. (9.2):

$$V(\hat{\theta}) = \int \left(f(x) - \int f(t)dt \right) dx.$$

An estimator of the variance is

$$\hat{V}(\hat{\theta}) = c^2 \frac{\sum (f_i - f)^2}{m - 1}.$$

This estimator is appropriate only if the elements of the set of random variables F_i, on which we have observations $\{f_i\}$, are assumed to be independent and thus have zero correlations.

9.3 Bootstrapping Method

The bootstrap is a broad class of usually non-parametric resampling methods for estimating the sampling distribution of an estimator. The method was described by Efron (1979, 1982, 1992), and was inspired by the previous success of the Jackknife procedure.

Imagine that a sample of n independent, identically distributed observations from an unknown distribution have been gathered, and a mean of the sample, $\overline{Y}$, has been calculated. To make inferences about the population mean we need to know the variability of the sample mean, which we know from basic statistical theory is $V[\overline{Y}] = V[Y]/n$. Here, since the distribution is unknown, we do not know the value of $V[Y] = \sigma^2$. The central limit theorem (CLT) states that the standardized sample mean converges in distribution to a standard normal Z as the sample size grows large, and we can invoke Slutsky's theorem to demonstrate that the sample standard deviation is an adequate estimator for standard deviation σ when the distribution is unknown. However, for other statistics of interest that do not admit the CLT, and for small sample sizes, the bootstrap is a viable alternative.

Briefly, the bootstrap method specifies that B samples be generated from the data by sampling with replacement from the original sample, with each sample set being of identical size as the original sample (here, n). The larger B is, the closer the set of samples will be to the ideal exact bootstrap sample, which is of the order of an n-dimensional simplex: $|C_n| = (2n - 1)C(n)$. The computation of this number, never mind the actual sample, is generally unfeasible for all but the smallest sample sizes (for example, a sample size of 12 has about 1.3 million with replacement subsamples). Furthermore, the bootstrap follows a multinomial distribution, and the most likely sample is in fact the original sample, hence it is almost certain that there will be random bootstrap samples that are replicates of the original sample. This means that the computation of the exact bootstrap is all but impossible in practice. However, Efron and Tibshirani (1994) have argued that in some instances, as few as 25 bootstrap samples can be large enough to form a reliable estimate. The next step in the process is to perform the action that derived the initial statistic here the mean: so we sum each bootstrap sample and divide the total by n, and use those quantities to generate an estimate of the variance of $V(\overline{Y})$ as follows:

$$V(\overline{Y})_B = \left(\frac{1}{B} \sum_{b=1}^{B} (\overline{Y}_b - \overline{Y})^2 \right).$$

The empirical distribution function (EDF) used to generate the bootstrap samples can be shown to be a consistent, unbiased estimator for the actual cumulative distribution function (CDF) from which the samples were drawn, F. In fact, the bootstrap performs well because it has a faster rate of convergence than the CLT: $O(1/n)$ vs. $O(1/\sqrt{n})$, as the bootstrap relies on the strong law of large numbers (SLLN), a more robust condition than the CLT.

9.4 Jackknife Method

The Jackknife method was proposed by Quenouille (1949) and later refined and given its current name by Tukey (1958). Quenouille (1949) originally developed the method as a procedure for correcting bias. Later, Tukey (1958) described its use in constructing confidence limits for a large class of estimators. It is similar to the bootstrap in that it involves resampling, but instead of sampling with replacement, the method is based on samples without replacement.

Many situations arise where it is impractical or even impossible to calculate good estimators or find those estimators' standard errors. The situation may be one where there is no theoretical basis to fall back on, or it may be that in estimating the variance of a difficult function of a statistic, say $g(\overline{X})$ for some functions with no closed-form integral, making use of the usual route of estimation the delta method theorem is impossible. In these situations, the Jackknife method can be used to derive an estimate of bias and standard error. Knight (2000) has noted, in his book Mathematical Statistics, that the Jackknife estimate of the standard error is roughly equivalent to the delta method for large samples.

Estimation Procedure

Suppose a sample consists of sample of size n observations. Select a sample by deleting one observation from the sample n observations, we get the sample of $(n-1)$ observations. Thus, there are n unique Jackknife samples, and the ith Jackknife sample vector is defined as

$$\mathbf{X}_{(i)} = \{X_1, X_2, ..., X_{i-1}, X_{i+1}, ...X_{n-1}, X_n\}.$$

This procedure is generated to k deletions. The ith Jackknife replicate is defined as the value of the estimator $s(.)$ evaluated at the ith Jackknife sample.

$$\hat{\theta}_{(i)} = s(\mathbf{X}_{(i)}).$$

The Jackknife variance estimate is defined as

$$V(\hat{\theta}_{jack}) = \frac{n-1}{n} \sum_{i=1}^{n} (\hat{\theta}_{(i)} - \hat{\theta}_{(.)})^2,$$

where $\hat{\theta}_{(.)}$ is the empirical average of the Jackknife replicates as given by

$$\hat{\theta}_{(.)} = \frac{1}{n} \sum_{i=1}^{n} \hat{\theta}_{(i)}.$$

The $(n-1)/n$ factor in the formula above looks similar to the formula for the standard error of the sample mean, except that there is a quantity $(n-1)$ included in the numerator. As motivation for this estimator, we consider the case that does not actually need any resampling methods: that of the sample mean. Here, the Jackknife estimator above is an unbiased estimator of the variance of the sample mean.

9.5 Importance Sampling

It is clear intuitively that we must get some samples from the interesting or important region. This can be done by the sampling from a distribution that have over weights the important region, hence the name importance sampling. Having over sampled the important region, we have to adjust our estimate somehow to account for having sampled from this other distribution. Importance sampling can bring enormous gains, making an otherwise infeasible problem amenable to Monte Carlo. It can also backfire, yielding an estimate with infinite variance when simple Monte Carlo would have had a finite variance. It is the hardest variance reduction method to use well. Importance sampling is more than just a variance reduction method. It can be used to study one distribution while sampling from another. As a result, we can use importance sampling as an alternative to acceptance-rejection sampling, as a method for sensitivity analysis, and as the foundation for some methods of computing normalizing constants of probability densities. Importance sampling is also an important prerequisite for sequential Monte Carlo.

The accurate estimation of probabilities of rare events through fast simulation is a primary concern of importance sampling. Rare events are almost always defined on the tails of probability density functions. They have small probabilities and occur infrequently in real applications or in a simulation. This makes it difficult to generate them in sufficiently large numbers that statistically significant conclusions may be drawn. However, these events can be made to occur more often by deliberately introducing changes in the probability distributions that govern their behavior. Results obtained from such simulations are then altered to compensate for or undo the effects of these changes. In this section, the concept of importance sampling is motivated by examining the estimation of tail probabilities (Srinivasan 2002). It is a problem frequently encountered in applications and forms a good starting point for the study of importance sampling (IS) theory.

Basics of Importance Sampling

Owen and Zhou (2000) discussed in detail the importance sampling with examples. To motivate our discussion consider the following situation. We want to use Monte Carlo method to compute $\mu = E[X]$. There is an event E such that $P(E)$ is small but X is small outside of E. When we run the usual Monte Carlo algorithm the vast majority of our samples of X will be outside E. But outside of E, X is close to zero. We will get only rare sample in E where X is not small.

Most of the time we think of our problem as trying to compute the mean of some random variable X. For importance sampling, we need a little more structure. We assume that the random variable we want to compute the mean of is of the form $f(\mathbf{X})$ where $\mathbf{X}$ is a random vector. We will assume that the joint distribution of $\mathbf{X}$ is absolutely continous and let $p(\mathbf{x})$ be the density. So, we focus on computing

$$Ef(\mathbf{X}) = \int f(\mathbf{x})p(\mathbf{x}). \tag{9.4}$$

Sometimes people restrict the region of integration to some subset D of R^d. We can instead just take $p(x) = 0$ outside of D and take the region of integration to be R^d. The idea of importance sampling is to rewrite the mean as follows. Let $q(\mathbf{x})$ be another probability density on R^d such that $q(\mathbf{x}) = 0$ implies $f(\mathbf{x})p(\mathbf{x}) = 0$. Then,

$$\mu = \int f(\mathbf{x})p(\mathbf{x})dx = \int \frac{f(\mathbf{x})p(\mathbf{x})}{q(\mathbf{x})}q(\mathbf{x})dx. \tag{9.5}$$

We can write the (9.5) as

$$E_q\left[\frac{f(\mathbf{X})p(\mathbf{X})}{q(\mathbf{X})}\right], \tag{9.6}$$

where E_q is the expectation for a probability measure for which the distribution of $\mathbf{X}$ is $q(\mathbf{x})$ rather than $p(\mathbf{x})$. The density $p(\mathbf{x})$ is called the nominal or target distribution, $q(\mathbf{x})$ the importance or proposal distribution and $p(\mathbf{x})/q(\mathbf{x})$ the likelihood ratio. Note that we assumed that $f(\mathbf{x})p(\mathbf{x}) = 0$ whenever $q(\mathbf{x}) = 0$. Note that we do not have to have $p(\mathbf{x}) = 0$ for all $\mathbf{x}$ where $q(\mathbf{x}) = 0$. The importance sampling algorithm is then as follows. Generate samples $\mathbf{X_1}, \ldots, \mathbf{X_n}$ according to the distribution $q(\mathbf{x})$. Then, the estimator for μ is

$$\hat{\mu}_q = \frac{1}{n}\sum_{i=1}^n \frac{f(\mathbf{X}_i)p(\mathbf{X}_i))}{q(\mathbf{X}_i)} \tag{9.7}$$

provided iff $f(\mathbf{x})p(\mathbf{x})/q(\mathbf{x})$ is computable. The unbiased estimator of μ is $\hat{\mu}_q$. Its variance is σ^2/n where

$$\sigma_q^2 = \int \frac{f^2(\mathbf{x})p^2(\mathbf{x})}{q(\mathbf{x})}dx - \mu^2 = \int \frac{(f(\mathbf{x})p(\mathbf{x}) - \mu q(\mathbf{x}))^2}{q(\mathbf{x})}dx. \tag{9.8}$$

We can think of this importance sampling Monte Carlo algorithm as just ordinary Monte Carlo applied to $E_q[f(\mathbf{X})p(\mathbf{X})/q(\mathbf{X})]$. So, a natural estimator for the variance is

$$\hat{\sigma}_q^2 = \frac{1}{n}\sum_{i=1}^n \left[\frac{f(\mathbf{X}_i)p(\mathbf{X}_i)}{q(\mathbf{X}_i)} - \hat{\mu}_q\right]^2. \tag{9.9}$$

Even if the original $f(\mathbf{X})$ has finite variance, there is no guarantee that σ_q will be finite. How the sampling distribution should be chosen depends very much on the particular problem. Nonetheless there are some general ideas which we illustrate with some trivial examples. If the function $f(\mathbf{x})$ is unbounded then ordinary Monte Carlo may have a large variance, possibly even infinite. We may be able to use importance sampling to turn a problem with an unbounded random variable into a problem with a bounded random variable in the following examples.

9.6 Examples

Example 9.1 We want to compute the integral

$$I = \int_0^1 x^{-\alpha} e^{-x} dx,$$

where $0 < \alpha < 1$. So the integral is finite, but the integrand is unbounded. We take $f(x) = x^{-\alpha} e^{-x}$ and the nominal distribution is the uniform distribution on $[0, 1]$. Note that f will have infinite variance if $\alpha \le -1/2$. We take the sampling distribution to be

$$q(x) = \frac{1}{1-\alpha} x^{-\alpha}$$

on $[0, 1]$. This can be sampled using inversion. We have

$$f(x)\frac{p(x)}{q(x)} = e^{-x}(1-\alpha).$$

So we do a Monte Carlo simulation of $E_q[e^{-X}(1-\alpha)]$ where X has distribution $q(x)$. Note that $e^{-X}(1-\alpha)$ is a bounded random variable. The second general idea we illustrate involves rare-event simulation. This refers to the situation where you want to compute the probability of an event when that probability is very small.

Example 9.2 Let Z have a standard normal distribution. We want to compute $P(Z \ge 4)$. We could do this by a Monte Carlo simulation. We generate a bunch of samples of Z and count how many satisfy $Z \ge 4$. The problem is that there won't be very many (probably zero). If $p = P(Z \ge 4)$, then the variance of $1_{Z \ge 4}$ is $p(1-p) \approx p$. So the error with n samples is of order $\sqrt{p/n}$. So this is small, but it will be small compared to p only if n is huge. Our nominal distribution is

$$\mu(x) = \frac{1}{\sqrt{2\pi}} exp(-x^2/2).$$

We take the sampling distribution to be

$$q(x) = \begin{cases} e^{-(x-4)}, \\ x \ge 4, \\ 0. \quad\quad x < 4. \end{cases}$$

The sampling distribution is an exponential shifted to the right by 4. In other words, if Y has an exponential distribution with mean 1, then $Y + 4$ has an exponential distribution q. The probability we want to compute is

$$p = \int_{1_{x \geq 4}} p(x)dx$$

$$= \int_{1_{x \geq 4}} \frac{p(x)}{q(x)} q(x)dx.$$

The likelihood ratio is

$$w(x) = \frac{p(x)}{q(x)} = \frac{1}{\sqrt{2\pi}} exp(-\frac{1}{2}x^2 + x - 4).$$

On $[4, 1)$ this function is decreasing. So its maximum is at 4 where its value is $exp(-8)/\sqrt{2\pi}$ which is really small. The variance is no bigger than the second moment which is bounded by this number squared. This is $exp(-16)/2\pi$. Compare this with the variance of ordinary MC which is of the order of $exp(-8)$. So the decrease in the variance is huge.

Example 9.3 Let $U_1, U_2, ..., U_5$ be independent and uniform on $[0, 1]$. Let Ti be U_i multiplied by the appropriate constant to give the desired distribution for the times T_i. We want to estimate the mean of $f(U_1, \ldots, U_5)$ where f is the minimum time. The nominal density is $p(u) = 1$ on $[0, 1]^5$. For our sampling density we take $g(u) = \prod_{i=1}^{5} \nu_i u_i^{\nu_i - 1}$, where ν_i are the parameters (This is a special case of the beta distribution). Note that $\nu_i = 1$ gives the nominal distribution p. There is no obvious choice for the ν_i. Note that with $\nu = (1.3, 1.1, 1.1, 1.3, 1.1)$, the variance is reduced by roughly a factor of 2.

We have discussed importance sampling in the setting, where we want to estimate $E[f(\mathbf{X})]$ and $\mathbf{X}$ is jointly absolutely continuous. Everything we have done works if $\mathbf{X}$ is a discrete RV. For this discussion, we will drop the vector notation. So, suppose we want to compute $\mu = E[f(X)]$ where X is discrete with probability mass function $p(x)$, i.e., $p(x) = P(X = x)$. If $q(x)$ is another discrete distribution such that $q(x) = 0$ implies $f(x)p(x) = 0$, then we have

$$\mu = E[f(X)] = \sum_x f(x)p(x) = \sum_x \frac{f(x)p(x)}{q(x)} q(x) = E_q[\frac{f(x)p(x)}{q(x)}],$$

where E_q means expectation with respect to $q(x)$.

Example 9.4 Union counting problem

We have a finite set which we will take to just be $1, 2, ..., r$ and will call Ω. We also have a collection $S_j, j = 1, \ldots, m$ of subsets of Ω. We know r, the cardinality of and the cardinalities $|S_j|$ of all the given subsets. Throughout this example, we use $|\ |$ to denote the cardinality of a set. We want to compute $l = |U|$ where $U = \cup_{j=1}^{m} S_j$. We assume that r and l are huge so that we cannot do this explicitly by finding all the elements in the union. We can do this by a straightforward Monte Carlo if two conditions are met. First, we can sample from the uniform distribution on Ω. Second, given an $w \in \Omega$ we can determine if $w \in S_j$ in a reasonable amount of time. The MC algorithm is then to generate a large number, n, of samples w_i from the uniform distribution on Ω and let X be the number that are in the union U. Our estimator is then rX/n. We are computing $E_p[f(w)]$, where $f(w) = r1_{w \in U}$. We are assuming r and n are both large, but suppose r/n is small. Then, this will be an inefficient MC method. For our importance sampling algorithm, define $s(w)$ to be the number of subsets S_j that contain w, i.e.

$$s(w) = |\{j : w \in S_j\}|$$

and let $s = \sum_w s(w)$. Note that $s = \sum_j |S_j|$. The importance distribution is taken to be

$$q(w) = \frac{s(w)}{s}.$$

The likelihood ratio is

$$\frac{p(w)}{q(w)} = \frac{s}{rs(w)}.$$

Note that $q(w)$ is zero when $f(w)$ is zero. So $f(w)p(w)/q(w)$ is $\frac{s}{s(w)}$. We then do a Monte Carlo to estimate

$$R_q \left[\frac{s}{s(w)} \right].$$

However, is it really feasible to sample from the $q(w)$ distribution? Since l is huge a direct attempt to sample from it may be impossible. We make two assumptions. We assume we know $|S_j|$ for all the subsets, and we assume that for each j, we can sample from the uniform distribution on S_j. Then, we can sample from $q(w)$ as follows. First generate a random $J \in \{1, 2, ..., m\}$ with

$$P(J = j) = \frac{|S_j|}{\sum_{i=1}^{m} |S_i|}.$$

Then sample w from the uniform distribution on S_J. To see that this gives the desired density $q(w)$, first note that if w is not in $\cup_i S_i$, then there is no chance of picking w. If w is in the union, then

$$P(w) = \sum_{j=1}^{m} P(w|J = j)P(J = j) = \sum_{j:w \in S_j} \frac{|S_j|}{|S_j| \sum_{i=1}^{m} |S_i|} = \frac{s(w)}{s}.$$

The variance will not depend on n. So if n, r are huge but r/n is small then the importance sampling algorithm will certainly do better than the simple Monte Carlo of just sampling uniformly from Ω.

9.7 Exercises

1. Write a short note on Jackknife variance estimator for regression estimator.
2. Write a short note on Jackknife variance estimator for ratio estimator.
3. Write a short note on Bootstrap variance estimator for regression estimator.
4. Write a short note on Bootstrap variance estimator for ratio estimator.
5. Use Monte Carlo methods to estimate the expected value of the fifth-order statistic from a sample of size 25 from a $N(0, 1)$ distribution. As with any estimate, you should also compute an estimate of the variance of your estimate. Compare your estimate of the expected value with the true expected value of 0.90501. Is your estimate reasonable? (Your answer must take into account your estimate of the variance.)
6. Obtain Monte Carlo estimates of the base of the natural logarithm. (a) For Monte Carlo sample sizes n = 8, 64, 512, 4096, compute estimates and the errors of the estimates using the known value of e. Plot the errors versus the sample sizes on log-log axes. What is the order of the error? (b) For your estimate with a sample of size 512, compute 95% confidence bounds.
7. What is self-normalized importance sampling?
8. Explain how the variance is minimized in importance sampling.
9. How do you choose the sample size in order to minimize the variance of importance sampling?

References

Gentle, J.E.: Random Number Generation and Monte Carlo Methods, 2nd edn. Springer, New York (2003)

Efron, B.: Bootstrap methods: another look at the jackknife. Ann. Stat. **7**(1), 1–26 (1979)

Efron, B.: The jackknife, the bootstrap, and other resampling plans. SIAM, Philadelphia (1982)

Efron, B.: Jackknife after bootstrap Bootstrap methods: (with discussions). J. R. Stat. Soc. B **54**, 1–127 (1992)

Quenouille, M.H.: Problems in plane sampling. Ann. Math. Stat. **20**(3), 355–375 (1949)

Tukey, J.W.: Bias and confidence in not quite large samples (abstract). Ann. Math. Stat. **29**(2), 614–623 (1958)

Srinivasan, R.: Importance Sampling. Springer, Berlin (2002)

Owen, A.B., Zhou, Y.: Safe and effective importance sampling. J. Am. Stat. Assoc. **95**(449), 135–143 (2000)

Knight, K.: Mathematical Statistics. Chapman and Hall/CRC, New York (2000)

Efron, B., Tibshirani, R.J.: An Introduction to the Bootstrap. CRC press, New York (1994)

Chapter 10
Adaptive Cluster Sampling

10.1 Introduction

While conducting a sample survey, a number of difficult sampling problems are encountered. One of them is the problem in estimating the population mean/total when it is rare or geographically uneven. If the population of interest is hidden or elusive, then it becomes difficult to identify it for sampling.

Adaptive sampling designs are the designs in which the additional units are selected depending on the observations made on the initial sampling units. The conventional sampling designs such as simple random sampling (SRS) lead to estimates with large variances and biases when they are used for the rare, hidden, or hard to reach populations. With a sufficient prior knowledge of the population, precision of the estimates can be increased by using stratification, systematic designs and using the auxiliary information in the design. Many times the uneven patterns in the populations can only be identified after sampling. For such populations, adaptive sampling designs can be useful.

The researcher may find the conventional sampling design such as SRS and stratified sampling as inadequate for producing sufficient data and information from the sampling units while studying such type of population. If conventional sampling designs are applied to population that are rare and clustered then usually very few units possessing the characteristic of interest are selected in the sample. Even a very large conventional sample would be inadequate in such cases. Due to these reasons, researchers have thought about the other sampling designs such as adaptive sampling.

In conventional sampling, sampling design is fixed before the study begins. It is based entirely on a prior information about the population. In this case, the sampling units can be identified before commencing the sampling. It does not add any sampling units satisfying the condition of interest discovered during the course of sampling.

On the other hand, in adaptive sampling, the sampling units that are found to satisfy the condition of interest during the survey are added in the sample. The conventional sampling designs are associated with well-established statistical procedures for making inferences about the population. On the other hand, in adaptive sampling, there have been very few guidelines about obtaining the estimates of population parameters.

There has been a significant development during the last few years in statistical theory for certain adaptive sampling designs. In some cases, the estimates based on data from adaptive sampling designs can be more precise than those obtained by conventional designs based on the same amount of sampling effort.

In adaptive sampling, the sampler specifies
(i) the initial sampling design
(ii) the initial sample size
(iii) the description of the neighborhood for a sampling unit. Neighborhood can be defined by social or institutional connection as well as geographically.
(iv) the pre-specified condition that initiates adaptive sampling at a sample unit selected during the initial sampling.

Thompson (1990) presented designs in which, whenever the observed value of a selected unit in sample satisfies a condition of interest, additional units are added to the sample from the neighborhood of that unit are called as adaptive design. In this sampling design, an initial sample is selected with a conventional sampling design. Whenever the variable of interest satisfies a specified condition for an individual unit in the sample, units in the neighborhood of that unit are added in the sample. Further, if any of the added units satisfy the condition, still more units in its neighborhood are also added in the sample. This is continued till no further unit in the neighborhood satisfies the specified condition.

Collection of all units that are observed under the design as a result of initial selection of a unit is termed as a 'cluster'. A sub-collection of units within each cluster excluding edge units is called a 'network'. It has the property that selection of any unit within network would lead to inclusion in the sample of every other unit in the network. Any unit not satisfying the condition but in the neighborhood of one that does is termed as an 'edge unit'. The selection of edge unit does not result in the inclusion of any other units. Any unit that does not satisfy the condition is considered as a network of size one. Thus, the population may be uniquely partitioned into networks. Thompson called this design as adaptive cluster sampling design.

Section 10.2 introduces the notations and terminology used in adaptive cluster sampling (ACS). Section 10.3 gives an illustration of the methodology used in ACS. In Sect. 10.4, the different types of sampling units in ACS are discussed. The computation of inclusion probabilities of the different types of sampling units in ACS is illustrated in Sect. 10.5. The different approaches of estimation in adaptive sampling are discussed in Sect. 10.6. Section 10.7 discusses the problem of varying probability estimation. Section 10.8 deals with the problem of estimation of the population mean/total in ACS without using the edge units of the networks. Section 10.9 discusses the estimation of the population mean/total by using the edge units of the networks. Sections 10.10 discuss the estimation of upper and lower bound, on population total by using the new method, restricted adaptive sampling (RAS).

In adaptive sampling, final sample size is a variable quantity. Considering it as a random variable, its expected value is derived in Sect. 10.11. Section 10.12 discusses a pilot study and illustrates the computation of the expected final sample size in ACS.

10.2 Notations and Terminology

The population under study is assumed to have N units. Y is the variable of interest. $\{y_1, y_2, \ldots y_N\}$ are values of Y associated with the population units $\{U_1, U_2, \ldots U_N\}$ respectively.

Neighborhood of the ith unit in the population is the collection of units that includes unit i. If unit j is in the neighborhood of unit i, then unit i is also in neighborhood of unit j.

A network is defined as a group of units in the neighborhood of an unit that satisfies the condition C which is specified well in advance before the survey. If ith unit in the population satisfies condition C, then the group of units in the neighborhood of unit i which also satisfy C constitute a network. It is denoted as A_i. Each unit included in A_i is called as a network unit. Selection of any unit in A_i leads to selection of all units in A_i. An unit that does not satisfy the condition C but is included in the neighborhood of an unit that does, such unit is called as an edge unit. Let m_i denote the number of sampling units in A_i. Let a_i denote total number of sampling units in the networks of which sampling unit i is an edge unit; If i satisfies C then $a_i = 0$ else $m_i = 1$. After sampling is terminated, we have clusters of units. Each cluster contains units satisfying C and edge units.

If the initial sample drawn by conventional design contains n_1 units then the final sample consists of at most n_1 clusters because two or more units satisfying C might have been selected from the same clusters in the initial sample.

If a unit in the initial sample does not satisfy C, then it is considered to be a network of size 1. Unit to the right, to the left, above and below an unit define neighbors of that unit. There are two common ways to define a neighborhood:

(i) First-order neighborhood: It consists of the unit itself and the four adjacent units sharing a common boundary.

(ii) Second-order neighborhood: It consists of eight new units and the first-order neighborhood units.

It is important to note the distinction between clusters and networks. Clusters are not necessarily disjoint because they may have overlapping edge units.

The entire population can be partitioned into a set of disjoint and exhaustive networks.

Any multi-unit cluster can be decomposed into one network whose units satisfy C and individual networks (edge units) of size 1 that do not satisfy C.

Probability of selection of the unit i in a given draw:

$$P_i = \frac{m_i + a_i}{N}$$

The probability that unit i is included in the initial sample is

$$\pi_i = 1 - \frac{\binom{N - m_i - a_i}{n_1}}{\binom{N}{n_1}}.$$

If the initial sample is selected by SRSWR, then

$$\pi_i = 1 - (1 - P_i)^{n_1}.$$

Since some of the a_i may not be known, P_i and π_i can not be calculated. In that case we use the partial inclusion probabilities:

$$\pi_i = \frac{m_i}{N}.$$

The values of m_i's are known for the units in the sample, but some of the a_i's are unknown. That requires sampling around the empty sampling units which is not done. Hence, we compute the partial inclusion probability:

$$\pi_i' = 1 - \frac{\binom{N-m_i}{n_1}}{\binom{N}{n_1}}.$$

It can be seen that π_i' is in fact the probability that the initial sample intersects A_i.

If there are m_k sampling units in the kth network, then

probability that kth network is included in the sample $= \alpha_k = 1 - \frac{\binom{N-m_k}{n_1}}{\binom{N}{n_1}}$.

When each $m_k = 1$ then $\alpha_k = \frac{n_1}{N}$.

Further,

probability that the jth and kth networks are not intersected is given by

$$P_{jk} = \frac{\binom{N-m_j-m_k}{n_1}}{\binom{N}{n_1}}$$

Hence, the joint probability that networks j and k are both intersected is given by

$$\alpha_{jk} = \alpha_j + \alpha_k - (1 - P_{jk}) \quad = 1 - \left[\frac{\binom{N-m_j}{n_1} + \binom{N-m_k}{n_1} - \binom{N-m_j-m_k}{n_1}}{\binom{N}{n_1}} \right].$$

Using these probabilities, estimate of the population total and its variance can be obtained.

10.3   Illustration of Methodology

It is easier to understand any methodology with the help of an illustrative example. We consider an example by Edgar Barry Moser to illustrate methodology of ACS.

Suppose that a scientist is studying a particular weed that grows in strawberry fields. The weed is not particularly abundant but serves as a host plant for a disease of strawberries. The scientist would like to estimate the total (or average) number of weeds in the field using ACS.

The scientist divides the field using a grid system to produce contiguous sampling units. The following Fig. 10.1 shows the hypothetical strawberry field divided into sampling units using a grid. The numbers in the squares of the grid indicate the number of weeds identified in the corresponding sampling unit. The absence of any number in the square indicates the absence of weeds in that sampling unit.

The scientist initially takes a sample by using SRSWOR of 10 sampling units. These are shown in Fig. 10.2. In adaptive sampling, it is required to define a neighborhood of a sampling unit. It is some pre-specified rule for associating other sampling units with each sampling unit in the frame. In general, the neighborhood will be defined the same way for each sampling unit. But the definition of neighborhood may vary from one experiment to the other. Here, the scientist defines the neighborhood to be the 4 contiguous sampling units for a given unit. It is illustrated in Fig. 10.3.

It is required to specify a condition C for searching the neighborhood. Here, the specified condition C be $\{y_i > 0\}$; where y_i is the number of weeds present in the ith sampling unit. By using the procedure of ACS, the final sample of units is obtained and shown in Fig. 10.4. After obtaining the final sample, the sampled networks are identified. A network consists of the initial sampling unit and any other sampling units in its neighborhood meeting the specified condition that are identified through the adaptive sampling process.

The initial units that do not have the weed present are considered as networks of size 1.

y_i^* denotes the total number of weeds observed in the ith network.

From the initial sample of $n_1 = 10$ units; 9 distinct networks are identified.

One network has $m_1 = 7$ units with $y_1^* = 30$ weeds.

				1					
	2	15	8	1					
	1	2							
						2			
					7	10			
					7	3			
					1				
		5	2						
	1	9							

Fig. 10.1 Hypothetical strawberry field divided into sampling units using a grid

				1					
	2	**15**	**8**	**1**		*		*	
	1*	**2**	*						
						2			
*		*			**7**	***10**			
					7	***3**			
				*	**1**				
		5	**2**						
	1	**9**						*	

Fig. 10.2 Field map with initial sampling units represented by * in those units

Fig. 10.3 Neighborhood of sampling unit

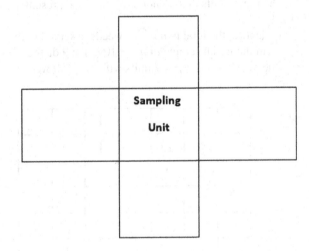

The second has $m_2 = 6$ with $y_2^* = 30$. The remaining 7 networks have one unit each with zero weeds.

That is, $m_i = 1$ and $y_i^* = 0$, for $i = 3, 4 \ldots 9$. Note that the numbers in squares in Fig. 10.1 indicate the number of weeds observed in that sampling unit.

Using these values, the inclusion and intersection probabilities are computed as follows:

$$\alpha_1 = 1 - \frac{\binom{100-7}{10}}{\binom{100}{10}} = 0.533$$

				1					
	2	15	8	1		*		*	
	1*	2	*						
						2			
*		*			7	*10			
					7	*3			
				*	1				
		5	2						
	1	9						*	

(* in a square indicates unit selected in the initial sample)

Fig. 10.4 Strawberry field showing the final sample, obtained by using the adaptive sampling process from the initial SRS

$$\alpha_2 = 1 - \frac{\binom{100-6}{10}}{\binom{100}{10}} = 0.478$$

$$\alpha_i = 1 - \frac{\binom{100-1}{10}}{\binom{100}{10}} = 0.10 \text{ , for } i = 3, 4, \ldots 9.$$

$$P_{12} = \frac{\binom{100-7-6}{10}}{\binom{100}{10}} = 0.231$$

$$\alpha_{12} = \alpha_1 + \alpha_2 - (1 - P_{12}) = 0.242.$$

These probabilities further can be used to obtain the estimate of the total/average number of weeds and variance of this estimate.

10.4 Sampling Units and their Inclusion Probabilities

Thompson and Seber (1996) described ACS in detail and provided two unbiased estimators of the population mean: Hansen–Hurwitz (HH) estimator and Horvitz–Thompson (HT) estimator. The HT estimator is based on the inclusion probabilities of the distinct networks in the population. They have not discussed the different types of units in a population with reference to ACS. It can be seen easily that in ACS, the final adaptive sample contains three types of sampling units:
(i) Sampling units that are in initial sample.
(ii) Sampling units that are added as a result of adaptive sampling and caused more units to be added.

(iii) Sampling units that are added as a result of adaptive sampling but did not cause more units to be added in the sample.

Among the sampling units that are added as a result of adaptive sampling, some units are network units and some are non-network units. Network units cause the addition of sampling units, whereas non-network units do not cause such an addition.

At this stage, it is necessary to make clear which neighbors of network units get added to the sample. For that purpose, let us number the rows and columns of the population grid. Let (i, j) denote the sampling unit in the ith row and jth column of the population grid. If (i, j) is a network unit then $(i - 1, j)$, $(i + 1, j)$, $(i, j - 1)$ and $(i, j + 1)$ are its neighbors and will be included in the sample.

We assume that there are K distinct networks in the population.

If the initial sample of size n does not include a sampling unit from any of the K networks, then that sample itself will be the final sample and no information will be available. In this case, the inclusion probability of a sampling unit $\frac{n}{N}$ remains unaffected. If the initial sample includes $1, 2, \ldots n$ network units then the inclusion probabilities of these number of network units and hence non-network units in the sample get affected. These probabilities depend upon the distribution of the network units in the different networks. Since each possible distribution of the network units in the sample is equally likely, it is required to average out all the inclusion probabilities.

10.5   Illustration of Computing Inclusion Probabilities

Consider the Fig. 10.5. It represents a clustered population divided into 100 sampling units (quadrats). The numbers in the quadrats in the grid are the observed values of the variable characteristic under study. The corresponding units are the network units.

There are three distinct networks. Network number 1 (N_1) shown in orange color, network number 2 (N_2) in blue color, and network number 3 (N_3) in green color. Among the 100 units, 22 are the network units and 78 are the non-network units.

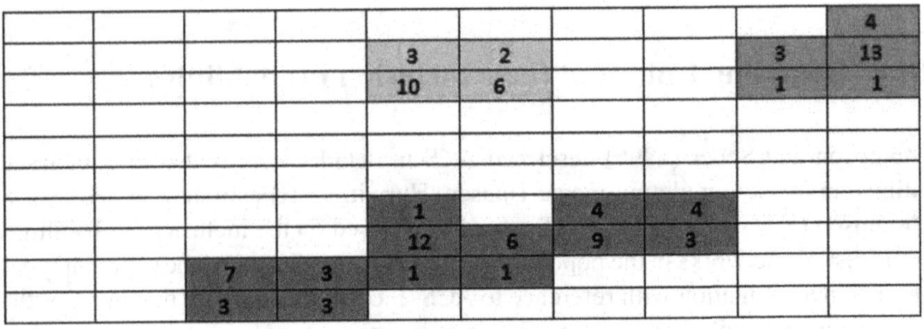

Fig. 10.5   Clustered population divided into 100 sampling units

Suppose a random sample of 5 units is drawn from this population. Then every unit in the population has a constant probability 0.05 of getting included in this sample.

If one of the network units from N_1 is selected, then all other network units in N_1 will also be included in the sample. The size of N_1 is 4. Hence, by additive law of probability, the probability of selecting a network unit from N_1 is 0.2. Similarly, the probability of selecting a network unit from N_2 is 0.25 and that from N_3 is 0.65. If the initial sample does not include a sampling unit from any of the networks, then this sample itself will be the final sample.

The probability that the initial sample has 1 network unit from N_1

$$= 0.2 \left[\frac{\binom{4}{1}\binom{5}{0}\binom{13}{0}\binom{78}{4}}{\binom{100}{5}} \right] = 0.0152.$$

The probability that the initial sample has 1 network unit from N_2

$$= 0.25 \left[\frac{\binom{4}{0}\binom{5}{1}\binom{13}{0}\binom{78}{4}}{\binom{100}{5}} \right] = 0.0237.$$

The probability that the initial sample has 1 network unit from N_3

$$= 0.65 \left[\frac{\binom{4}{0}\binom{5}{0}\binom{13}{1}\binom{78}{4}}{\binom{100}{5}} \right] = 0.1601.$$

Since all these events are equally likely, we average out these probabilities.

Thus, probability that the initial sample has exactly one network unit

$$= \frac{0.0152 + 0.0237 + 0.1601}{3}$$

$$= 0.0663.$$

Continuing in this fashion the probabilities of having 2, 3, 4, and 5 network units in the initial sample can be worked out.

These inclusion probabilities can be used to estimate the population total/mean and the variance of that estimator.

10.6 Different Approaches of Estimation

The usual unbiased estimators in ACS are developed by using the design-based approach of inference. A general problem with these estimators is that they depend on the design being used. If the sampling is not carried out according to the design, the estimation of the parameters may get affected. Most of the design-based unbiased estimators require the entire networks to be observed. The researcher may not have enough resources for that. For these reasons, it is important to study the model-based estimators. These estimators can be useful to make a choice of estimator for incorporating the auxiliary information (Cochran 1977; Elliot and Little 2000).

In situations where HT estimator is not reasonable (Little and Rubin 2002), a model-assisted modification is used to predict the non-sample values and then the HT estimator is applied to the residuals from that model.

When the population under study is rare and clustered, the inclusion probabilities of the units vary. In that case, one has to use the varying probability approach to estimate the population parameters.

10.6.1 Model-Based Approach

This approach of estimation uses conceptual statistical and mathematical models to estimate the population quantities based on detailed information from a limited number of sampling units. The model may not be explicit. The set of criteria used to select sample units may be based on this model. The model specifies the relationship of the sampling units to the population. The validity of the population estimates depends upon the validity of the model. This approach estimates the parameters of the model that generates the data and not the population parameters.

The assumed statistical model describes the error distribution (i.e. the distribution of the discrepancies between reality and model results). This approach of estimation relies on this model. So the expected values are the averages over possible error realization.

This approach uses information from units in the population. It ignores how they are selected. The estimates of population quantities depend upon the assumption of the behavior of the population. Even though this approach does not make explicit use of the probability structure of the sample, the model-based estimates are strengthened by the characteristics of a probability sample. Model-based parameter estimates can be either biased or unbiased under a Judgemental Sampling Design.

10.6.2 Design-Based Approach

Design-based estimation is essentially an empirical approach. The design specifies what information and from where it is to be collected and the design stipulates the population estimates. The validity of the estimates depends upon the ability of the design to produce representative information. In general, design-based program depends on the methods of statistical survey sampling and is valid only with a probability sample. This approach also obtains the estimates of population quantities and relies on statistical theory to describe the properties of estimators. This approach assumes that a random sample of units is selected from the population. The estimates obtained by this approach are based on all possible samples.

The design-based approach has a number of strengths that makes it popular among the researchers. It automatically takes into account features of survey design and provides reliable estimates in large samples without using any strong modeling assumption. It is essentially asymptotic. It gives limited guidance for small sample adjustments.

The model-based approach leads to efficient inference based on Likelihood or Bayesian principles. The design-based approach is not prescriptive for the choice of estimator. It lacks a theory for optimal estimation. The estimates obtained by this approach are potentially inefficient.

10.6.3 Model Assisted-Cum-Design-Based Approach

All design-based unbiased estimators face a common problem that is they depend upon the design being carried out. If the sampling is not carried out according to the design, it affects the estimation of the parameters of interest to a great extent. The sampling problems may be correlated with the parameter of interest. For example, a researcher may not have enough resources to sample the entire network, if it is too large and many design-based unbiased adaptive sampling estimators require the entire network to be observed. So it is important to study model-based estimators in combination with an adaptive sampling design being used.

10.7 Variable Probability Estimation

In a probability sample, it is essential to know the probability of inclusion in the sample for every element of the population.

In a continuous population, this knowledge is contained in an inclusion probability density function (pdf) $\pi(s)$. Thus, the requirement is that $\pi(s)$ must be known and it should be positive for all s. This function specifies the density of sample points; therefore, it has units with dimension like the number of sample points per unit area. Conversely, the reciprocal of this function has units of area per sample point. Thus, it represents the area that each sample point represents.

A design-based analysis of a sample with varying inclusion pdf needs to attach a different weight to each observation (i.e. the different amount of area represented by each point). So the proper way to give each observation its correct weight is to multiply each observation by the reciprocal of the inclusion density function at that observation site.

If we have samples at locations $S_1, S_2 \ldots, S_n$ with corresponding observations $Z_1, Z_2 \ldots, Z_n$ then the estimate of the mean value of Z is

$$\hat{\mu}_z = \frac{\sum_{i=1}^n \frac{z(s_i)}{\pi(s_i)}}{\sum_{i=1}^n \frac{1}{\pi(s_i)}} = \frac{\sum_{i=1}^n z(s_i) w(s_i)}{\sum_{i=1}^n w(s_i)},$$

where

$$w(s_i) = \frac{1}{\pi(s_i)} \quad i = 1, 2 \ldots, n.$$

In general, the variance of a quantity estimated from a probability sample depends upon inclusion pdf and also on the pair-wise probabilities of only two points being included. If we assume that the sample observations are independent and if the population has the characteristic that values measured on sites that are close together tend to be more similar than the values on the sites far apart, then we get

$$V(\hat{\mu}_z) = \frac{ns^2(d)}{[\sum_{i=1}^{n} w(s_i)]^2},$$

where

$$s^2(d) = \frac{\sum_{i=1}^{n}(d_i - \bar{d})^2}{n}$$

and

$$d_i = [z(s_i) - \hat{\mu}_z]w(s_i).$$

In the above discussion, we have dealt with estimation of the population mean and its variance.

The same technique can be used to estimate the proportion of a population that meets the condition specified. In that case, we define a new response variable that assumes the value 1 if a sample site meets the condition and 0 otherwise. This new response variable is called the indicator variable for the condition specified. The mean value of this indicator variable gives the proportion of the population that meets the specified condition.

We estimate it using the formula:

$$\hat{P} = \frac{\sum_{i=1}^{n} I(s_i)w(s_i)}{\sum_{i=1}^{n} w(s_i)},$$

where the indicator function,

$$I(s_i) = \begin{cases} 1 & ; \text{if } s_i \text{ satisfies the condition} \\ 0 & ; otherwise. \end{cases} \quad for \ i = 1, 2 \dots, n.$$

10.8   Estimation of Population Mean/Total without Using the Edge Units

We will discuss the problem of estimation of population mean/total using ACS; since it is more frequently used in practice. In ACS, an initial random sample of n_1 units is selected either with or without replacement. We consider without replacement method. All units in the neighborhood of the initial sample units satisfying the condition are added and so on. We describe two estimators. The first one is Hansen–Hurwitz (HH) estimator based on draw by draw selection probabilities. The second one is Horvitz–Thompson (HT) estimator based on inclusion probabilities. Both of these estimators indicate that the inclusion probability of an unit depends upon the size of the network to which it belongs.

There are two important elements in the estimation. One is to consider the network as a sampling unit. That is because selection of a unit in the initial sample leads to selection of all units in the corresponding network. The initial unit may be a network of size 1. It is possible to determine the inclusion probability for a network.

The second important element is to exclude the edge units from the estimator which were not part of initial sample. In the following process of estimation, the edge units do not play any role.

10.8.1 Different Estimators in ACS

In this case, the sample mean is a biased estimator of the population mean.

We describe below three estimators suitable for adaptive sampling:

(i) Initial Sample Mean:

If the initial sample in adaptive cluster sampling design is selected either by SRSWR or SRSWOR, then the mean of the initial observations is unbiased for the population mean.

This estimator completely ignores the observations on the units, other than those initially selected.

(ii) Modified HH Estimator:

In conventional sampling, HH estimator, in which each value of y is divided by the number of times the unit is selected, is an unbiased estimator of population mean.

But in adaptive cluster sampling, selection probabilities are not known for every unit in the sample. An unbiased estimator can be formed by modifying HH estimator. Due to this modification, the observations not satisfying the condition are used only when they are selected in the sample.

Let ψ_k denote the network which consists of the unit U_k and m_k be the number of units in that network.

Let $\bar{y}_k^* = \frac{\sum_{i \in \psi_k} y_i}{m_k}$.

Modified HH estimator is defined as

$$\bar{y}_{HH}^* = \frac{\sum_{k=1}^{n_1} \bar{y}_k^*}{n_1}.$$

Below we state a result already proved by Thompson and Seber (1996), related to the mean and variance of $\bar{y}_{HH}^*$.

Result 1: The estimator $\bar{y}_{HH}^*$ is unbiased for the population mean.

That is, $E(\bar{y}_{HH}^*) = \bar{Y}_N$

(a) If the initial sample is selected by SRSWOR, then

$V(\bar{y}^*_{HH}) = \frac{N-n_1}{Nn_1(N-1)} \sum_{k=1}^{N} (\bar{y}^*_k - \mu)^2.$

Unbiased estimator of $Var(\bar{y}^*_{HH})$ is given by

$\hat{V}(\bar{y}^*_{HH}) = \frac{N-n_1}{Nn_1(n_1-1)} \sum_{k=1}^{n_1} (\bar{y}^*_k - \hat{\mu})^2.$

(b) If the initial sample is selected by SRSWR, then

$V(\bar{y}^*_{HH}) = \frac{\sum_{k=1}^{N}(\bar{y}^*_k - \mu)^2}{N \, n_1},$

where $\bar{y}^*_k$ is the average of observations in the network that includes the kth unit in sample and $\mu = \frac{\sum_{i=1}^{N} Y_i}{N}.$

(iii) Modified HT Estimator:

Using the first-order inclusion probabilities π_k, we can construct HT estimator for estimating population mean. With adaptive designs, the inclusion probabilities are not known for all units included in the sample. Hence, it cannot be used to estimate the mean unbiasedly. An unbiased estimator can be obtained by modifying the HT estimator. This estimator uses the units not satisfying the condition only when they are included in the initial sample. In this case, the probability that a unit is used in the estimator can be computed, even though its actual probability of inclusion in the sample may be unknown.

Define the indicator variable:

$$Z_k = \begin{cases} 1 & ; \text{if } k\text{th unit is included in the sample and satisfies} \\ & \quad condition \ C \\ 0 & ; otherwise \end{cases}$$

for $k = 1, 2 \ldots, N.$

The modified estimator is

$\bar{y}^*_{HT} = \frac{1}{N} \sum_{k=1}^{\nu} \frac{y_k z_k}{\pi_k},$

where ν is the number of distinct units in the final sample.

π_k is the probability that unit k is included in the estimator $= 1 - \frac{\binom{N-m_k}{n_1}}{\binom{N}{n_1}}$ Thompson and Seber (1996) proved a result given below related to the mean and variance of $\bar{y}^*_{HT}.$

Result 2: The estimator $\bar{y}^*_{HT}$ is unbiased for the population mean.

That is $E(\bar{y}^*_{HT}) = \bar{Y}$

and

$V(\bar{y}^*_{HT}) = \frac{1}{N^2} \sum_{h=1}^{K} \sum_{j=1}^{K} \bar{y}^*_h \bar{y}^*_j [\frac{\pi_{hj} - \pi_h \pi_j}{\pi_h \pi_j}],$

where K is the number of networks in the population and π_{hj} is the probability that the initial sample contains at least one unit each from the networks h and j.

$\hat{V}(\bar{y}^*_{HT}) = \frac{1}{N^2} \left[\sum_{k=1}^{\nu} (\frac{1-\pi_k}{\pi_k^2})(\bar{y}^*_k)^2 \, Z_k + \sum_{h=1}^{\nu} \sum_{j \neq h=1}^{K} (\frac{\pi_{hj} - \pi_h \pi_j}{\pi_{hj} \pi_h \pi_j}) \bar{y}^\star_h Z_h \bar{y}^\star_j Z_j \right].$

Illustrative Example:

Suppose it is desired to estimate the abundance of a particular species of plants in a

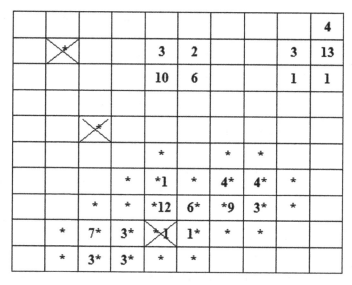

Fig. 10.6 Number of plants on the different plots of a square region

square region which is a clustered region. Let this region be divided into 100 plots which include 100 plants.

Figure 10.6 shows the number of plants on each plot.

Suppose we define the neighborhood of a plot which consists of the four plots directly above, below, to the left, and to the right of the individual plot.

The neighborhood of a plot in the initial sample will be searched if the plot has at least one plant on it, i.e. the neighborhood of ith plot will be searched if the condition $C=\{y_i > 0\}$ is satisfied for it. Where y_i is the observed value of Y on plot i. Suppose that a random sample without replacement of 3 plots is selected from the region. The plots selected in this sample are indicated by a X in Fig. 10.6. Following the adaptive procedure, we get the final sample. Plots included in this final sample are represented by $\star$ (asterisk) in Fig. 10.6.

If a unit in the initial sample satisfies the condition C, then the neighboring units of that unit are included in the sample. If any of these units satisfy the condition C, then its neighbors are also added to the sample. This is continued till the neighbors not satisfying condition C are obtained. The initial unit satisfying C along with its neighbors satisfying C form a network. The units that are added as neighbors but do not satisfy condition C are called edge units. The set of network units along with the edge units is called a cluster. The entire cluster gets selected in the final sample if any of the units satisfying C in it is selected in the initial sample. Thus, a network is a subset of a cluster. An unit that is selected in the initial sample but does not satisfy C is considered as a network of size 1.

The networks in the region form a partition of the plots in the grid.

Suppose that the initial sample of 3 plots is drawn without replacement as shown in Fig. 10.6. The 27 plots (with asterisk) are added to the sample.

In the above example, there are 3 networks of sizes $m_1 = 1, m_2 = 1$ and $m_3 = 13$ included in the final sample.

We have $\bar{y}_1^\star = \frac{0}{1} = 0$; $\bar{y}_2^\star = \frac{0}{1} = 0$ and $\bar{y}_3^\star = \frac{57}{13}$.

Therefore,

$$\hat{\mu} = \bar{y}_{HH}^\star = \frac{0 + 0 + \frac{57}{13}}{3} = \frac{57}{39} = 1.46.$$

Hence, $\hat{V}ar(\bar{y}_{HH}^\star) = \frac{N - n_1}{N n_1 (n_1 - 1)} \sum_{k=1}^{n_1} (\bar{y}_k^\star - \hat{\mu})^2$

$$= \frac{1}{n_1(n_1 - 1)} \sum_{k=1}^{n_1} (\bar{y}_k^\star - \hat{\mu})^2 \quad \text{(Approximately)}$$

$$= 2.14.$$

Hence, $\hat{S}E(\bar{y}_{HH}^\star) = \sqrt{2.14} = 1.46.$

Thus, the estimated total number of plants in the whole region $= N\bar{y}_{HH}^\star = 146$ with SE $= 146$.

We can also calculate the HT estimator of population mean and total.

HT estimate of population mean $= \bar{y}_{HT}^\star = \frac{1}{N} \sum_{k=1}^{\nu} \frac{\bar{y}_k^\star}{\pi_k}$.

There are three networks of sizes $x_1 = 1$; $x_2 = 1$; and $x_3 = 13$.

With $y_1^\star = 0$; $y_2^\star = 0$; $y_3^\star = 57$.

The inclusion probabilities of these networks are given as

$$\pi_1 = 1 - \frac{\binom{100-1}{3}}{\binom{100}{3}} = 0.03,$$

$$\pi_2 = 1 - \frac{\binom{100-1}{3}}{\binom{100}{3}} = 0.03,$$

$$\pi_3 = 1 - \frac{\binom{100-13}{3}}{\binom{100}{3}} = 0.3445.$$

The joint inclusion probabilities of these networks are given as

$$\pi_{12} = 1 - \left[\frac{\binom{100-1}{3} + \binom{100-1}{3} - \binom{100-1-1}{3}}{\binom{100}{3}} \right] = 0.000606,$$

$$\pi_{13} = 1 - \left[\frac{\binom{100-1}{3} + \binom{100-13}{3} - \binom{100-14}{3}}{\binom{100}{3}} \right] = 0.007396 = \pi_{23}.$$

Therefore, $\bar{y}_{HT}^\star = \frac{1}{100} \left[\frac{0}{.03} + \frac{0}{.03} + \frac{57}{0.3445} \right] = 1.655$ with SE$=1.34$.

Hence, estimated total number of plants $= N\bar{y}_{HT}^\star = 165.5$ with $SE = 134$.

10.8.2 Improved Unbiased Estimators in ACS

Dryver and Thompson (1998a, b) have obtained two new unbiased estimators by applying Rao–Blackwell theorem to $\hat{\mu}_1$ and $\hat{\mu}_2$ (i.e. $\bar{y}_{HH}^\star$ and $\bar{y}_{HT}^\star$, respectively) discussed in the previous section. These estimators are easy to compute. While computing the estimators $\hat{\mu}_1$ and $\hat{\mu}_2$, we use only those edge units which were in the initial sample. The new estimators $\hat{\mu}_{1+}$ and $\hat{\mu}_{2+}$ are developed considering only how many edge units were initially selected but not which ones.

Let S_0 denote the set of units selected in the initial sample and S denote the set of units selected in the final sample. Let n denote the initial sample size and v be the final sample size.

The final sample S can be partitioned into two parts as a core part S_c and the remaining part $S_{\bar{c}}$. The core part S_c is the set of all distinct units in the sample for which the condition C is satisfied. The remaining part $S_{\bar{c}}$ consists of all distinct units in the sample for which C is not satisfied.

For unit i, let f_i be the number of times the network to which unit i belongs, which is intersected by the initial sample. That is, f_i is the number of units in the initial sample that are in the network to which unit i belongs.

Let the statistic d^+ be defined as

$d^+ = [(i, y_i, f_i) : i \in S_c, (j, y_j) : j \in S_{\bar{c}}]$.

In d^+, the intersection frequency f_i is included only for $i \in S_c$. Let D^+ denote a r.v. that takes on possible values of d^+. D^+ denote sample space for d^+.

For $i \in S$, define indicator variable e_i as

$$e_i = \begin{cases} 1 & ; if \ y_i < C \ and \\ & \quad i \ is \ in \ the \ neighborhood \ of \ some \ j \in S_c \\ 0 & ; otherwise \end{cases}$$

$e_i = 1$ if i is an edge unit and the network that makes it an edge unit is selected in the initial sample. Other units in the initial sample may be edge units but sample units are the edge units whose network that classifies them as an edge unit was intersected in the initial sample.

The number of edge units in the sample is

$e_s = \sum_{i=1}^{v} e_i = \sum_{i \in S} e_i$.

The number of sample edge units selected in the initial sample S_0 is

$e_{S_0} = \sum_{i=1}^{n} e_i = \sum_{i \in S_0} e_i$.

The average y value for the sample edge units in the final sample is

$\bar{y}_e = \sum_{i=1}^{v} \frac{e_i y_i}{e_s}$.

For the ith unit in the sample, define a new variable W_i' as

$W_i' = \bar{y}_i^{\star}(1 - e_i) + \bar{y}_e e_i$.

The variable W_i' is the value $\bar{y}_i^{\star}$ when it is not related to sample edge units. While dealing with sample edge units W_i' equals the average of the sample edge units $\bar{y}_e$.

The new estimator $\hat{\mu}_{1+}$ (based on HH type estimator) is defined by

$\hat{\mu}_{1+} = E[\hat{\mu}_1 | D^+ = d^+] = E[\bar{y}_{HH}^{\star} | D^+ = d^+]$.

By Rao–Blackwell theorem $\hat{\mu}_{1+}$ is unbiased for μ and $V(\hat{\mu}_{1+}) \le V(\hat{\mu}_1)$.

Dryver and Thompson (1998a, b) have shown that

$\hat{\mu}_{1+} = \frac{\sum_{i=1}^{n} W_i'}{n}$

and

$V(\hat{\mu}_{1+}) = \frac{N-n}{Nn(N-1)} \sum_{i=1}^{N} (\bar{y}_i^{\star} - \mu)^2 - \frac{1}{n^2} \sum_{d^+ \in D^+} \frac{P(d^+)}{L(d^+)} \sum_{s_0' \in s} I(s_0', d^+)(\sum_{i \in s_0', e_i=1}$

$y_i - e_{s_0'} y_e)^2$,

where the indicator variable,

$$I(s'_0, d^+) = \begin{cases} 1 & ; \quad \text{if } g(S'_0) = d^+ \\ 0 & ; \quad otherwise. \end{cases}$$

$g(S'_0)$ denotes the function that maps an initial sample into a value of d^+ resulting from its selection.

$L(d^+)$= Number of initial samples compatible with d^+.

$P(d^+) = Prob(D^+ = d^+)$.

An unbiased estimator of $V(\hat{\mu}_{1+})$ when the initial sample is drawn by SRSWOR is given by

$$\hat{V}(\hat{\mu}_{1+}) = \frac{N-n}{Nn(n-1)} \sum_{i=1}^{N} (\bar{y}^\star_i - \hat{\mu}_1)^2 - \frac{1}{L(d^+)\, n^2} \sum_{s'_0 \in s} I(s'_0, d^+)$$

$(\sum_{i \in s'_0, e_i=1} y_i - e_{s'_0} \bar{y}_e)^2.$

A more efficient estimator is

$\hat{V}(\hat{\mu}_{1+}) = E[V(\hat{\mu}_{1+})|d^+]$

$$= \frac{1}{L(d^+)} \sum_{s'_0 \in s} I(s'_0, d^+) \frac{N-n}{Nn(n-1)} \sum_{i=1}^{n} (\bar{y}^\star_i - \hat{\mu}_1)^2 - \frac{1}{L(d^+)\, n^2} \sum_{s'_0 \in s} I(s'_0, d^+)$$

$(\sum_{i \in s'_0, e_i=1} y_i - e_{s'_0} \bar{y}_e)^2.$

Another estimator $\hat{\mu}_{2+}$ based on HT type estimator is developed as follows:

For the kth network in the sample, define the indicator variable:

$$e'_k = \begin{cases} 1 & ; \quad \textit{if the sum of y values in network } k \textit{ satisfies} \\ & \quad C \textit{ and } k \textit{ is in the neighborhood of some } k' \in S_c \\ 0 & ; \quad otherwise. \end{cases}$$

The variable e'_i for $i = 1, 2 \ldots, k$ is same as e_i but it is indexed by the network instead of individual unit.

All networks of size greater than 1 have $e'_k = 0$

Also $e'_k = 0$ for those units not in S.

$$y'_k = \begin{cases} y^\star_k & ; \quad \text{if } e'_k = 0 \\ \bar{y}_e & ; \quad \text{if } e'_k = 1. \end{cases}$$

Thus for a network of units satisfying the condition, y'_k is the total of y-values in that network while for a sample edge unit y'_k is the averages of y-values for all the sample edge units in the sample.

The new estimator based on $\hat{\mu}_2$ (i.e. based on HT type estimator) is defined as follows:

$\hat{\mu}_{2+} = E[\hat{\mu}_2 | D^+ = d^+]$

$$= \frac{1}{N} \sum_{k=1}^{K} \frac{y'_k Z_k}{\pi_k}$$

$$V(\hat{\mu}_{2+}) = \frac{1}{N^2} \sum_{k=1}^{K} \sum_{h=1}^{K} y'_k y'_h [\frac{\pi_{kh} - \pi_k \pi_h}{\pi_k \pi_h}] - \frac{1}{n^2} \sum_{d^+ \in D^+} \frac{P(d^+)}{L(d^+)} \sum_{s'_0 \in s} I(s'_0, d^+)$$

$(\sum_{i \in s'_0, e_i = 1} y_i - e_{s'_0} y_e)^2.$

An unbiased estimator of this variance is

$$\hat{V}(\hat{\mu}_{2+}) = \frac{1}{N^2} \sum_{k=1}^{K} \sum_{h=1}^{K} y_k^\star y_h^\star [\frac{\pi_{kh} - \pi_k \pi_h}{\pi_{kh} \pi_k \pi_h}] - \frac{1}{L(d^+) \, n^2} \sum_{s'_0 \in s} I(s'_0, d^+)$$

$(\sum_{i \in s'_0, e_i = 1} y_i - e_{s'_0} \bar{y}_e)^2.$

A more efficient estimator of the variance is

$$\hat{V}(\hat{\mu}_{2+}) = \frac{1}{L(d^+)} \sum_{s'_0 \in s} I(s'_0, d^+) \frac{1}{N^2} \sum_{k=1}^{K} \sum_{h=1}^{K} y_k^\star y_h^\star [\frac{\pi_{kh} - \pi_k \pi_h}{\pi_{kh} \pi_k \pi_h}] - \frac{1}{L(d^+) \, n^2}$$

$$\sum_{s'_0 \in s} I(s'_0, d^+)(\sum_{i \in s'_0, e_i = 1} y_i - e_{s'_0} \bar{y}_e)^2.$$

10.9   Estimation of Population Mean in ACS by using Edge Units

The unbiased estimators of population mean/total of rare and clustered population developed by Thompson and Seber (1996) are based on the networks that consider units not satisfying the predefined condition as the edge units.

These edge units are identified during the adaptation stage, when the initial sample is expanded around sampling units that satisfy the pre-defined condition. In this process, the population is divided into clusters that are of the convex type.

In this section, we discuss the two new estimators of the population mean based on the networks that consider the units satisfying the pre-defined condition as the edge units. These units are identified during the adaptation stage, where the initial sample is expanded around sampling units that do not satisfy the pre-defined condition. The new estimators use the edge units in the estimation process.

We consider a rare and clustered population divided into N rectangular grid points of equal sizes. Each grid point serves as a sampling unit. Let Y be the binary variable that takes value 1 if a particular grid point satisfies the defined condition C and 0 otherwise. An ACS design based for the above type of situation usually includes the following steps:

(i) An initial sample of size n_1 is selected from the population by using SRSWOR method.

(ii) The neighbors of an initial sampling unit are sampled if it does not satisfies the condition C. Further, the neighbors of these units are sampled if they do not satisfy the condition C. This is continued till desirable neighbors are sampled. Here, the groups of adjacent units do not satisfy the condition C. It is called a network. The adjacent units are the outermost units in the network that satisfy the condition C. These are now the edge units. They are used in the estimation. Based on the above method, we propose the following new unbiased estimators of the population mean.

(I) HH Type Estimator

$$\mu_1^* = \frac{\sum_{i=1}^{n_1} W_i}{n_1},$$

where $W_i = \frac{\sum_{j \in \psi_i} y_j}{e_i}$ for $i = 1, 2 \ldots, n$

ψ_i: ith network.

y_j: Value of the variable Y corresponding to jth edge unit in ψ_i .

e_i : Number of edge units in ψ_i.

n_1: Number of networks.

Theorem 10.1 $E(\hat{\mu}_1^*) = \mu_y$.

Unbiased estimator of the variance of $\hat{\mu}_1^*$ is given by

$\hat{V}(\hat{\mu}_1^*) = \frac{N-n_1}{N \, n_1 \, (n_1-1)} \sum_{i=1}^{n_1} (w_i - \mu_1^*)^2$.

Proof Using Result 1 above an unbiased estimator of the variance of $\hat{\mu}_1^*$ can be obtained as given above and left for exercise.   □

(II) HT Type Estimator

Let K = Number of distinct networks in the population under study.

ψ_k= Set of edge units in the kth network.

x_k= Number of interior units that make up the kth network.

e_k = Number of units in ψ_k.

y_k^* = Sum of y-values in the edge units of the kth network.

$= \sum_{j \in \psi_k} y_i$

α_k^*= The inclusion probability of kth network.

$= 1 - \frac{\binom{N-x_k-e_k}{n_1}}{\binom{N}{n_1}}$.

Define

$$z_k = \begin{cases} 1 & \text{; if any unit of the kth network is an initial} \\ & \quad \textit{sampling unit} \\ 0 & \text{; otherwise.} \end{cases}$$

Then, HT type estimator of the population mean is given by

$\mu_2^* = \frac{1}{N} \sum_{k=1}^{K} \frac{z_k y_k^*}{\alpha_k^*}$.

Theorem 10.2 $E(\hat{\mu}_2^*) = \mu_y$.

An unbiased estimator of the variance of $\hat{\mu}_2^*$ is given by

$\hat{V}(\hat{\mu}_2^*) = \frac{1}{N^2} \sum_{k=1}^{K} \sum_{h=1}^{K} y_k^* y_h^* z_k z_h \left[\frac{(\alpha_{kh}^* - \alpha_k^* \alpha_h^*)}{\alpha_{kh}^* \alpha_k^* \alpha_h^*} \right]$,

where

$\alpha_{kh}^* = 1 - \left[\frac{\binom{N-x_k-e_k}{n_1} + \binom{N-x_h-e_h}{n_1} - \binom{N-x_h-x_k-e_k-e_h}{n_1}}{\binom{N}{n_1}} \right]$.

Note that $\alpha_{kk}^* = \alpha_k^*$.

Proof Using Result 2 above an unbiased estimator of the variance of this $\hat{\mu}_2^*$ can be obtained as given above.   □

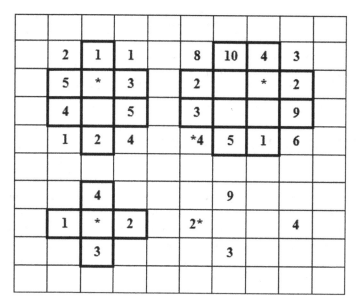

Fig. 10.7 Number of animals observed on the different quadrats. ⋆ in a quadrat indicates its selection in the initial sample of ACS. Empty square with thick borders indicates a sampling unit included in the network of the respective initial sampling units that do not satisfy the condition C

Illustrative Example

Figure 10.7 shows that area is divided into 100 square quadrats of equal sizes for estimating the mean abundance of particular species of animal.

In this example

$x_1 = 2, x_2 = 4, x_3 = x_4 = x_5 = 1$

$e_1 = 6, e_2 = 8, e_3 = 4, e_4 = e_5 = 1$

$y_1 = 20, y_2 = 36, y_3 = 10, y_4 = 4, y_5 = 2.$

Hence, $W_1 = 3.67, W_2 = 4.5, W_3 = 2.5, W_4 = 4.0, W_5 = 2.0.$

Thus, $\hat{\mu}_1^* = 3.33$

$\hat{V}(\hat{\mu}_1^*) = 0.21; \quad \widehat{SE}(\hat{\mu}_1^*) = 0.46.$

Further,

$\alpha_1^* = 0.35, \alpha_2^* = 0.48, \alpha_3^* = 0.23, \alpha_4^* = \alpha_5^* = 0.10.$

Hence, $\hat{\mu}_2^* = 2.35.$

The joint inclusion probabilities are

$\alpha_{12}^* = 0.15, \alpha_{13}^* = 0.07, \alpha_{14}^* = 0.02, \alpha_{15}^* = 0.03, \alpha_{23}^* = 0.10, \alpha_{24}^* = 0.04,$

$\alpha_{25}^* = 0.04, \alpha_{34}^* = 0.02, \alpha_{35}^* = 0.02, \alpha_{45}^* = 0.01.$

It gives

$\hat{V}(\hat{\mu}_2^*) = -0.28.$

If the usual ACS design is used in the same example, then we get more precise estimate of the population mean. But the final sample size in that case is 34. Whereas if we use the above proposed new design the final sample size is 27. As the size of the initial sample increases this difference in the final sample size is expected to grow much faster. If the cost of acquisition, collection, and measurement is high then the above new design of ACS is preferable to the regular.

The variance of the estimator proposed in the new design faces the drawback of being negative in some situations as seen above. This is a drawback of HT estimator.

10.10   Restricted Adaptive Sampling Design

There are practical difficulties for applying ACS in practice. Such as, how to determine appropriate critical values, how to schedule survey times, the indefinite sampling problem due to a low critical value and costly edge unit issues. In ACS, sometimes the final sample size may be very large. That in turn may increase the cost of selection, acquisition, and taking observations. In such situations instead of obtaining the exact estimate of the population total, it may be desirable to estimate simply an upper bound for the population total.

10.10.1   Estimation of an Upper Bound for the Population Total

We have proposed the restricted adaptive sampling (RAS) design to obtain such an estimate.

The steps involved in RAS are as follows:

(i) Draw a sample of size n by SRSWOR from the rare and clustered population. (First stage)

(ii) From these units, locate the units not satisfying the condition C.

(iii) Discard these units along with their first-order neighbors.

(iv) Draw another sample of size n by SRSWOR from the reduced population (second stage).

(v) Identify the units satisfying condition C from the first-stage and second-stage sampling units.

(vi) Estimate an upper bound for the population total Y_t based on the values of the variables of interest for the units identified in (v).

RAS works opposite to ACS. In ACS, adaptation is based on the units satisfying the condition C, whereas in RAS, the adaptation is based on the units not satisfying the condition C. The population size N gets reduced as shown in Table 10.1, depending upon the number of units satisfying C in the first sample before drawing the second sample.

Thus, the maximum possible reduction in the population size would be $5n$.

Hence, the minimum possible reduced population size would be $N - 5n$.

Table 10.1 Maximum reduction in the population size according to the number of units satisfying C in the initial sample

Number of units satisfying condition C	n	n − 1	n − 2.......0
Maximum possible reduction in population size	n	n + 4	n + 8......5n

Probability that unit i is selected in the first-stage sample $= P_{i1} = \frac{1}{N}$.

The probability of selection of unit i in the second stage sample P_{i2} depends upon the number of units found in the first sample that do not satisfy the condition C.

These probabilities will be $\frac{1}{N-n}, \frac{1}{N-n-4}, \ldots, \frac{1}{N-5n}$, respectively.

Thus, the probability of selection of an unit in the second stage sample can be at least $\frac{1}{N-n}$.

The probabilities of selection of units being different, probability proportional sampling (PPS) estimator is recommended to estimate the population total.

Let y_{i1} and y_{i2} denote values of the variable observed for the ith unit in the first-stage and second-stage samples, respectively. Define an estimator of upper bound for the population total:

$$\hat{Y}_u = \frac{1}{2}[\hat{Y}_1 + \hat{Y}_2] = \frac{\sum_{i=1}^{n}(\frac{y_{i1}}{P_{i1}} + \frac{y_{i2}}{P_{i2}})}{2n}$$

Consider $E[\hat{Y}_1] = \frac{1}{n} \sum_{i=1}^{n} E(\frac{y_{i1}}{P_{i1}})$

$$= \frac{1}{n} \sum_{i=1}^{n} \sum_{i=1}^{N} (\frac{Y_j}{P_j}) P_j$$
$$= Y.$$

Similarly, $E[\hat{Y}_2] = Y$.

Hence, $E[\hat{Y}_u] = Y$.

Thus, $\hat{Y}_u$ is unbiased for Y

Variance of $\hat{Y}_u$ is given by

$$V(\hat{Y}_u) = \frac{1}{4}[V(\hat{Y}_1) + V(\hat{Y}_2)],$$

where

$\hat{Y}_1$ and $\hat{Y}_2$ are the unbiased estimators of the population total obtained from first-stage and second-stage sampling units, respectively.

$$V(\hat{Y}_1) = V[\frac{1}{n} \sum_{i=1}^{n} \frac{y_{i1}}{P_{i1}}]$$
$$= \frac{1}{n^2} \sum_{i=1}^{n} V(\frac{y_{i1}}{P_{i1}})$$
$$= \frac{1}{n^2} \sum_{i=1}^{n} E(\frac{y_{i1}}{P_{i1}} - E(\frac{y_{i1}}{P_{i1}}))^2$$
$$= \frac{1}{n^2} \sum_{i=1}^{n} E(\frac{y_{i1}}{P_{i1}} - Y)^2$$
$$= \frac{1}{n^2} \sum_{i=1}^{n} \sum_{i=1}^{N} (\frac{Y_j}{P_j} - Y)^2 P_j$$
$$= \frac{1}{n} \sum_{i=1}^{N} (\frac{Y_j}{P_j} - Y)^2 P_j.$$

Similarly,

$$V(\hat{Y}_2) = \frac{\sum_{j=1}^{N'} (\frac{Y_j}{P_j} - Y)^2 P_j}{n}$$

N' = Reduced population size.

An unbiased estimator of variance of $\hat{Y}_{ut}$ is given by

$$\hat{V}(\hat{Y}_{ut}) = \frac{1}{4}\left[\hat{V}(\hat{Y}_1) + \hat{V}(\hat{Y}_2)\right]. \tag{10.1}$$

We will first obtain an unbiased estimator of $V(\hat{Y}_1)$.

In probability proportional sampling, we have

$V(\hat{Y}_1) = \frac{1}{n}\sum_{i=1}^{N}(\frac{Y_i}{P_i} - Y)^2 P_i$

$= \frac{1}{n}\sum_{i=1}^{N}\frac{Y_i^2}{P_i} - Y^2.$

Also, $V(\hat{Y}_1) = E[\hat{Y}_1^2] - Y^2.$

If $E[\hat{V}(\hat{Y}_1)] = V(\hat{Y}_1)$, then we get

$E[\hat{V}(\hat{Y}_1)] = E[\hat{Y}_1^2] - Y^2.$

Hence, $Y_t^2 = E[\hat{Y}_1^2] - E[\hat{V}(\hat{Y}_1)]$

$= E[\hat{Y}_1^2 - \hat{V}(\hat{Y}_1)].$

Note that $E[\frac{1}{n}\sum_{i=1}^{n}\frac{y_{i_1}^2}{P_{i_1}}] = \sum_{i=1}^{N}\frac{Y_i^2}{P_i}.$

Hence, $E[\hat{V}(\hat{Y}_1)] = \frac{1}{n} E[\frac{1}{n}\sum_{i=1}^{n}\frac{y_{i_1}^2}{P_{i_1}}] - \frac{1}{n} E[\hat{Y}_1^2 - \hat{V}(\hat{Y}_1)].$

Hence, $\frac{n-1}{n} E[\hat{V}(\hat{Y}_1)] = E[\frac{1}{n^2}\sum_{i=1}^{n}\frac{y_{i_1}^2}{P_{i_1}} - \frac{1}{n}\hat{Y}_1^2].$

This gives, $E[\hat{V}(\hat{Y}_1)] = E\left[\frac{1}{n(n-1)}\left\{\sum_{i=1}^{n}(\frac{y_{i_1}}{P_{i_1}} - \hat{Y}_1)^2 P_{i_1}\right\}\right].$

Thus, $\frac{1}{n(n-1)}\left\{\sum_{i=1}^{n}(\frac{y_{i_1}}{P_{i_1}} - \hat{Y}_1)^2 P_{i_1}\right\}$ is unbiased for $V(\hat{Y}_1)$.

Similarly, it can be shown that $\frac{1}{n(n-1)}\left\{\sum_{i=1}^{n}(\frac{y_{i_2}}{P_{i_2}} - \hat{Y}_2)^2 P_{i_2}\right\}$ is unbiased for $V(\hat{Y}_2)$.

Using these expressions in (10.1), we get

$$\hat{V}(\hat{Y}_{ut}) = \frac{\sum_{i=1}^{n}(\frac{y_{i_1}}{P_{i_1}} - \hat{Y}_1)^2 P_{i_1} + \sum_{i=1}^{n}(\frac{y_{i_2}}{P_{i_2}} - \hat{Y}_2)^2 P_{i_2}}{4n(n-1)}.$$

10.10.2   *Estimation of a Lower Bound for the Population Total*

In Sect. 10.10.1, we have proposed an estimator of the upper bound for the population total. Sometimes the researcher may be interested in estimating the minimum possible value that the population total can assume in the prevailing conditions. For such situation, using RAS discussed above one can estimate a lower bound for the population total as follows:

As seen in Sect. 10.6, the probability that unit i is selected in the first-stage sample $= P_{i1} = \frac{1}{N}$, for $i = 1, 2, \ldots, n$.

The probability of selection of unit i in the second stage sample depends upon the number of units in the initial sample that do not satisfy the condition C. This number can vary from $0, 1, 2, \ldots, n$.

Accordingly the probabilities of selection of an unit in the second stage sample will be $\frac{1}{N-n}, \frac{1}{N-n-4}, \ldots, \frac{1}{N-5n}$, respectively.

Hence, Probability of selection of an unit in the second stage sample will be at most $\frac{1}{N-5n}$.

Thus, $P_{i2} = \frac{1}{N-5n}$ for $i = 1, 2, \ldots, n$.

The selection probabilities being different at the two stages, PPS estimator is proposed for estimating the population total.

We have proposed the following estimator for the lower bound of population total:

$$\hat{Y}_l = \frac{\frac{y_{i1}}{P_{i1}} + \frac{y_{i2}}{P_{i2}}}{2n}.$$

Variance of $\hat{Y}_{lt}$ is given by

$$V(\hat{Y}_l) = V(\hat{Y}_1) + V(\hat{Y}_2),$$

where

$\hat{Y}_1$ and $\hat{Y}_2$ are the unbiased estimators of the population total obtained from first-stage and second-stage sampling units, respectively.

$$V(\hat{Y}_1) = \frac{\sum_{j=1}^{N} (\frac{Y_j}{P_j} - Y)^2 P_j}{n}$$

$$V(\hat{Y}_2) = \frac{\sum_{j=1}^{N'} (\frac{Y_j}{P_j} - Y)^2 P_j}{n}$$

$N' =$ Reduced population size.

An unbiased estimator of variance of $\hat{Y}_{lt}$ is given by

$$\hat{V}(\hat{Y}_{lt}) = \frac{\sum_{i=1}^{n} (\frac{y_{i1}}{P_{i1}} - \hat{Y}_1)^2 P_{i1} + \sum_{i=1}^{n} (\frac{y_{i2}}{P_{i2}} - \hat{Y}_2)^2 P_{i2}}{4n(n-1)}.$$

10.10.3 Illustrative example of RAS

Figure 10.8 represents a population of 60 plants which is highly clumped and divided into 400 quadrats of equal size. The quadrats shown in blue color were selected at the first stage of RAS. The quadrats with a circle in it were selected at the second stage of RAS. The quadrats with a cross symbol were discarded from the population before selecting the second stage sample.

In order to obtain the estimate of upper bound for population total ($\hat{Y}_{ut}$), we take

$$P_{i1} = \frac{1}{N} = \frac{1}{400} \quad \text{and} \quad P_{i2} = \frac{1}{N-n} = \frac{1}{390}.$$

In the first-stage sample of 10 quadrats, 3 quadrats are occupied by 1 plant each and the remaining 7 quadrats are empty. Thus, $\sum_{i=1}^{n} y_{i1} = 3$.

In the second stage sample 3 quadrats are occupied by 2 plants each and the remaining quadrats are empty. Hence, $\sum_{i=1}^{n} y_{i2} = 6$.

Using the formula of $\hat{Y}_u$ and $\hat{V}(\hat{Y}_u)$ derived in Sect. 10.6, we get

$$\hat{Y}_u = 177$$

and

$$\hat{V}(\hat{Y}_u) = \frac{1}{4 \times 10 \times 9} \left[\frac{3 \times (400-177)^2}{400} + \frac{3 \times (780-177)^2}{390} \right]$$
$$= 8.80$$

$$\widehat{SE}(\hat{Y}_u) = 2.96.$$

In order to obtain the estimate of lower bound for population total ($\hat{Y}_l$), we take

$$P_{i1} = \frac{1}{N} = \frac{1}{400}$$

							X									1	1	1	
X						X	X	X									②	1	
X		1	②	1			X				X						②	1	
		1	1	1	1				1		X	X	X						
												X				X			
			X												X	X	X		
								1								X			
1	2					1	1	1				1	1						
1	2	1		◯		1	1	1		X		1		◯					
1	1	1				1	1x	X	X										
	1							X								1x	1		
	X											1							
X	X	X			1						②	1							
	X			◯	1					1	1	1	1						
												1							
◯									◯										
	1											1x							
		1			1x							1							
		◯	1	1	1	1								◯					

Fig. 10.8 Representation of sample selected by RAS

and
$$P_{i2} = \frac{1}{N-5n} = \frac{1}{350}.$$
Hence, $\hat{Y}_I = 165$
$$\hat{V}(\hat{Y}_I) = \frac{1}{4\times10\times9}\left[\frac{3\times(400-165)^2}{400} + \frac{3\times(780-165)^2}{390}\right]$$
$$= 9.23$$
$$\widehat{SE}(\hat{Y}_I) = 3.03.$$

10.11   Estimation of the Final Sample Size in ACS

In ACS, the final sample size depends upon the units selected in the initial sample. Since the initial sample is drawn by using SRSWOR, the final sample size is not fixed. It can be looked upon as a random variable. Its expected value can be calculated. In view of the expected cost of collecting the information, knowledge of the expected value of the final sample size is important. In this section, we have discussed the problem of estimation of the final sample size in ACS in the form a theorem and further have discussed a pilot survey. This illustrates the use of the theorem to estimate the final sample size in ACS and further the estimation of population total and its standard error.

Theorem 10.3 *The final sample size in ACS is given by*
$$n_s = n + \sum_{k=1}^{K}(m_k - X_k)\delta_k = \left(n - \sum_{k=1}^{K} X_k\delta_k\right) + \sum_{k=1}^{K} m_k\delta_k$$

and

$$E[n_s] = n + (1 - \tfrac{n}{N})N^\star - \sum_{k=1}^{K} m_k \frac{\binom{N-m_k}{n}}{\binom{N}{n}},$$

where

n_S: *Final sample size*

U_i: *Sampling units in the population,* $i = 1, 2 \ldots, N.$

K: *Number of networks in the population.*

m_k : *Size of the kth network,* $k = 1, 2 \ldots, K.$

X_k : *Number of units included in the initial sample from the kth network* $k = 1, 2, \ldots, K.$

π_{ki}: *Inclusion probability for sampling unit i from network k,* $i = 1, 2, \ldots, m_k.$

$$\delta_k = \begin{cases} 1 & \text{; if the initial sample includes a sampling unit} \\ & \quad from\ kth\ network. \\ 0 & ; otherwise. \end{cases}$$

$N^\star$: *Number of network units in the population.*

Proof The number of network units in the population is given by

$$N^\star = \sum_{k=1}^{K} m_k.$$

Hence, the number of non-network units in the population is equal to $N - N^\star$.

When a non-network unit is included in sample, no more sampling unit is added to the sample. Hence, the inclusion probability for every non-network unit to be in the final sample is same as its inclusion probability in the initial sample.

This probability that an unit in the population is included at a particular draw in the initial sample $= \frac{1}{N}$.

The non-network unit may get selected in the initial sample at any of the n draws.

Hence, probability that a non-network unit gets selected in the initial sample $= \frac{1}{N} + \frac{1}{N} \cdots + \frac{1}{N}$ (n terms)

$$= \frac{n}{N}.$$

If a unit from network k is included in the initial sample, then every other sampling unit from that network is included in the final adaptive sample. It is therefore of interest to note that the inclusion probability of a sampling unit in the sample is same for all sampling units in the population, but that for the adaptive sample depends on the network size.

If a sampling unit from the network k is included in the initial sample, then all m_k sampling units of the network k are included in the final adaptive sample. Hence, probability that the ith unit from the kth network is included in the final sample is given by

$$\pi_{ik} = P[\delta_k = 1] = \tfrac{m_k}{N}, \quad i = 1, 2 \ldots, m_k; \ k = 1, 2 \ldots, K.$$

Now, if a sampling unit from the network k is included in the initial sample then all the m_k sampling units of the network k are included in the final adaptive sample. Since there are X_k sampling units from the network k in the initial sample, an additional $(m_k - X_k)$ sampling units are added adaptively to the sample.

Hence, the final adaptive sample size,

$$n_s = n + \sum_{k=1}^{K}(m_k - X_k)\delta_k. \tag{10.2}$$

Alternatively, the initial sample has a total of $\sum_{k=1}^{K} X_k$ network units and hence there are $(n - \sum_{k=1}^{K} X_k)$ non-network units. These units do not cause any change in the sample size after adaptation. The network units in the final adaptive sample are from those networks that contain at least one sampling unit in the initial sample. It is obvious that this number is given by $\sum_{k=1}^{K} m_k \delta_k$.

The two terms together give

Final adaptive sample size$(n_s) = (n - \sum_{k=1}^{K} X_k) + \sum_{k=1}^{K} m_k \delta_k$.

Noting that $\sum_{k=1}^{K} X_k$ can also be written as $\sum_{k=1}^{K} X_k \delta_k$, we get

$$n_s = \left(n - \sum_{k=1}^{K} X_k \delta_k\right) + \sum_{k=1}^{K} m_k \delta_k. \tag{10.3}$$

This shows the equivalence between (10.2) and (10.3) above.

It should be noted that X_k, $k = 1, 2 \ldots, K$ are the only random variables in the expression for the final adaptive sample size n_s. The other random quantity δ_k is a function of X_k, $k = 1, 2 \ldots, K$.

More specifically,

$P[\delta_k = 1] = P[X_k > 0]$, $k = 1, 2 \ldots, K$.

If we express, the number of non-network units in the initial sample as $X_{k+1} = n - \sum_{k=1}^{K} X_k$, the vector $(X_1, X_2 \ldots, X_{k+1})$ follows the multivariate hypergeometric distribution.

The joint probability mass function of $(X_1, X_2 \ldots, X_{k+1})$ is given by

$$P[X_1 = x_1, X_2 = x_2 \ldots, X_{k+1} = x_{k+1}] = \frac{\binom{m_1}{x_1}\binom{m_2}{x_2}\ldots\binom{m_{k+1}}{x_{k+1}}}{\binom{N}{n}},$$

where $m_{k+1} = N - \sum_{k=1}^{K} m_k$. It is the number of non-network units in the population.

Hence,

$E[X_k] = n\frac{m_k}{N}$, for $k = 1, 2 \ldots, K + 1$

$V[X_k] = n\left(\frac{N-n}{N-1}\right)\left(\frac{m_k}{N}\right)\left(\frac{N-m_k}{N}\right)$, for $k = 1, 2 \ldots, K + 1$.

Note that, for $k = 1, 2 \ldots, K + 1$,

$P[X_k = 0] = \frac{\binom{N-m_k}{n}}{\binom{N}{n}}$.

Hence, $P[\delta_k = 1] = 1 - P[\delta_k = 0] = 1 - \frac{\binom{N-m_k}{n}}{\binom{N}{n}}$.

Now,

$E[n_s] = E\left[(n - \sum_{k=1}^{K} X_k \delta_k) + (\sum_{k=1}^{K} m_k \delta_k)\right]$

$$= E[X_{k+1}] + \sum_{k=1}^{K} m_k E[\delta_k]$$

$$= n\left(\frac{m_{k+1}}{N}\right) + \sum_{k=1}^{K} m_k \left[1 - \frac{\binom{N-m_k}{n}}{\binom{N}{n}}\right]$$

$$= n + (1 - \frac{n}{N})N^* - \sum_{k=1}^{K} \left[m_k \frac{\binom{N-m_k}{n}}{\binom{N}{n}}\right].$$

Hence the theorem.   □

10.12   Pilot Study for Estimating the Final Sample Size

A pilot study was conducted by using ACS. The area of 100 acres in Tamhini Ghat, Maharashtra was divided into 100 plots, each of size one acre. The interest was to estimate the total number of evergreen plants which are rare in that area due to the presence of Basalt rocks.

A random sample of 15 plots was drawn from this area using SRSWOR. The plots selected in the initial sample are shown by putting ⋆ in that plot (Fig. 10.9).

Further, it was checked which of the plots selected in the initial sample satisfy the condition $C = \{y > 0\}$, where y denotes the number of evergreen plants in a plot. The neighbors of the plots where C was satisfied were included in the sample by using the usual procedure. The plots selected in the initial sample which did not satisfy the condition C were treated as networks of size one.

In Fig. 10.9, there are in total 14 networks shown in blue color of which eleven are of size 1, one network of size 4, the other of size 5, and the last network is of size 13. The edge units of the last three networks are shown in pink color.

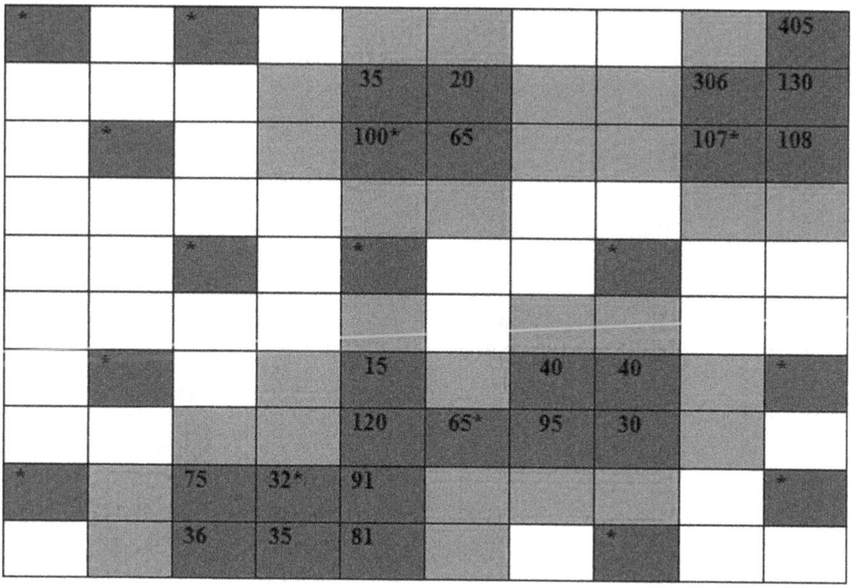

Fig. 10.9 Number of evergreen plants observed on the plots in the population

Expected value of the final sample size was calculated by using the formula derived above. For computational efficiency in the estimation, r number of repetitions were performed. The values of r varied as 5000, 10000, 20000, and 100000. For ACS, we required initial sample size(n). It was varied as 15, 20, 25, 35, 45. The samples were selected by using probability proportional to size (pps) without replacement method.

The expected final sample sizes were calculated in each case. The results are presented in Table 10.2.

The expected final sample size increased as the initial sample size increased.

10.13 Exercises

1. Explain the adaptive sampling procedure.
2. Compare adaptive sampling design with conventional sampling design.

Table 10.2 Estimated values of population total, standard error of its estimator and expected final sample size

Number of repetitions (r)	Initial Sample Size (n)	Expected Sample Size (n_s)	$\bar{y}_{HT}$	$\hat{SE}(\bar{y}_{HT})$
5000	10	21.32	2031.62	1316.18
	15	28.22	2033.26	973.95
	20	33.84	2032.51	756.07
	25	38.90	2031.08	599.86
	35	48.10	2032.12	383.25
	45	56.53	2030.19	244.15
10000	10	21.38	2029.31	1315.63
	15	28.21	2033.75	969.66
	20	33.87	2032.51	755.62
	25	38.90	2029.71	595.55
	35	48.05	2030.47	382.99
	45	56.58	2031.99	241.71
20000	10	21.40	2033.84	1314.15
	15	28.18	2030.97	967.08
	20	33.92	2032.02	755.00
	25	38.93	2032.05	598.03
	35	48.08	2031.53	383.49
	45	56.52	2031.71	237.25
100000	10	21.39	2029.16	1309.28
	15	28.22	2031.48	970.77
	20	33.88	2029.15	754.17
	25	38.92	2029.00	592.94
	35	48.06	2031.42	382.12
	45	56.53	2031.08	236.11

3. Which factors are required to be specified before using adaptive sampling design?
4. Give Thompson's definition of adaptive sampling design.
5. Define the following terms with reference to adaptive sampling: (i) neighborhood of a sampling unit (ii) cluster of sampling units (iii) network of sampling units (iv) edge unit of a network
6. Describe adaptive cluster sampling design with an example.
7. If the initial sample in ACS is selected by SRSWOR, then obtain the probability that ith unit in the population is included in the sample.
8. In ACS, find the probability of inclusion in the sample for a network.
9. In ACS, find the joint probability of inclusion in the sample for two distinct networks.
10. Write a note on the different approaches of estimation in adaptive sampling.
11. Explain the procedure of estimating the population mean and proportion in case of a probability sample.
12. Discuss various estimators of the population mean in ACS.
13. Show that $\bar{y}_{HH}^{\star}$ is unbiased for the population mean.
14. Derive variance of $\bar{y}_{HH}^{\star}$ when the initial sample is drawn by SRSWOR and SRSWR.
15. Show that $\bar{y}_{HT}^{\star}$ is unbiased for the population mean and derive its variance.
16. Explain how to obtain two new improved unbiased estimators from $\bar{y}_{HH}^{\star}$ and $\bar{y}_{HT}^{\star}$ by using Rao–Blackwell theorem.
17. Explain the procedure of estimating the population mean in ACS by using the edge units of the networks.
18. Show that $\hat{\mu}_1^{\star}$ is unbiased for the population mean. Obtain an unbiased estimate of the variance of $\hat{\mu}_1^{\star}$.
19. Show that $\hat{\mu}_2^{\star}$ is unbiased for the population mean. Obtain an unbiased estimate of its variance.
20. Describe RAS procedure.
21. State the estimator of the upper bound for the population total using RAS. Show that it is unbiased estimator of the population total. Obtain its variance.
22. State the estimator of the lower bound for the population total using RAS. Is it unbiased for the population total? Obtain its variance and expression for unbiased estimate of this variance.
23. State and prove an expression for the final sample size in ACS.

References

Cochran, W.G.: Sampling Techniques, 3rd edn. Wiley, New York (1977)
Dryver, A.L., Thompson, S.K.: Improving Unbiased Estimators in Adaptive Cluster Sampling. Technical Report, Department of Statistics, Pennsylvania State University (1998a)
Dryver, A.L., Thompson, S.K.: Adaptive Cluster Sampling Without Replacement of Clusters. Technical Report 9, Department of Statistics, Pennsylvania State University (1998b)

Elliott, M.R., Little, R.J.A.: Model-based alternatives to trimming survey weights. J. Off. Stat. **16**(3), 191–209 (2000)

Little, R.J.A., Rubin, D.B.: Statistical Analysis with Missing Data. Wiley Series in Probability and Statistics, 2nd edn. Wiley, New York (2002)

Thompson, S.K.: Adaptive cluster sampling. J. Am. Stat. Assoc. **85**(412), 1050–1058 (1990)

Thompson, S.K., Seber, G.A.F.: Adaptive Sampling. Wiley, New York (1996)

Chapter 11
Two-Stage Adaptive Cluster Sampling

11.1 Introduction

Adaptive sampling is a method in which selection of units at any stage of sampling depends upon the information collected from the already selected units in the initial sample. It means, if one finds what he/she is looking for at a particular location then he/she would sample in the vicinity of that location with the hope of obtaining more information. Thompson (1990, 1992) and Seber and Thompson (1994) surveyed briefly the research work made by others on this type of sampling. In 1996, they presented this subject in detail along with some new results in their book (Thompson and Seber 1996).

We have already studied some situations where an adaptive approach is preferable. Thompson and Seber (1996) discussed the different schemes other than SRS that can be used for obtaining the initial sample.

The essential features of ACS consist of an initial selection of units, a criterion C that decides when to add neighboring units to the initial sample and the definition of neighborhood of an unit. Thompson (1991a, b, 1992) has extended this method to the use of primary units and strata where a primary unit consists of a set of secondary units and can arise when one uses strip sampling or systematic sampling. Thompson suggested to choose a SRS of PSUs without replacement and then to add units adaptively.

In this chapter, we have discussed the method developed by Salehi and Seber (1997a; 1997b) which follows the traditional two-stage approach. One selects the sample of primary units and then draws a subsample of secondary units from each of the selected primary units.

Adaptive sampling encounters several problems. Firstly, the final sample size is random and therefore unknown. The appropriate theory has not been developed for using a pilot survey to establish a sampling experiment with a given precision of estimation.

Secondly, if criterion C is not appropriate then there may be a over or under-representation of the desired sampling units.

Thirdly, a lot of effort can be expended in locating initial units as one has to travel to the site of each unit.

Two-Stage ACS deals with all of the above problems in a reasonably optimal manner. We shall discuss Two-Stage ACS using two different sampling schemes.

In Sect. 11.2, notations and procedures of two-stage ACS are introduced. Sections 11.3 and 11.4 discuss the derivations of unbiased estimators for the two different designs. In the first design, clusters are allowed to overlap primary unit boundaries and in the second the clusters are truncated at the boundaries. In Sect. 11.5, the theory developed is applied to an example from Smith et al. (1995).

11.2   Notations and Selection Procedure

11.2.1   Notations

N: The size of a clustered and patchy population.

Let this population be partitioned into M primary units. Let $N_i, i = 1, 2, \ldots, M$ denote the size of the ith primary unit. Hence, we have $N = \sum_{i=1}^{M} N_i$.

Let Y be the characteristic under study. The interest is to estimate the population mean μ of Y values.

Let (i, j) denote the jth unit in the ith primary unit.

Let y_{ij} be the value of Y corresponding to (i, j).

Let τ_i = Sum of Y values in the ith primary unit.

$$= \sum_{j=1}^{N_i} y_{ij} \quad i = 1, 2, \ldots, M$$

τ = Sum of Y values in the entire population.

$$= \sum_{i=1}^{M} \tau_i.$$

Thus, we have $\mu = \frac{\tau}{N}$.

Let C denote the condition specified for adding the neighboring units. Thompson's (1990) procedure is followed to define neighborhood,network and cluster formed around a sampling unit.

11.2.2   Selection Procedure

In the first stage, m primary units by using SRSWOR from M is selected, and the condition of having SRS may be relaxed.

At the second stage, take an initial sample of n_i units from the ith primary unit $(i = 1, 2, \ldots, m)$ by using SRSWOR so that $n_0 = \sum_{i=1}^{m} n_i$ is the total initial sample size.

Then further add the neighborhood units following Thompson's procedure and build up the clusters. As in Thompson (1991b), we have two different designs. In the first design, the clusters are allowed to overlap the boundaries of primary units. That means clusters can grow across PSU boundaries and in the second design, the clusters are truncated at the boundaries.

When $m = M$, we get the stratified sampling scheme of Thompson (1991b).

11.3 Overlapping Clusters Scheme

In this scheme, it is assumed that the N population units are partitioned into K distinct networks ignoring PSU boundaries. To discover these K networks, the condition C and Thompson's definition of neighborhood are used. Two different estimators of population mean are developed by Salehi and Seber (1997a; 1997b).

11.3.1 Modified HT Estimator

A modified HT estimator of the population mean μ is given by
$$\hat{\mu} = \frac{1}{N} \sum_{k=1}^{K} y_k^\star J_k/\alpha_k,$$
where
$y_k^\star$ is the sum of Y values for the network k.
α_k is the inclusion probability for the network k.

$$J_k = \begin{cases} 1 \text{ ; if the initial sample contains atleast one unit from network k} \\ \\ 0 \text{ ; } otherwise. \end{cases}$$

$E(J_k) = 1.P(Initial\ sample\ contains\ atleast\ one\ unit\ from\ network\ k) + 0.P(Initial\ sample\ does\ not\ contain\ any\ unit\ from\ network\ k)$
$$= \alpha_k.$$
Hence, $E(\hat{\mu}) = \frac{1}{N} \sum_{k=1}^{K} y_k^\star E(J_k)/\alpha_k$
$$= \frac{\sum_{k=1}^{K} y_k^\star}{N}$$
$$= \mu.$$

Thus, $\hat{\mu}$ is an unbiased estimator of the population mean of Y. This estimator uses only those edge units which are included in the initial sample.
Variance of $\hat{\mu}$ is given by

$$V(\hat{\mu}) = \frac{1}{N^2} \sum_{k=1}^{K} \sum_{k'=1}^{K} y_k^\star y_{k'}^\star \left(\frac{\alpha_{kk'} - \alpha_k \alpha_{k'}}{\alpha_k \alpha_{k'}}\right),$$

where

$\alpha_{kk'}$ is the probability that initial sample contains at least one unit from network k and k' each. Note that $\alpha_{kk} = \alpha_k$.

An unbiased estimator of the above variance is given by

$$\hat{V}(\hat{\mu}) = \frac{1}{N^2} \sum_{k=1}^{K^*} \sum_{k'=1}^{K^*} y_k^* y_{k'}^* \left(\frac{\alpha_{kk'} - \alpha_k \alpha_{k'}}{\alpha_{kk'} \alpha_k \alpha_{k'}} \right),$$

where

K^* is the number of distinct networks observed in the sample.

11.3.1.1 Derivations of the Inclusion Probabilities

In order to derive the inclusion probabilities of the network k, we need to treat each primary unit separately.

Let n_{ik} be the number of units of network k that are located in primary unit i. Let B_k be the set of primary unit labels representing those PSU's included in the network k. Let g_k be the number of elements in B_k. Let C_{ik} denote the event that at least one of the n_{ik} units of network k lying in the primary unit i is selected in the initial sample.

Then,

$$C_k = \cup_{i \in B_k} C_{ik}$$

is the event that at least one of the units of network k is included in the initial sample.

Thus,

$$\alpha_k = P(C_k) = P(\cup_{i \in B_k} C_{ik}).$$

The above probability can be derived by 'inclusion- exclusion' formula (Billingsly, 1995, P. 24) as follows:

$$\alpha_k = \sum_{i \in B_k} P(C_{ik}) - \sum_i \sum_{i' < i} P(C_{ik} \cap C_{i'k}) + \cdots + (-1)^{g_k+1} P(\cap_{i \in B_k} C_{ik}$$

$$= \sum_{i \in B_k} \frac{m}{M} \left[1 - \frac{\binom{N_i - n_{ik}}{n_i}}{\binom{N_i}{n_i}} \right] - \sum_i \sum_{i' < i} \frac{m(m-1)}{M(M-1)} \left[1 - \frac{\binom{N_i - n_{ik}}{n_i}}{\binom{N_i}{n_i}} \right] \left[1 - \frac{\binom{N_{i'} - n_{i'k}}{n_{0i'}}}{\binom{N_i'}{n_{0i'}}} \right] +$$

$$\cdots + \frac{(-1)^{g_k+1} m(m-1)\dots(m-g_k+1)}{M(M-1)\dots M(M-1)\dots(M-g_k+1)} \prod_{i \in B_k} \left[1 - \frac{\binom{N_i - n_{ik}}{n_i}}{\binom{N_i}{n_i}} \right].$$

$$(11.1)$$

To see this consider the last term of α_k.

Let S_1 be the set of labels representing the primary units that are selected at the first stage.

Then, given S_1, the events C_{ik} are independent.

Further, $P(S_1) = \frac{1}{\binom{M}{m}}$

Hence,

$$P(\cap_{i \in B_k} C_{ik}) = \sum_{S_1} P(\cap_{i \in B_k} C_{ik} | S_1) P(S_1)$$
$$= P(S_1) \sum_{S_1 \in B_k} \prod_{i \in B_k} P(C_{ik} | S_1)$$
$$= \frac{\binom{M-g_k}{m-g_k}}{\binom{M}{m}} \prod i \in B_k P(C_{ik} | S_1).$$

Now,

$\alpha_{kk'} = P(Initial\ sample\ intersects\ both\ networks\ k\ and\ k\prime)$

$$= P\left[(\cup_{B_k} C_{ik}) \cap (\cup_{B_{k'}} C_{ik'})\right]$$
$$= P(\cup_{B_k} C_{ik}) + P(\cup_{B_{k'}} C_{ik'}) - P\left[(\cup_{B_k} C_{ik}) \cup (\cup_{B_{k'}} C_{ik'})\right]$$
$$= \alpha_k + \alpha_{k'} - P(\cup_{B_{kk'}} C_{ikk'}),$$

where

$B_{kk'}$ denotes the set of primary units in which network k or k' or both are located. Let $g_{kk'}$ denote the number of units included in $B_{kk'}$.

$C_{kk'}$ denotes the event that at least one of the initial sample units chosen from primary unit i intersects network k and k'.

Hence,

$$P(\cup_{B_{kk'}} C_{ikk'}) = \sum_{i \in B_{kk'}} P(C_{ikk'}) - \sum_i \sum_{i' < i} P(C_{ikk'} \cap C_{i'kk'}) + \cdots + (-1)^{g_{kk'}+1}$$
$$P(\cap_{i \in B_{kk'}} C_{ikk'}).$$

This expression is given by (11.1) by replacing n_{ik}, g_k and B_k by $n_{ik} + n_{ik'}$, $g_{kk'}$ and $B_{kk'}$, respectively.

Though the expressions for α_k and $\alpha_{kk'}$ look complicated, g_k and $g_{kk'}$ are usually small. So, only one or two terms in the above expression are needed.

11.3.2 Modified HH Estimator

Assume that the inclusion probability of the ith primary unit in the initial sample is P_i and the joint inclusion probability for the ith and i'th primary units is $P_{ii'}$.

Let A_{ij} denote the network containing unit (i, j).(That is, jth unit in the primary unit i).

Let A_{ijl} denote the part of A_{ij} in primary unit l. It is to be noted that A_{ij} will be same for all units (i, j) in any network k.

Let f_{ijl} be the number of units from the initial sample in primary unit l that belongs to A_{ijl} and let α_{ijl} denote the number of units in A_{ijl}. Then the number of units from an initial sample of n_0 units that belongs to A_{ij} is

$$f_{ij.} = \sum_{l=1}^{M} f_{ijl}.$$

Note that $f_{ij.}$ can be zero.

The modified HH estimator of μ is given by

$$\mu^\star = \frac{1}{N} \sum_{i=1}^{M} \sum_{j=1}^{N_i} \frac{y_{ij} f_{ij.}}{E(f_{ij.})}. \tag{11.2}$$

It is easy to see that $E(\mu^\star) = \mu$.
Thus, $\mu^\star$ is an unbiased estimator of μ.
Now define I_l as follows:

$$I_l = \begin{cases} 1 \; ; with \; probability \; P_l \; if \; the \; primary \; unit \; l \; has \; been \; chosen \; initially \\ \\ 0 \; ; otherwise. \end{cases}$$

Then,

$f_{ijl}|I_l = 0$ is zero and $f_{ijl}|I_l = 1$ has hypergeometric distribution with parameters (N_l, a_{ijl}, n_l).

Hence,

$E(f_{ij.}) = \sum_{l=1}^{M} E(f_{ijl}) = \sum_{l=1}^{M} E_1 E(f_{ijl}|I_l)$,

where

E_1 denotes expectation with respect to first stage of sampling.

So,

$E(f_{ij.}) = \sum_{l=1}^{M} E(f_{ijl}|I_l = 1) P_l = \sum_{l=1}^{M} \frac{n_l\, a_{ijl}\, P_l}{N_l}$

The term $y_{ij} f_{ij.}$ in (11.2) tells that A_{ij} is intersected $f_{ij.}$ times by the initial sample, so that μ^* is a weighted sum of all units in all the networks corresponding to initial sample. Some networks are possibly repeated. Since $E(f_{ij.})$ is same for each unit in A_{ij} and $n_i = 0$ for PSUs not initially selected, we have

$\mu^* = \frac{1}{N} \sum_{i=1}^{M} (\sum_{j=1}^{n_i} \frac{1}{E(f_{ij.})} \sum_{i',j' \in A_{ij}} y_{i'j'})$

$= \frac{1}{N} \sum_{i=1}^{m} \sum_{j=1}^{n_i} Y_{ij}/E(f_{ij.})$

where Y_{ij} is the sum of y values in A_{ij}

One can alternatively express μ^* as follows:

$\mu^* = \frac{1}{N} \sum_{i=1}^{m} N_i\, \bar{w}_i/P_i$,

where

$\bar{w}_i = \sum_{j=1}^{n_i} w_{ij}/n_i$ and $w_{ij} = \frac{P_i\, n_i\, Y_{ij}}{N_i\, E(f_{ij.})}$.

Further, let $\hat{t}_{i\pi} = N\, \bar{w}_i$ and $V(N_i\, \bar{w}_i) = V_i$

Then, from Särndal et al. (1992, P. 137), we get

$$V(\mu^*) = \frac{1}{N^2} \sum_{i=1}^{M} \sum_{i'=1}^{M} W_i\, W_{i'} \left(\frac{P_{ii'} - P_i\, P_{i'}}{P_i\, P_{i'}} \right) + \frac{1}{N^2} \sum_{i=1}^{M} V_i/P_i, \qquad (11.3)$$

where

$P_{ii} = P_i$, $V_i = \frac{N_i\, (N_i - n_i)\, \sigma_i^2}{n_i}$, $W_i = \sum_{j=1}^{N_i} W_{ij}$

$\bar{W}_i = \frac{W_i}{N_i}$, $\sigma_i^2 = \sum_{j=1}^{N_i} \frac{(W_{ij} - \bar{W}_i)^2}{N_i - 1}$.

An unbiased estimator of $V(\mu^*)$ is given by

$$\hat{V}(\mu^*) = \frac{1}{N^2} \sum_{i=1}^{m} \sum_{i'=1}^{m} N_i\, N_{i'}\, \bar{W}_i\, \bar{W}_{i'} \left(\frac{P_{ii'} - P_i\, P_{i'}}{P_i\, P_{i'}\, P_{ii'}} \right) + \frac{1}{N^2} \sum_{i=1}^{m} \hat{V}_i/P_i, \qquad (11.4)$$

where

$\hat{V}_i = \frac{N_i\, (N_i - n_i)\, S_i^2}{n_i}$ and $S_i^2 = \sum_{i=1}^{n_i} \left[(w_{ij} - \bar{w}_i)^2/(n_i - 1) \right]$.

If the first stage of sampling is by SRSWOR then

$P_i = \frac{m}{M}$, and $P_{ii'} = \frac{m(m-1)}{M(M-1)}$ then (11.3) reduces to

$$V(\mu^*) = \frac{M(M-m)\, \sigma_M^2}{N^2\, m} + \frac{M}{N^2\, m} \sum_{i=1}^{M} V_i, \qquad (11.5)$$

where
$$\sigma_M^2 = \sum_{i=1}^{M}(W_i - \bar{W})^2/(M-1) \quad \text{and} \quad \bar{W} = \sum_{i=1}^{M} W_i/M.$$
The expression (11.4) reduces to

$$\hat{V}(\mu^{\star}) = \frac{M(M-m)}{N^2} \frac{S_M^2}{m} + \frac{M}{N^2 m} \sum_{i=1}^{m} \hat{V}_i, \tag{11.6}$$

where
$$S_M^2 = \sum_{i=1}^{m}(N_i \, \bar{w}_i - \sum_{i=1}^{m} N_i \, \bar{w}_i/m)^2/(m-1).$$

11.4 Non-overlapping Clusters Scheme

In this scheme, the clusters are truncated at the primary unit boundaries. Using a SRS of primary units and a SRS of secondary units from each primary unit, the HT estimator is given by
$\hat{\mu}_{\star} = \frac{M}{N} \sum_{i=1}^{m} \hat{\tau}_i/m,$
where
$\hat{\tau}_i = \sum_{k=1}^{K_i} y_{ik}^{\star} I_{ik}/\alpha_{ik}.$
Here, K_i, $y_{ik}^{\star}$ and α_{ik} are the number of networks in the primary unit i, the sum of y values in the network k and probability that the initial sample of units in primary unit i intersects network k, respectively.

Let $\alpha_{ikk'}$ denote the probability that the initial sample of units in primary unit i intersects both k and k' networks.
Then,
$$\alpha_{ikk'} = \alpha_{ik} + \alpha_{ik'} - \left[1 - \binom{N_i - x_{ik} - x_{ik'}}{n_i}/\binom{N_i}{n_i}\right],$$
where
$\alpha_{ik} = 1 - \binom{N_i - x_{ik}}{n_i}/\binom{N_i}{n_i}.$
We have, $E(I_{ik}) = \alpha_{ik}$, $E(\hat{\tau}_i) = \tau_i$ and $\hat{\mu}_{\star}$ unbiased.

Variance of $\hat{\mu}_{\star}$ is given by (11.5) with
$\sigma_M^2 = \sum_{i=1}^{M}(\tau_i - \bar{\tau})^2/(M-1)$, $\bar{\tau} = \sum_{i=1}^{M} \tau_i/M$ and $V_i = \sum_{k=1}^{k_i} \sum_{k'=1}^{k_i} y_{ik}^{\star} y_{ik'}^{\star}$
$\left(\frac{\alpha_{ikk'} - \alpha_{ik}.\alpha_{ik'}}{\alpha_{ikk'}.\alpha_{ik}.\alpha_{ik'}}\right),$

where k_i is the number of distinct networks intersected in primary unit i.

The modified HH estimator $\mu^{\star\star}$ is similar to $\mu^{\star}$ except that the dot subscript of f_{ij}. drops out.

Thus, if A_{ij} containing a_{ij} units is truncated at the boundary of ith PSU and f_{ij} is the number of units from the initial sample that fall in A_{ij} then
$E(f_{ij}) = \frac{n_i.a_{ij}.p_i}{N_i}$ and $w_{ij}^{\star} = \frac{n_i.Y_{ij}}{N_i.E(f_{ij})}.$
Thus, replacing w_{ij} by $w_{ij}^{\star}$ one can use all earlier developed formulae in Sect. 11.3.

When p_i and $\frac{n_i}{N_i}$ are same for all primary units then $w_{ij}^\star = \frac{y_{ij}}{a_{ij}}$. That is the mean of the units in truncated network of unit (i, j).

The distinct unordered units and their labels form a minimal sufficient statistic. The Rao–Blackwell theorem can be used to obtain unbiased estimators with smaller variances.

11.4.1 Choice of a Scheme

It is not clear which of the two schemes as discussed in Sects. 11.3 and 11.4 is better. They both end up with a random final sample size(n). Hence, they should be compared using the same value of $E(n)$.

Initially, the overlapping scheme appears to be more efficient because of additional network information. But in that case the expected sample size will also be larger. Also, various costs must be taken into account. Ignoring costs, some preliminary simulations show that the non-overlapping scheme is more efficient.

The next question is which of the two unbiased estimators should be used because neither is uniformly better than the other. However, various examples in literature and some unpublished theoretical works by Salehi and Seber suggest that the modified HT estimator tends to be more efficient in many situations. So, $\hat{\mu}_\star$ is considered as the preferred estimator. One disadvantage of this estimator is that the PSU variance estimates $\hat{V}_i$ may be negative. In order to overcome this disadvantage, Salehi and Seber (1997a; 1997b) recommend to keep n_i's constant and small.

11.5 Illustrative Example

To demonstrate the calculations, we use an example discussed in Salehi and Seber (1997a; 1997b). In that example, the authors used the data related to a blue-winged teal population given by Smith et al. (1995, P. 787).

An extremely clustered population of $N = 200$ units is divided into $M = 8$ PSU's as shown in Fig. 11.1, each containing $N_i = 25$ units.

A random sample of $m = 4$ PSUs was selected using SRSWOR (labeled 1, 2, 3, and 8) and selected $n_i = 3$ units from each of these PSUs using SRSWOR. Further, the neighborhoods of the units selected in this sample were determined by using criterion of Thompson (1990) with the condition $C = (Number\ of\ blue\ winged\ teals\ observed\ in\ an\ unit > 0)$. It led to Fig. 11.2.

Estimates in Overlapping Scheme

In the overlapping clusters case, two networks containing 7 and 4 units, respectively, are discovered. The corresponding network totals are $y_1^\star = 13753$ and $y_2^\star = 38$.

Since x_{ik} is the number of units of network k that are located in the primary unit i, we get

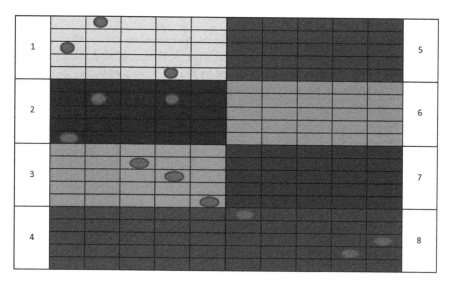

Fig. 11.1 Population is divided into eight PSUs showing the SSUs having blue circles selected from four selected PSUs

1		*							60		5
					1			122	114	3	
	*				7144	6339		14			
				103	150	6					
				10*							
2		*		*		2			2		6
					3						
	*										
3				12							7
			*	2							
				4*							
		5		20							
			3		*						
4						*					8
										*	
									*		

Fig. 11.2 The networks formed in white color and edge units in blue color. The units having * denotes the SSUs selected from the different PSUs. Number in an unit is the corresponding y-value

$x_{11} = 5, \quad x_{51} = 2, \quad x_{32} = 4.$

g_1 = Number of primary units intersected by the first network.

    The first network intersects primary units 2 and 5. Hence, $g_1 = 2$.

    g_2 = Number of primary units intersected by the second network.

    The second network intersects primary unit 2 only.

    Hence, $g_2 = 1$.

Then

$$p_1 = \left[1 - \frac{\binom{N_1 - x_{11}}{n_1}}{\binom{N_1}{n_1}} \right]$$

$$= \left[1 - \frac{\binom{25-5}{3}}{\binom{25}{3}} \right]$$

$$= 0.5043$$

$$p_3 = \left[1 - \frac{\binom{N_3 - x_{32}}{n_3}}{\binom{N_3}{n_3}} \right]$$

$$= \left[1 - \frac{\binom{25-4}{3}}{\binom{25}{3}} \right]$$

$$= 0.4217$$

$$p_5 = \left[1 - \frac{\binom{N_5 - x_{51}}{n_5}}{\binom{N_5}{n_5}} \right]$$

$$= \left[1 - \frac{\binom{25-2}{3}}{\binom{25}{3}} \right]$$

$$= 0.23.$$

Hence,

$\alpha_1 = \frac{4}{8}(p_1 + p_5) - \frac{4}{8} \cdot \frac{3}{7}(p_1 \cdot p_5) = 0.3423$

$\alpha_2 = \frac{4}{8}(p_3) = 0.2109$

The probability of selecting one of the other networks of size 1 in the initial sample= 12/200=0.06.

Estimate of HT estimator

$\hat{\mu} = \frac{1}{200}(13753/0.3423 + 38/0.2109 + 0/0.06 + \cdots + 0/0.06)$

   $= 201.80.$

To find the estimate of variance of $\hat{\mu}$, we need α_{12}. No PSU overlaps both networks. Hence, $x_{ik} + x_{ik'}$ reduces to either x_{ik} or $x_{ik'}$.

Further,

g_{12}= Number of primary units in which networks 1 or 2 are located.=3.

Hence,

$P(\cup_{i=1,3,5} C_{i12}) = \frac{4}{8}(p_1 + p_3 + p_5) - \frac{4}{8} \cdot \frac{3}{7}(p_1 p_3 + p_1 p_5 + p_3 p_5) + \frac{4}{8} \cdot \frac{3}{7} \cdot \frac{2}{6}(p_1 \cdot p_3 \cdot p_5)$

   $= 0.4903.$

So

$\alpha_{12} = \alpha_1 + \alpha_2 - P(\cup_i C_{i12}) = 0.0629.$

Using the values of α_1, α_2 and α_{12} in the formula, we get

$\hat{V}(\hat{\mu}) = (13753^2 \times 5.6125 - 13753 \times 38 \times 4.1045 + 38^2 \times 17.7468)/40000$

   $= 26486.68$

$\hat{SD}(\hat{\mu}) = 162.7.$

Estimate of HH estimator

Recall that

w_{ij} = The network mean for the jth initial unit in the ith PSU.

So that we have

$w_{11} = w_{12} = 0, \quad w_{13} = \frac{13753}{7},$ and $\bar{w}_1 = \sum w_{ij}/3 = 654.9048.$

Also,

$w_{31} = w_{33} = 0, \quad w_{32} = \frac{38}{4}$ and $\bar{w}_3 = 3.1667.$

All w_{2j} and w_{8j} are zero.

Hence,

$\mu^{\star} = \frac{1}{200} \cdot \frac{8}{4} \cdot 25(654.90 + 3.17 + 0 + 0) = 164.5.$

To calculate $\hat{V}(\mu^{\star})$ we need to find the two terms of (11.6).

The first term is $\frac{8.4.25^2}{200^2.4}$ times the sample variance of $\bar{w}_1, \bar{w}_3, \bar{w}_2,$ and $\bar{w}_8$ in which $\bar{w}_2 = \bar{w}_8 = 0.$

So the first term is 13360.24.

Now, $S_1^2 = 1286701, \quad S_3^2 = 30.083, \quad S_5^2 = S_8^2 = 0.$

So, the second term is $\frac{25.22.8}{200^2.3.4} \cdot (S_1^2 + S_3^2) = 11795.03.$

The sum of these two terms gives $\hat{V}(\mu^{\star}) = 25155.$

Hence, the estimated standard deviation = 158.6.

The population standard deviation is $\sqrt{V(\mu^{\star})} = 105.5.$

Thus, the HH estimator is slightly more efficient than the HT estimator.

Estimates in Non-overlapping Scheme

Estimate of HT Estimator

Recall that $y_{ik}^{\star}$ denotes the sum of the y-values for the truncated network k in the ith PSU and x_{ik} refers to truncated networks.

Here, $y_{11}^{\star} = 7408, \quad y_{32}^{\star} = 38, \quad x_{11} = x_{32} = 4..$

Then, $\alpha_{11} = p_1, \quad \alpha_{32} = p_3, \quad \hat{\tau}_1 = \frac{7408}{p_1} = 14688.28, \quad \hat{\tau}_3 = \frac{38}{p_3} = 90.10, \quad \hat{\tau}_2 = \hat{\tau}_8 = 0.$

Hence,

$\hat{\mu}_{\star} = \frac{2.(14688.28+90.10+0+0)}{200} = 147.8$

$SE(\hat{\mu}_{\star}) = 107.3$ and $E(v) = 20.705.$

The first term of (11.6) is $\frac{8}{200^2}$ times the sample variance of the $\hat{\tau}_i$'s. That is equal to 10743.57.

To find the second term we take $\hat{V}_2 = \hat{V}_8 = 0$ because $y_{ik}^{\star} = 0$ for all six networks of size 1 selected in primary units 2 and 8.

We have $\hat{V}_1 = 7408^2 \cdot \frac{(1-p_1)}{p_1^2}, \quad \hat{V}_3 = 38^2 \cdot \frac{(1-p_3)}{p_3^2}.$

Hence, the second term = $\frac{2(\hat{V}_1 + \hat{V}_3)}{200^2} = 5346.97.$

Adding the two terms finally we get

$\hat{V}(\hat{\mu}_{\star}) = 16090.50$

$\hat{SE}(\hat{\mu}_{\star}) = 126.80.$

Estimate of HH estimator

For the HH estimator, we have w_{ij}^* as the truncated network mean for the jth initial unit in the ith PSU.

Here $w_{11}^* = w_{12}^* = 0$, $w_{13}^* = 7408/5$ and $\bar{w}_1^* = 493.8667$.

Also $w_{31}^* = w_{33}^* = 0$, $w_{32}^* = 38/4$, $\bar{w}_3^* = 3.1667$ and all w_{2j}^* and w_{8j}^* are zero.

The HH estimate is given by μ^* on page 6 but with $\bar{w}_i$ replaced by $\bar{w}_i^*$. We have denoted it by μ^{**}

Hence,

$\mu^{**} = \frac{1}{200} \cdot \frac{8}{4} .25 (493.87 + 3.17 + 0 + 0) = 124.3$.

To calculate the variance estimate , it is required to find the two terms of (11.6).

The first term is $\frac{8.4.25^2}{200^2.4} = 1/8$ times sample variance of $\bar{w}_1^*, \bar{w}_2^*, \bar{w}_3^*,$ and $\bar{w}_8^*$(with $\bar{w}_2^*$ and $\bar{w}_8^*$ equal to zero)

Hence,

first term = 7589.741.

Now,

$S_1^{*^2} = 731712.9$ $S_3^{*^2} = 30.08$, $S_2^{*^2} = S_8^{*^2} = 0$

Hence, the second term = $\frac{1.25.22.8}{200^2.3.4} (S_1^{*^2} + S_3^{*^2}) = 6707.64$.

The sum of these two terms gives

$\hat{V}(\mu^{**}) = 14297.38$ with $\hat{S}E(\mu^{**}) = 119.6$.

So, HH estimator is less efficient than HT estimator. Here in this example, in the overlapping case, the HH and HT estimators require a smaller sample size. They are more efficient than the corresponding estimators in the overlapping case. This is not true in general.

11.6 Exercises

1. Discuss the problems faced in ACS.
2. Explain the procedure of two-stage ACS.
3. Show that the modified HT estimator($\hat{\mu}$) used in two-stage ACS with overlapping clusters is unbiased for the population mean.
4. Derive the expressions for the inclusion probability of the kth network(α_k) and the joint inclusion probability of networks k and k' ($\alpha_{kk'}$) in two-stage ACS with overlapping clusters.
5. Derive an expression for the variance of $\hat{\mu}$ and also obtain its unbiased estimator.
6. Show that the modified HH estimator(μ^*) used in two-stage ACS with overlapping clusters is unbiased for the population mean.
7. Derive variance of μ^* and obtain its unbiased estimator.
8. Show that the modified HT estimator ($\hat{\mu}_*$) used in two-stage ACS with non-overlapping clusters is unbiased for the population mean. Derive variance of $\hat{\mu}_*$.
9. Discus the efficiency of overlapping cluster scheme as compared to the non-overlapping clusters scheme in two-stage ACS.

References

Billingsley, P.: Probability and Measure, 3rd edn, p. 25. John Wiley and Sons (1995)

Salehi, M.M., Seber, G.A.F.: Adaptive cluster samplingwith networks selected without replacement. Biometrika **84**, 209–219 (1997a)

Salehi, M.M., Seber, G.A.F.: Two stage adaptive clustersampling. Biometrics **53**, 959–70 (1997b)

Särndal, C.E., Swensson, B., Wretman, J.: Model Assisted Survey Sampling. Springer, New York Inc (1992)

Seber, G.A.F., Thompson, S.K.: Environmental adaptive sampling. Handb. Stat. **12**, 201–220 (1994)

Smith, D.R., Conroy, M.J., Brakhage, D.H.: Efficieny of adaptive cluster sampling for estimating density of wintering waterfowl. Biometrics **51**, 777–788 (1995)

Thompson, S.K.: Adaptive cluster sampling. J. Am. Stat. Assoc. **85**(412), 1050–1058 (1990)

Thompson, S.K.: Adaptive cluster sampling: Designs with primary and secondary units. Biometrics **47**(3), 1103–1115 (1991a)

Thompson, S.K.: Stratified adaptive cluster sampling. Biometrika **78**(2), 389–397 (1991b)

Thompson, S.K.: Sampling. Wiley, New York (1992)

Thompson, S.K., Seber, G.A.F.: Adaptive Sampling. Wiley, New York (1996)

Chapter 12
Adaptive Cluster Double Sampling

12.1 Introduction

ACS introduced by Thompson (1990) has been found appropriate for sampling of rare and clustered populations. But it suffers from drawback of losing control of the final sample size. There have been several suggestions for limiting this final sample size of adaptive cluster samples. In this design, traveling costs are increased because the second phase sample is selected after the first phase sample is completed. In the second phase of the sampling design, the sampler cannot allocate the subsample near the places of interest. The proposed unbiased estimators of the population mean do not take the advantage of the relation between the variable of interest and the auxiliary variable. Medina and Thompson (2004) proposed a multiphase variant of ACS. It is obtained by combining the ideas of double sampling and ACS. It is called adaptive cluster double sampling (ACDS). In this design, an auxiliary variable which is easy to measure and is inexpensive is considered. This variable is used to select the first phase of ACS. The network structure of this first phase sample is used to select the subsequent subsamples, which are selected using conventional design. Values of the variable of interest associated with the units selected in the final phase subsample only are recorded and the population mean is estimated by a regression type estimator. The ACDS allows the sampler to overcome the drawbacks of ACS.

Brown (1994) proposed a method where the initial sample is selected sequentially until the final sample size reaches a specified value or exceeds that value for the first time. That method controls the final sample size well. Salehi and Seber (2002) proposed a promising estimator of the population mean by using this method. However, further research is necessary to observe the performance of that estimator in situations ordinarily found in applications.

Salehi and Seber (1997a, b) proposed a two-stage version in which primary units are selected using a conventional design and the secondary units are selected from the selected primary units, using ACS. The secondary units are not allowed to cross the borders of the primary units even though the secondary units with interesting values of the survey variable can be found beyond the borders of the primary units.

© The Author(s), under exclusive license to Springer Nature Singapore Pte Ltd. 2021   171
R. Latpate et al., *Advanced Sampling Methods*,
https://doi.org/10.1007/978-981-16-0622-9_12

These boundaries do not correspond to natural patterns in the population. Hence, it imposes an artificial constraint on the adaptive response to encountered patterns. Consequently, it reduces the efficiency of the design.

Lee (1998) proposed a two-phase version of ACS in which the first phase sample is an ACS based on an auxiliary variable and the second phase sample is selected from the first phase units using a probability proportional to size with replacement sampling design or an inclusion probability proportional to size sampling design (see Särndal et al. 1992 for the definitions of these designs). This design allows the sampler to control the number of measurements of the survey variable and not that of the auxiliary variable. Further, traveling costs are increased because the second phase sample is selected after the first phase sample is completed. The second phase sampling design does not allow the sampler to allocate the subsample near interesting places and the proposed unbiased estimators of the population mean do not take the advantage of the relationship between the survey variable and the auxiliary variable.

In this chapter, we present a multiphase variant of ACS obtained by combining the ideas of the double sampling and ACS. This sampling design ACDS was developed by Medina and Thompson (2004). This design involves the availability of an inexpensive and easy to measure auxiliary variable. It is used to select a first phase ACS. The network structure of this first phase sample is used to select the subsequent subsamples that are selected using some conventional design. The values of the survey variable associated with the units in the final phase subsample are only recorded. The population mean is estimated by a regression type estimator. This design allows the sampler to control the number of measurements of the survey variable; to allocate the final phase subsample near interesting places and to start the second phase sampling before the first phase sampling is completed. It also allows the use of auxiliary variable at the estimation stage.

12.2 Notations and Sampling Design

Let $U = \{U_1, U_2, \ldots, U_N\}$ be a finite population of N units. Let X and Y be the auxiliary and survey variables, respectively. Let (X_i, Y_i), $i = 1, 2, \ldots, N$ denotes the values of these two variables corresponding to U_i.

It is assumed that no information about the values of the auxiliary variable is available before starting the sampling stage. Here, goal is to estimate the population mean of the Y values given by

$$\mu_Y = \frac{\sum_{i=1}^{N} Y_i}{N}.$$

The first phase of this design consists of selecting an ordinary ACS S_1 based on the values of X. We assume that a condition C_X for additional sampling and a set of neighboring units for an unit in the population U have been defined. These definitions induce a partition of U into K networks $A_1, A_2, \ldots, A_K$ as indicated by Thompson (1990). Let S_0 denote the initial sample that is used to select the ACS S_1. Let n be the size of S_0. Let the k distinct networks intersected by S_0 be denoted as $A_1, A_2, \ldots, A_k$.

In the second phase, a conventional sample S_2 of k_1 networks is selected from the k different networks intersected by S_0. Note that if S_2 is selected with replacement, then the number of distinct networks in S_2 might be less than k_1. Let $A_1, A_2, \ldots, A_{k_2}$, $k_2 \leq k_1$, denote the distinct networks in S_2.

Finally, the third phase consists of selecting a conventional subsample of units from each of the distinct networks in S_2. Y-values associated with unit in those subsamples are recorded. Let S_{3i}, $i = 1, 2, \ldots, k_2$ denote the k_2 subsamples of units. It is assumed that they are selected independently.

A wide variety of practical sampling procedures is possibly based on ACDS. One possibility is to omit the second-phase sampling and subsample every network intersected by S_0. Each network might be sampled as soon as the sampler knows which units belong to that network and even before the sampler has any information about the other networks. This procedure would help to save the traveling costs. But it would not help to control the number of measurements of the survey variable. To regulate the sample size, completion of the first phase sample before starting the other sampling phases seems to be necessary.

Other possible sampling procedures are obtained by combining different designs to select S_2 and S_{3i}, $i = 1, 2, \ldots, k_2$.

Most combinations will allow the sampler to control the costs and the number of measurements of the survey variable by fixing the total number of units in which that variable is to be measured. This is possible because the Y-values are measured only for the units in S_{3i}, $i = 1, 2, \ldots, k_2$. Therefore, the size or at least upper bound for the size of $S_3 = \cup_{i=1}^{k_2} S_{3i}$ can be fixed before starting the sampling procedure.

Furthermore, different types of auxiliary variables can be used that lead the sampler to the most promising areas where exact observations of the survey variable can be made.

In Medina and Thompson's procedure, the X-value associated with every unit in ACS S_1 has to be measured. Therefore, the procedure does not control the number of measurements of the auxiliary variable. It controls only the number of measurements of the survey variable. The measurement of auxiliary variable is easy and inexpensive. Hence, a relatively large initial sample can be used. This would increase the probability of intersecting the networks with units satisfying the condition C_X and this would improve the efficiency of the estimators.

12.3 Estimators Under ACDS

12.3.1 Regression Type Estimator

Medina and Thompson (2004) constructed a regression type estimator of the population mean μ_Y. This estimator is a function of HT type estimator. It is constructed under the assumption that the relationship between X and Y can be modeled by means of a regression model ζ having

$E_\zeta(Y_i|X_i) = X_i'\beta$ and $V_\zeta(Y_i|X_i) = \nu_i\sigma^2$ for $i = 1, 2, \ldots, N$ where $X_i = (X_{i1}, X_{i2}, \ldots, X_{ip})'$ is a p-dimensional vector whose elements are functions of the auxiliary variable X_i, such as $X_i = (1, x_i, x_i^2)'$, $\beta = (\beta_1, \beta_2, \ldots, \beta_p)'$ and $\nu_i = \phi(x_i)$, where the function ϕ is assumed to be known.

Let μ_X denote the population mean of the vectors X_j. That is,

$$\mu_X = \frac{\sum_{j=1}^N X_j}{N}.$$

The size of the network A_i is denoted by m_i. Let X_{i+} and Y_{i+} be the sums of X vector values and Y-values associated with the elements in A_i. That is,

$X_{i+} = \sum_{u_j \in A_i} X_j, Y_{i+} = \sum_{u_j \in A_i} Y_j$ for $i = 1, 2, \ldots, K$.

The general form of the regression estimator is given by

$$\hat{\mu}_R = \hat{\mu}_Y^\star + (\hat{\mu}_X - \hat{\mu}_X^\star)' \hat{B}_{S3}. \tag{12.1}$$

It is same as that presented in Särandal et al. (1992, P. 364) for the regression estimator used in two-phase sampling. In Eq. (12.1) above, $\hat{\mu}_X$ is the HT estimator of μ_X. That is,

$$\hat{\mu}_X = \frac{\sum_{i=1}^K \frac{X_{i+}}{\pi_i}}{N},$$

where π_i is the probability of intersecting A_i by S_0, for $i = 1, 2, \ldots, K$.

The estimators $\hat{\mu}_X^\star$ and $\hat{\mu}_Y^\star$ are the HT type estimators of μ_X and μ_Y, respectively, based on the sample S_3.
Thus,

$$\hat{\mu}_X^\star = \frac{\sum_{i=1}^{k_2} \frac{\hat{X}_{i+}}{\pi_i \, \pi_{i|S_1}}}{N} = \frac{\sum_{i=1}^{k_2} \sum_{u_j \in S_{3i}} \frac{X_j}{\pi_j^\star}}{N}$$

and

$$\hat{\mu}_Y^\star = \frac{\sum_{i=1}^{k_2} \frac{\hat{Y}_{i+}}{\pi_i \, \pi_{i|S_1}}}{N} = \frac{\sum_{i=1}^{k_2} \sum_{u_j \in S_{3i}} \frac{Y_j}{\pi_j^\star}}{N}, \tag{12.2}$$

where $\pi_{i|S_1}$ is the conditional probability, given S_1 of including the network A_i ($i = 1, 2, \ldots, K$) in the second phase sample S_2; $\pi_j^\star = \pi_i \, \pi_{i|S_1} \, \pi_{j|i}$; $\pi_{j|i}$ is the conditional probability of including u_j in S_{3i} given that A_i is in S_2 and $\hat{X}_{i+} = \sum_{u_j \in S_{3i}} \frac{X_j}{\pi_{j|i}}$, $\hat{Y}_{i+} = \sum_{u_j \in S_{3i}} \frac{Y_j}{\pi_{j|i}}$ ($i = 1, 2, \ldots, k_2$) are the HT type estimators of X_{i+} and Y_{i+}, respectively.

$$\hat{B}_{S3} = \left(\sum_{i=1}^{k_2} \sum_{u_j \in S_{3i}} \frac{X_j \, X_{j'}}{\pi_j^\star \, \nu_j} \right)^{-1} \left(\sum_{i=1}^{k_2} \sum_{u_j \in S_{3i}} \frac{X_j \, Y_j}{\pi_j^\star \, \nu_j} \right).$$

By using Särandal et al.(1992, Ch.7), variance estimator of $\hat{\mu}_R$ can be obtained. Using their results, we get
$V(\hat{\mu}_R) = V_{FP}(\hat{\mu}_R) + V_{SP}(\hat{\mu}_R) + V_{TP}(\hat{\mu}_R),$
where

$$V_{FP}(\hat{\mu}_R) = \frac{\sum_{i=1}^K \sum_{j=1}^K \frac{\delta_{ij} \, Y_{i+} \, Y_{j+}}{\pi_i \, \pi_j}}{N^2}, \tag{12.3}$$

$$V_{SP}(\hat{\mu}_R) = E_1 \left(\frac{\sum_{i=1}^{K} \sum_{j=1}^{K} \frac{\delta_{ij|S_1} \, E_{i+} \, E_{j+}}{\pi_i \, \pi_j \, \pi_{i|S_1} \, \pi_{j|S_1}}}{N^2} \right), \tag{12.4}$$

$$V_{TP}(\hat{\mu}_R) = E_1 E_2 \left\{ \frac{1}{N^2} \sum_{i=1}^{k_2} \frac{1}{(\pi_i \, \pi_j | S_1)^2} \sum_{u_j \in A_i} \sum_{u_{j'} \in A_i} \left(\frac{\delta_{jj'|i} \, e_j \, e_{j'}}{\pi_{j|i} \, \pi_{j'|i}} \right) \right\}. \tag{12.5}$$

In the above expression, the subscript i of the expectation operator E indicates that the expectation is taken over all the phase i sample selections.

In Eq. (12.3), $\delta_{ij} = \pi_{ij} - \pi_i \, \pi_j$, where π_{ij} is the probability that the networks A_i and A_j for $i \neq j$ are intersected by S_0 and $\pi_{ii} = \pi_i$ for $i = 1, 2, \ldots K$.

In Eq. (12.4), $E_{i+} = \sum_{u_j \in A_i} e_j$,
where $e_j = Y_j - X'_j \, B_U$, e_j is the population regression residual associated with the unit $u_j \in U$.
$\delta_{ij|S_1} = \pi_{ij|S_1} - \pi_{i|S_1} \, \pi_{j|S_1}$,
where $\pi_{ij|S_1}$ is the conditional probability of including the networks A_i and A_j given S_1 for $i \neq j$ in the second-stage sample S_2 and $\pi_{ii|S_1} = \pi|S_1$ for $i = 1, 2, \ldots k$.

In Eq. (12.5), $\delta_{jj'|i} = \pi_{jj'|i} - \pi_{j|i} \, \pi_{j'|i}$,
where $\pi_{jj'|i}$ for $j \neq j'$ is the conditional probability that u_j and $u_{j'}$ are included in $S_{3i}|S_2$.

Also, $\pi_{jj|i} = \pi_{j|i}$; for $j = 1, 2, \ldots m_i$ and $i = 1, 2, \ldots k_2$.

An estimator of $Var(\hat{\mu}_R)$ is given by
$\hat{V}(\hat{\mu}_R) = \hat{V}_{FP}(\hat{\mu}_R) + \hat{V}_{SP}(\hat{\mu}_R) + \hat{V}_{TP}(\hat{\mu}_R)$,
where

$$\hat{V}_{FP}(\hat{\mu}_R) = \frac{1}{N^2} \left(\sum_{i=1}^{k_2} \sum_{j=1}^{k_2} \frac{\delta_{ij} \, \hat{Y}_{i+} \, \hat{Y}_{j+}}{\pi_{ij} \, \pi_{ij|S_1} \, \pi_i \, \pi_j} - \sum_{i=1}^{k_2} \frac{1 - \pi_i}{\pi_i^2 \, \pi_{i|S_1}} \sum_{u_j \in S_{3i}} \sum_{u_{j'} \in S_{3i}} \frac{\delta_{jj'|i} \, Y_j \, Y_{j'}}{\pi_{jj'|i} \, \pi_{j|i} \, \pi_{j'|i}} \right), \tag{12.6}$$

$$\hat{V}_{SP}(\hat{\mu}_R) = \frac{1}{N^2} \left(\sum_{i=1}^{k_2} \sum_{j=1}^{k_2} \frac{\delta_{ij|S_1} \, \hat{E}_{i+} \, \hat{E}_{j+}}{\pi_{ij|S_1} \, \pi_{i|S_1} \, \pi_i \, \pi_j \, \pi_{j|S_1}} - \sum_{i=1}^{k_2} \frac{1 - \pi_{i|S_1}}{\pi_{i|S_1}^2} \frac{V_{S_{3i}}}{\pi_i^2} \right), \tag{12.7}$$

$$\hat{V}_{TP}(\hat{\mu}_R) = \frac{1}{N^2} \sum_{i=1}^{k_2} \frac{\hat{V}_{S_{3i}}}{(\pi_i \, \pi_{i|S_1})^2}, \tag{12.8}$$

where
$\hat{V}_{S_{3i}} = \sum_{u_j \in S_{3i}} \sum_{u_{j'} \in S_{3i}} \frac{\delta_{jj'|i} \, g_{S_j} \, \hat{e}_j \, g_{S_{j'}} \, \hat{e}_{j'}}{\pi_{jj'|i} \, \pi_{j|i} \, \pi_{j'|i}}$
$\hat{E}_{i+} = \sum_{u_j \in S_{3i}} \frac{g_{S_j} \, \hat{e}_j}{\pi_{j|i}}$.
$\hat{e}_j = Y_j - X'_j \, \hat{B}_{S_3}$ is the sample regression residual and
$g_{S_j} = 1 + (\hat{\mu}_X - \hat{\mu}^\star)' \, (\frac{1}{N} \sum_{i=1}^{k_2} \sum_{u_j \in S_{3i}} \frac{X_j \, X_{j'}}{\pi_j^\star \, v_j})^{-1} \frac{X_j}{v_j}$
is the g-weight associated with the unit $u_j \in S_3$.

12.3.2 HT Type Estimator

Though the regression type estimator discussed above is expected to be very efficient, Y and X are highly correlated and the assumed model is a reasonable approximation to the relationship between X and Y; two problems are faced.

First, the requirement to specify an appropriate regression model and the second, possibility that in some real applications of ACDS the matrix of the sampled X vector values, X_s, is not of full rank. In some situations, the sampler might not be able to carry out the analysis required to specify the regression model.

With respect to the second problem, if X_s is not of full rank, one cannot use $\hat{\mu}_R$ to estimate μ_Y. This type of situation arises when the condition for additional sampling is of the form $C_X = \{X > 0\}$ and a relatively small number of units in U satisfy C_X. Then, one may get $X_i = 0$ for every unit $u_i \in S_0$.

In view of the above-mentioned problems with the regression estimator, the HT type estimator given in (12.2) above can be used.

It can be seen that
$$V(\hat{\mu}_Y^*) = V_{FP}(\hat{\mu}_Y^*) + V_{SP}(\hat{\mu}_Y^*) + V_{TP}(\hat{\mu}_Y^*),$$
where $V_{FP}(\hat{\mu}_Y^*)$ is given by the right-hand side of (12.3). $V_{SP}(\hat{\mu}_Y^*)$ is given by the right-hand side of (12.4) replacing E_{i+} by Y_{i+} and $V_{TP}(\hat{\mu}_Y^*)$ is given by the right hand side of (12.5) by replacing e_j by Y_j.

A design-based estimator of $V(\hat{\mu}_Y^*)$ is given by
$$\hat{V}(\hat{\mu}_Y^*) = \hat{V}_{FP}(\hat{\mu}_Y^*) + \hat{V}_{SP}(\hat{\mu}_Y^*) + \hat{V}_{TP}(\hat{\mu}_Y^*),$$
where first term in the right-hand side is given by (12.6), second term by (12.7) by replacing $\hat{E}_{i+}$ by $\hat{Y}_{i+}$ and computing $V_{S_{3i}}$ by using Y_j instead of $g_{s_j} \hat{e}_j$ and the third term by (12.8) by replacing $e_{s_j} \hat{e}_j$ by Y_j in the expression for $V_{S_{3i}}$.

12.4 Exercises

1. Bring out the difference between ACS and ACDS.
2. Explain the selection procedure of ACDS.
3. Define regression type estimator of population mean proposed by Mediana and Thompson ($\hat{\mu}_R$).
4. Derive an expression for the variance of $\hat{\mu}_R$. Obtain an estimate of the variance of $\hat{\mu}_R$.
5. Discuss the problems encountered while using $\hat{\mu}_R$ to estimate the population mean.
6. Define HT type estimator($\hat{\mu}_Y^*$) of the population mean used in ACDS. Obtain the variance of $\hat{\mu}_Y^*$. Also give its estimate.

References

Brown J.A: The applications of adaptive cluster sampling to ecological studies. In: Fletcher, Manley, B.F.J. (eds.) Statistics in Ecological and Environmental Monitoring. Otago Conference Series 2, pp. 86–97. University of Otago Press, New Zealand (1994)

Lee, K.: Two phase adaptive cluster sampling with unequal probabilities selection. J. Korean Stat. Soc. **27**, 265–278 (1998)

Medina, M.H.F., Thompson, S.K.: Adaptive cluster double sampling. Biometrika **91**, 877–891 (2004)

Salehi, M.M., Seber, G.A.F.: Adaptive cluster samplingwith networks selected without replacement. Biometrika **84**, 209–219 (1997a)

Salehi, M.M., Seber, G.A.F.: Two stage adaptive clustersampling. Biometrics **53**, 959–70 (1997b)

Salehi, M.M., Seber, G.A.F.: Unbiased estimators for restricted adaptive cluster sampling. Aust. N. Z. J. Stat. **44**(1), 63–74 (2002)

Särndal, C.E., Swensson, B., Wretman, J.: Model Assisted Survey Sampling. Springer, New York Inc (1992)

Thompson, S.K.: Adaptive cluster sampling. J. Am. Stat. Assoc. **85**(412), 1050–1058 (1990)

Chapter 13
Inverse Adaptive Cluster Sampling

13.1 Introduction

When the population under study is homogeneous with respect to the characteristic of interest then the traditional method such as SRSWR/SRSWOR can be used to estimate the population parameters. But if the population under study is rare and clustered then the use of these sampling methods may lead to poor estimates of the population parameters. Since the population is rare, the sample drawn may not include the desired proportion of units satisfying the condition of interest.

In such cases, in order to have desired proportion of units, one may use inverse sampling. In inverse sampling, one continues to choose units until an event occurs specified number of times. It is often used when one does not know the exact sample size. The idea of inverse sampling for attributes was introduced by Haldane (1945). He obtained an unbiased estimator of the population proportion of units possessing the attribute under study. This estimator considered equal probabilities of selection of units. Sampford (1962a, b) introduced inverse sampling with unequal probabilities of selection of units. Pathak (1964) showed that this method is equivalent to sampling with unequal probabilities without replacement in some sense. Lan (1999) showed that estimators based on fixed sample size are biased when inverse sampling is used. Section 13.2 discusses the inverse sampling by using two different stopping rules. The unbiased estimators of the population total and their variances are obtained under these rules.

In Sect. 13.3, inverse sampling with unequal probabilities of different groups in the population is discussed. An unbiased estimator of the population total is given along with the derivation of its variance. Further, unbiased estimators of the population total using the two stopping rules proposed by Greco and Naddeo (2007) are discussed along with their variance estimators of these unbiased estimators. Under the set up of equal probabilities of selection of the groups in the population, estimator of the population total is given along with the estimator of its variance.

© The Author(s), under exclusive license to Springer Nature Singapore Pte Ltd. 2021
R. Latpate et al., *Advanced Sampling Methods*,
https://doi.org/10.1007/978-981-16-0622-9_13

In Sect. 13.4, general inverse sampling is discussed. Section 13.5 discusses regression estimators under general inverse sampling and inverse sampling with unequal selection probabilities. Improved estimator and its variance of the population total in inverse adaptive cluster sampling proposed by Pochai (2008) and are given in Sect. 13.6. In Section, general unequal probability inverse adaptive cluster sampling is discussed.

13.2   Inverse Sampling

Let Y be the characteristic under study. Let $U = \{U_1, U_2, \ldots, U_N\}$ be the finite population. Let $\{Y_1, Y_2, \ldots, Y_N\}$ denote the y-values corresponding to the population units $U_1, U_2, \ldots, U_N$, respectively.

$\tau_Y = \sum_{i=1}^{N} Y_i$ denotes the population total. It is unknown and is required to be estimated. The population is divided into two sub-populations P_M and P_{N-M}, where $P_M = \{Y_i | Y_i \ satisfy$

$C, i = 1, 2, \ldots, N\}$ and $P_{N-M} = \{Y_i | Y_i \ does \ not \ satisfy \ C, i = 1, 2, \ldots, N\}$. M and $N - M$ denotes the cardinality of P_M and P_{N-M}, respectively.

C denotes the condition on Y with respect to which the units in the population are classified. A common form of condition C is $C = (Y | Y > c)$ where c is some specified constant.

It is assumed that the population is rare and clustered with respect to Y. It means, there are a few elements in P_M. It is to be noted that the sub-population to which a unit belongs is not known until the unit is sampled.

With this background, inverse sampling is performed by using any of the following two stopping rules.

13.2.1   Sampling Until k Units of P_M are Observed

Suppose SRSWR or SRSWOR is undertaken and it stops when $1 < k \leq M$ units from P_M are sampled, where k is specified. Lan (1999) has shown that the conventional sampling estimators such as sample mean or Horvitz–Thompson estimator are biased because the sample size is random. So, an alternative unbiased estimator of τ_Y is derived under this stopping rule.

If M is known, an unbiased estimator of τ_Y is given by

$T = M \ \bar{y}_M + (N - M) \ \bar{y}_{N-M},$

where $\bar{y}_M = \frac{\sum_{i \in S_M} y_i}{k}$, S_M is the index set of units in the sequential sample that are included in P_M.

$\bar{y}_{N-M} = \frac{\sum_{i \in S_{N-M}} y_i}{n-k}$, S_{N-M} is the index set of units in the sequential sample that are included in P_{N-M}.

n is the total sequential sample size.

Usually, M is unknown and hence it is replaced by its unbiased estimator $\hat{M} = \frac{N(k-1)}{n-1}$ assuming $k > 1$. Then, the estimator of τ_Y is given by

$$T_I = \hat{M}\, \bar{y}_M + (N - \hat{M})\, \bar{y}_{N-M}.$$

The unbiasedness of T_I can be shown as follows.

Let f_i be the number of times $y_i \in P_{N-M}$ appears in the sample; $i \in U_{N-M}$. Then, $f_i|n$ has binomial distribution with parameters $(n-k; P = \frac{1}{N-M})$, if the sampling is with replacement.

Similarly, let z_i be the number of times $y_i \in P_M$ appears in the sample, $i \in U_M$. Then, $z_i|n$ has binomial distribution with parameters $(k, P = \frac{1}{M})$.

Hence, $E(T_I) = E_n \left[E[\hat{M}\, \bar{y}_M + (N - \hat{M})\, \bar{y}_{N-M}|n] \right]$

$$= E_n \left[\frac{\hat{M}}{M} \sum_{i \in U_M} y_i + \frac{N-\hat{M}}{N-M} \sum_{i \in U_{N-M}} y_i \right]$$

$$= \tau_Y.$$

Variance of T_I is given by

$$V_w(T_I) = E_n[V(T_I|n)] + V_n[E(T_I|n)]$$

$$= E_n \left[\hat{M}^2\, V(\bar{y}_M|n) + (N - \hat{M})^2\, V(\bar{y}_{N-M}|n) \right] + V_n \left[\hat{M}\, \mu_M + (N - \hat{M})\, \mu_{N-M} \right].$$

It reduces to

$$V_w(T_I) = \frac{\sigma_M^2}{k}\, E_n \left[\hat{M}^2 \right] + \sigma_{N-M}^2\, E_n \left[\frac{(N-\hat{M})^2}{n-k} \right] + (\mu_M - \mu_{N-M})^2\, V_n(\hat{M})$$

$$= \left[\frac{\sigma_M^2}{k} + (\mu_M - \mu_{N-M})^2 \right]\, V_n(\hat{M}) + \frac{\sigma_M^2\, \hat{M}^2}{k} + \sigma_{N-M}^2\, E_n \left[\frac{(N-\hat{M})^2}{n-k} \right].$$

If the sampling is without replacement then the distribution of the sample size n is negative hypergeometric with parameters (k, N, M).

The unbiasedness of T_I can be proved as above except that, instead of using binomial distribution, negative hypergeometric is used.

Variance of T_I in this case is given by

$$V_{wo}(T_I) = E_n[V(T_I|n)] + V_n[E(T_I|n)]$$

$$= E_n \left[\hat{M}^2\, V(\bar{y}_M|n) + (N - \hat{M})^2\, V(\bar{y}_{N-M}|n) \right] + V_n \left[\hat{M}\, \mu_M + (N - \hat{M})\, \mu_{N-M} \right]$$

$$= \frac{\sigma_M^2}{k}\, E_n \left[\hat{M}^2 (1 - \frac{k}{M}) \right] + \sigma_{N-M}^2\, E_n \left[\frac{(N-\hat{M})^2}{n-k} (1 - \frac{n-k}{N-M}) \right] + (\mu_M - \mu_{N-M})^2\, V_n(\hat{M}).$$

It is not easy to determine $V_n(\hat{M})$. Mikulski and Smith (1976) have shown it to be bounded as

$$\frac{M^2 (1-\frac{M}{N})}{k} \leq V_n(\hat{M}) \leq \frac{M^2 (1-\frac{M}{N})}{k+\frac{M}{N}-2}.$$

13.2.2 Mixed Design Stopping Rule

In this case, a more common approach to sampling rare populations is considered.

Suppose a random sample of fixed size n' is selected according to SRSWR or SRSWOR. Sampling is stopped after n' selections if any unit of P_M are selected in this sample; otherwise sampling is continued sequentially till k units from P_M are sampled.

Let S denote the number of units from P_M selected in the first n' units. In this stopping rule, the final sample size n is such that $n = n'$ if $S > 0$ and $n > n'$ if $S = 0$. Lan (1999) defined the estimator of τ_Y as

$$T_{mix} = \begin{cases} \frac{N}{n'} \sum_{i=1}^{n'} y_i & ; \text{if } n = n' \\ \hat{M}\, \bar{y}_M + (N - \hat{M})\, \bar{y}_{N-M} & ; if \ n > n'. \end{cases}$$

It is positively biased for τ_Y under this stopping rule. If the stopping rule is modified a little bit, the above estimator becomes unbiased. The modified stopping rule is given as follows.

If less than k units satisfying C are observed in the random sample of n' units, sampling continues in a sequential manner until exactly k units of P_M are selected.

Thus, in this case, $n = n'$ if $S \geq k$ and $n > n'$ if $S < k$.

The variance of the estimator under this unbiased stopping rule has been derived by Christman and Lan (2001).

It is given by

$$V(T_{mix}) = \sigma_M^2 \left\{ \frac{1}{k}\ E_n \left[\hat{M}^2\ I[n > n'] \right] + \frac{M\,N}{k}\ P(S^\star \geq k - 1) \right\}$$

$$+ \sigma_{N-M}^2 \left\{ E_n \left[\frac{N - \hat{M}^2}{n - k}\ I[n > n'] \right] + \frac{N^2}{n'} \left[P(S \geq k) - \frac{M}{N}\ P(S^\star \geq k - 1) \right] \right\}$$

$$+ (\mu_M - \mu_{N-M})^2 \left\{ \frac{M\,N}{n'}(1 - \frac{M\,n'}{N})\ P(S^\star \geq k - 1) + \frac{n_0 - 1}{n_0}\ M^2\ P(S^{\star\star} \geq k - 2) \right\}$$

$$+ (\mu_M - \mu_{N-M})^2 \left\{ V_n(\hat{M}\ I[n > n']) - 2\,M^2\ P(S^\star \geq k - 1)\ P(S^{\star\star} \geq k - 2) \right\}$$

$$- 2MN(\mu_M - \mu_{N-M})\,\mu_M \left\{ 1 - P(S < k) - P(S^\star \geq k - 1) \right\},$$

where S is binomially distributed with parameters $(n', P = \frac{M}{N})$; I is an indicator function equal to one; if the inequality is true and zero; otherwise. $S^\star$ and $S^{\star\star}$ are binomially distributed with parameters $(n' - 1, P)$ and $(n' - 2, P)$, respectively.

13.3   Inverse Sampling with Unequal Selection Probabilities

In many practical situations, population can be subdivided into two groups. The group to which an unit belongs is not known until the unit is sampled and observed. In Sect. 13.2, inverse sampling introduced by Christman and Lan (2001) was discussed. They considered all population having equal probability of selection. In this section, we discuss the derivation of unbiased estimators of the totals of the two groups, their variances and corresponding variance estimators in inverse sampling with replacement (ISWR) when the units have unequal selection probabilities. This type of situations is common in environmental surveys.

Consider a population of N units, divided into two groups containing N_1 and N_2 units, respectively, so that $N_1 + N_2 = N$. Let Y_{i1}, $(i = 1, 2, \ldots \cdots, N_1)$ and Y_{i2}, $(i = 1, 2, \ldots \cdots, N_2)$ be the values of the variable of interest Y in the first and second group, respectively. Let C be the condition with respect to which the population units are divided into two groups according to whether the Y-values satisfy the condition C.

Let T_1 and T_2 be the totals of the two groups and $T_1 + T_2 = T$ be the population total. P_{i1} be the probability of selection of the ith unit of the first group, and P_{i2} be the probability of selection of the ith unit of the second group. Let $P = \sum_{i=1}^{N_1} P_{i1}$ denote the selection probability of the first group and $1 - P = \sum_{i=1}^{N_2} P_{i2}$ denote the selection probability of the second group.

Assume that the units are selected sequentially till k units of the first group (rare group) are observed in the sample. Let n^* denote the sample size, which is a random variable assuming values $k, k + 1, \ldots$. Consider the event $n^* = n$. Among them, only k units are coming from the first group and the remaining $n - k$ from the second group. If we assume that this inverse sampling is performed with replacement then

$$P(n^* = n) = P^k (1 - P)^{n-k}.$$

Hence, the probability that a sample of n units is selected in a given order is

$$\frac{\prod_{i=1}^{N_1} P_{i1}^{m_{i1}} \; \prod_{i=1}^{N_2} P_{i2}^{m_{i2}}}{P^k (1-P)^{n-k}} = \prod_{i=1}^{N_1} \left(\frac{P_{i1}}{P}\right)^{m_{i1}} \prod_{i=1}^{N_2} \left(\frac{P_{i2}}{1-P}\right)^{m_{i2}},$$

where m_{i1} and m_{i2} denote the number of times the ith unit from the first group and the second group are selected in the sample such that $\sum_{i=1}^{N_1} m_{i1} = k$ and $\sum_{i=1}^{N_2} m_{i2} = n - k$.

Thus, the selection probability of the ith unit in the first group$= \frac{P_{i1}}{P}$ and that of the second group is$= \frac{P_{i2}}{1-P}$. Obviously, the selection of units from the groups is independent.

Let Y_{ij} denote the Y value on the ith sample unit from the jth group. Hence, the population total can be written as

$$T_Y = T_{1Y} + T_{2Y} = \frac{P}{k} \sum_{i=1}^{k} \frac{Y_{i1}}{P_{i1}} + \frac{1-P}{\nu-k} \sum_{i=1}^{\nu-k} \frac{Y_{i2}}{P_{i2}}.$$

Usually, P is unknown. ν has negative binomial distribution with parameters (k, P). Hence, $\hat{P} = \frac{k-1}{\nu-1}$ is unbiased for P.

If we consider $W_{ij} = \frac{Y_{ij}}{P_{ij}}$ as independent and identically distributed random variables, then estimator of T_Y can be given by

$$\hat{T}_Y = \hat{T}_{1Y} + \hat{T}_{2Y} = \frac{\hat{P}}{k} \sum_{i=1}^{k} W_{i1} + \frac{1-\hat{P}}{\nu-k} \sum_{i=1}^{\nu-k} W_{i2}.$$

That is, $\hat{T}_Y = \hat{P} \, \bar{W}_1 + (1 - \hat{P}) \, \bar{W}_2$.

Greco and Naddeo (2007) have shown that $\hat{T}_Y$ is unbiased for T_Y.

They have also shown that

$$V(\hat{T}_Y) = (W_1 - W_2)^2 \, V_\nu(\hat{P}) + \frac{\sigma_{W_1}^2}{k} E_\nu(\hat{P}^2) + \frac{\sigma_{W_2}^2}{k-1} E_\nu\left[\hat{P}(1-\hat{P})\right],$$

where $\sigma_{wj}^2 = \sum_{i=1}^{N_j} (w_{ij} - w_j)^2 P_{ij}$; $j = 1, 2$. Greco and Naddeo (2007) have also derived unbiased estimator of $V(\hat{T}_Y)$.

In inverse sampling, it is possible to run out of resources (such as money, time, and labor) prior to selecting k units from the first group. Greco and Naddeo (2007) showed that it is still possible to get unbiased estimators of the group totals and their variances.

They considered the sampling design in which the selection procedure stops if (i) k units of the first group are observed in the sample and (ii) the maximum sample size M is obtained without getting k units of the first group.

In the first case, the unbiased estimator of the population total is given by

$$\hat{T}_Y = \hat{P}\,\bar{W}_1 + (1 - \hat{P})\,\bar{W}_2.$$

In the second case, the estimator is based on the estimator of the population totals of the two groups used in fixed sample size designs when M units are randomly selected from the whole population. Hence, the number of selected units from the first group is a random variable X having binomial distribution with parameters (M, P). In this case, the unbiased estimator of T_Y is given by

$$T_Y^\star = \hat{P}_\star\,\bar{W}_1 + (1 - \hat{P}_\star)\,\bar{W}_2,$$

where $\hat{P}_\star = \frac{X}{M}$.

Greco and Naddeo (2007) have derived the expression for $V(T_Y^\star)$ and its unbiased estimator.

Salehi and Seber (2004) proved that Murthy's estimator can be applied for inverse sampling. Using this approach, (Sangngam 2012) obtained the following unbiased estimator of the population total and its variance estimator. These estimators are derived with the following background.

A finite population of N units is studied with respect to characteristic Y. It is divided into two classes. P_M the class of units satisfying the specified condition C and P_{N-M} the class of remaining units.

In unequal probability inverse sampling, units are selected one at a time with unequal probability with replacement until a given m number of units of the class P_M are obtained. The sample size n is a random variable. The sample (S) can be partitioned into two parts. S_M is the set of units from P_M and S_{N-M} is the set of units from P_{N-M} with cardinalities m and $n - m$, respectively. Let k and g be the number of distinct units in the sets S_M and S_{N-m} indexed by $i = 1, 2.\cdots, k$ and $i = k + 1, k + 2, \ldots \cdots, \nu$, respectively.

Let r_i be the number of times unit i appears in the sample.

The probability of getting an ordered sample $(S^\star) = P(S^\star) = \prod_{i=1}^{\nu} (z_i)^{r_i}$ where the last sample unit belongs to S_M.

An unordered sample S can be obtained in $\sum_{i=1}^{k} \binom{n-1}{r_1, r_1-1, \ldots, r_\nu}$ ways. Hence, the probability of getting an unordered sample $S = \sum_{i=1}^{k} \binom{n-1}{r_1, r_1-1, \ldots, r_\nu} \prod_{i=1}^{\nu} (z_i)^{r_i}$ Sangngam (2012) has shown that

$$T_I = \hat{P}\,\bar{y}_M + (1 - \hat{P})\,\bar{y}_{N-M} \text{ is unbiased for } \tau_Y,$$

where $\hat{P} = \frac{m-1}{n-1}$, $\quad \bar{y}_M = \frac{1}{m}\sum_{i \in S_M}\frac{y_i}{z_i}$, $\quad \bar{y}_{N-M} = \frac{1}{n-m}\sum_{i \in S_{N-M}}\frac{y_i}{z_i}$

and

$$\hat{V}(T_I) = (\bar{y}_M - \bar{y}_{N-M})^2\,\hat{V}(\hat{P}) + \frac{\hat{\sigma}_M^2}{m}\left[(m-1)\,\hat{V}(\hat{P}) - \hat{P}_\star\,\hat{P}^2\right] + \hat{\sigma}_{N-M}^2\,\frac{(n-m-1)}{(n-1)(n-2)},$$

where $\hat{P}_\star = \frac{(m-1)(m-2)}{(n-1)(n-2)}$.

13.4 General Inverse Sampling

By continuing the use of Lan's (1999) notations, Salehi and Seber (2004) introduced this design. In this design, at the beginning, a simple random sample of n_0 units is selected. If at least k units from P_M are selected in this sample, then sampling is stopped. Otherwise, it is continued till k units from P_M are selected or final sample size n_2 is reached. The estimator of the population total τ_Y is given by

$$
T_{gI} =
\begin{cases}
\frac{N}{n_0} \sum_{i=1}^{n_0} y_i & ; \text{if } n_1 = n_0 \\[2mm]
\hat{M}\, \bar{y}_M + (N - \hat{M})\, \bar{y}_{N-M} & ; \text{if } n_0 < n_1 < n_2 \text{ or } n_1 = n_2 \text{ and } |S_M| = k \\[2mm]
\frac{N}{n_2} \sum_{i=1}^{n_2} y_i & ; \text{if } n_1 = n_2 \text{ and } |S_M| < k.
\end{cases}
$$

An unbiased estimator of the variance of this estimator is given by

$$
\hat{V}(T_{gI}) =
\begin{cases}
N^2 \left(1 - \frac{n_0}{N}\right) \frac{S_o^2}{n_0} & ; \text{if } n_1 = n_0 \\[2mm]
a\, S_M^2 + \hat{V}(\hat{M})\,(\bar{y}_M - \bar{y}_{-N-M})^2 + b\, S_{N-M}^2 & ; \text{if } n_0 < n_1 < n_2 \text{ or } n_1 = n_2 \\
& \quad \text{and } |S_M| = k \\[2mm]
N^2 \left(1 - \frac{n_2}{N}\right) \frac{S_2^2}{n_2} & ; \text{if } n_1 = n_2 \text{ and } |S_M| < k,
\end{cases}
$$

where

$$S_M^2 = \frac{\sum_{i \in S_M} (\bar{y}_i - \bar{y}_M)^2}{k-1}$$

$$S_{N-M}^2 = \frac{\sum_{i \in S_{N-M}} (\bar{y}_i - \bar{y}_{N-M})^2}{n_1 - k}$$

$$\hat{V}(\hat{M}) = \left(1 - \frac{n_1 - 1}{N}\right) \frac{\hat{M}\,(N - \hat{M})}{(n_1 - 2)}$$

$$a = \frac{\hat{M}^2}{k}\, \frac{(N - n_1 + 1)\,(n_1 k - n_1 - k) - N\,(n_1 - 2)}{N\,(n_1 - 2)(k - 1)}$$

$$b = \frac{N\,(N - n_1 + 1)\,(n_1 - k - 1)}{(n_1 - 1)\,(n_1 - 2)}$$

$$S_o^2 = \frac{\sum_{i=1}^{n_0} (\bar{y}_i - \bar{y}_0)^2}{n_0 - 1}$$

$$\bar{y}_0 = \frac{\sum_{i=1}^{n_0} \bar{y}_i}{n_0}$$

$$S_2^2 = \frac{\sum_{i=1}^{n_2} (\bar{y}_i - \bar{y}_2)^2}{(n_2 - 1)}$$

$$\bar{y}_2 = \frac{\sum_{i=1}^{n_2} \bar{y}_i}{n_2}.$$

It has been shown that in case of general unequal probability inverse sampling, an unbiased estimator of τ_Y is given by

$$
T_{gII} =
\begin{cases}
\frac{1}{n_0} \sum_{i=1}^{n_0} \frac{y_i}{z_i} & ; \text{if } n_1 = n_0 \\[2mm]
\hat{P}\, \bar{y}_M + (1 - \hat{P})\, \bar{y}_{N-M} & ; \text{if } n_0 < n_1 < n_2 \text{ or } n_1 = n_2 \text{ and } |S_M| = k \\[2mm]
\frac{1}{n_2} \sum_{i=1}^{n_2} \frac{y_i}{z_i} & ; \text{if } n_1 = n_2 \text{ and } |S_M| < k,
\end{cases}
$$

where $\hat{P} = \frac{k-1}{n-1}$, $\bar{y}_M = \frac{1}{k}\sum_{i \in S_M} \frac{y_i}{z_i}$, $\bar{y}_{N-M} = \frac{1}{n-k}\sum_{i \in S_{N-M}} \frac{y_i}{z_i}$.

An unbiased estimator of the variance of T_{gI1} is given by

$$
\hat{V}(T_{gI1}) = \begin{cases} \frac{1}{n_0\,(n_0-1)} \sum_{i=1}^{n_0} (\frac{y_i}{z_i} - T_I)^2 & ; \text{if } n_1 = n_0 \\[2em] (\bar{y}_M - \bar{y}_{N-M})^2\, \hat{V}(\hat{P}) + \frac{\hat{\sigma}_M^2}{m} \\ \left[(k-1)\,\hat{V}(\hat{P}) - \hat{P}_\star\,\hat{P}^2 \right] + \hat{\sigma}_{N-M}^2\, \frac{(n-k-1)}{(n-1)(n-2)} & ; \text{if } n_0 < n_1 < n_2 \text{ or } n_1 = n_2 \\ & \quad\text{and } |S_M| = k \\[2em] \frac{1}{n_2\,(n_2-1)} \sum_{i=1}^{n_2} (\frac{y_i}{z_i} - T_I)^2 & ; \text{if } n_1 = n_2 \text{ and } |S_M| < k, \end{cases}
$$

where $\hat{P}_\star = \frac{(k-1)\,(k-2)}{(n-1)\,(n-2)}$, $\hat{\sigma}_M^2 = \frac{1}{k-1}\sum_{i \in S_M} (\frac{y_i}{z_i} - \bar{y}_M)^2$, $\hat{\sigma}_{N-M}^2 = \frac{1}{n-k-1}\sum_{i \in S_{N-M}}$ $(\frac{y_i}{z_i} - \bar{y}_{N-M})^2$ and $\hat{V}(\hat{P}) = \frac{\hat{P}\,(1-\hat{P})}{n-2}$.

13.5   Regression Estimator Under Inverse Sampling

Sometimes the population under study is rare and clustered with respect to the characteristic under study. This population has many zero or low values and very few units having high values of the characteristic under study. In simple random sampling, regression and ratio estimators are undefined for those samples containing information from only non-rare units having zero or low values. In such cases, the use of auxiliary variable improves the sampling design along with the improvement in estimator. Moradi et al. (2011) introduced the modified regression estimators and their associated variance estimators for general inverse sampling and inverse sampling with unequal selection probabilities.

13.5.1   General Inverse Sampling

Considering a finite population of N units, X as the auxiliary variable and Y as the interest variable. Salehi and Seber (2004) proposed the following estimator of the population mean of Y:

$$\hat{\mu}_Y = \begin{cases} \frac{1}{n_0} \sum_{i=1}^{n_0} y_i & ; \text{if } n_s = n_0 \ , |S_M| \geq m \\[2mm] \hat{P}\ \bar{y}_M^{\star} + (1 - \hat{P})\ \bar{y}_{N-M}^{\star} & ; \text{if } n_0 < n_s < n_1 \ \text{and } |S_M| = m \\[2mm] \frac{1}{n_1} \sum_{i=1}^{n_1} y_i & ; \text{if } n_s = n_1 \ \text{and } |S_M| < m, \end{cases}$$

where m is the pre-decided number of rare units in the general inverse sampling design. Estimator of μ_X can also be calculated by defining $\hat{\mu}_X$ using the form of $\hat{\mu}_Y$ above. By using Murthy's (1957) estimator, estimators of $\sum_{i=1}^{N} X_i Y_i$ and $\sum_{i=1}^{N} X_i^2$ are given by

$$\hat{\tau}_{XY} = \sum_{i=1}^{n} \frac{P(s|i)}{P(s)} x_i y_i \quad ; \hat{\tau}_{X^2} = \sum_{i=1}^{n} \frac{P(s|i)}{P(s)} x_i^2.$$

Using the form of $\hat{\mu}_Y$ above, $\hat{\tau}_{X^2}$ and $\hat{\tau}_{XY}$ in general inverse sampling are given as follows:

$$\hat{\tau}_X^2 = \begin{cases} \frac{N}{n_0} \sum_{i=1}^{n_0} y_i^2 & ; \text{if } n_s = n_0 \ ; |S_M| \geq m \\[2mm] N\ (\hat{P}\ \bar{x}^2{}_M + (1 - \hat{P})\ \bar{x}^2{}_{N-M}) & ; \text{if } n_0 < n_s <= n_1 \ ; |S_M| = m \\[2mm] \frac{N}{n_1} \sum_{i=1}^{n_1} x_i^2 & ; \text{if } n_s = n_1 \ ; |S_M| < m, \end{cases}$$

$$\hat{\tau}_{XY} = \begin{cases} \frac{N}{n_0} \sum_{i=1}^{n_0} x_i y_i & ; \text{if } n_s = n_0 \ ; |S_M| \geq m \\[2mm] N\ (\hat{P}\ \bar{xy}_M + (1 - \hat{P})\ \bar{xy}_{N-M}) & ; \text{if } n_0 < n_s <= n_1 \ ; |S_M| = m \\[2mm] \frac{N}{n_1} \sum_{i=1}^{n_1} x_i y_i & ; \text{if } n_s = n_1 \ ; |S_M| < m. \end{cases}$$

The modified regression estimator of population mean in general inverse sampling is given by

$$\hat{\mu}_{mr1} = \hat{\mu}_Y + \hat{B}(\mu_x - \hat{\mu}_x)$$

where, $\hat{B} = \frac{\hat{\tau}_{XY} - N\ \hat{\mu}_x\ \hat{\mu}_Y}{\hat{\tau}_X^2 - N\ \hat{\mu}_x^2}$.

The variance estimator of this modified regression estimator is given by

$$\hat{V}(\hat{\mu}_{mr1}) = \begin{cases} \left(\frac{1}{n_0} - \frac{1}{N}\right) \frac{1}{n_0 - 1} \sum_{i=1}^{n_0} (y_i - \bar{y})^2 & ; \text{if } n_s = n_0 \ ; |S_M| \geq m \\[3mm] \hat{P}^2 \left(\frac{N\ (N-n+1)\ (n\ m-n-m) - N\ (n-2)}{(n-2)\ (m-1)} \right) \frac{S_M^2}{m} & \\ +N^2\ \hat{V}(\hat{P})\ (\bar{y}_M - \bar{y}_{N-M})^2 + \left(\frac{N\ (N-n+1)\ (n-m-1)}{(n-1)\ (n-2)} \right) S_{N_M}^2 & ; \text{if } n_0 < n_s <= n_1 \\ & |S_M| = m \\[3mm] \left(\frac{1}{n_1} - \frac{1}{N}\right) \frac{1}{n_1 - 1} \sum_{i=1}^{n_1} (y_i - \bar{y})^2 & ; \text{if } n_s = n_1 \ ; |S_M| < m, \end{cases}$$

where $S_M^2 = \frac{\sum_{i=1}^{m}(y_i - \bar{y})^2}{m - 1}$; $S_{N-M}^2 = \frac{\sum_{i=1}^{n_s - m}(y_i - \bar{y})^2}{n_s - m - 1}$; $\hat{V}(\hat{P}) = \frac{(1 - \frac{n_s - 1}{N})\ \hat{P}\ (1 - \hat{P})}{n_s - 2}$.

13.5.2   *Inverse Sampling with Unequal Selection Probabilities*

Greco and Naddeo (2007) proposed an unbiased estimator of the population total as

$$\tau_Y^\star = \frac{\hat{P}}{m} \sum_{i=1}^m \frac{y_{1i}}{P_{1i}} + \frac{1-\hat{P}}{\nu-m} \sum_{i=1}^{\nu-m} \frac{y_{2i}}{P_{2i}},$$

where $\hat{P} = \frac{m-1}{\nu-1}$.

On the same lines, Moradi et al. (2011) proposed modified regression estimator of μ_Y in inverse sampling with unequal selection probabilities as

$$\hat{\mu}_{mr2} = \mu_Y^\star + B^\star(\mu_x - \mu_x^\star),$$

where $\tau_{xy}^\star = \frac{\hat{P}}{m} \sum_{i=1}^m \frac{x_{1i}\, y_{1i}}{P_{1i}} + \frac{1-\hat{P}}{\nu-m} \sum_{i=1}^{\nu-m} \frac{x_{2i}\, y_{2i}}{P_{2i}}$

$$\tau_{x2}^\star = \frac{\hat{P}}{m} \sum_{i=1}^m \frac{x_{1i}^2}{P_{1i}} + \frac{1-\hat{P}}{\nu-m} \sum_{i=1}^{\nu-m} \frac{x_{2i}^2}{P_{2i}}.$$

Variance of the above estimator $\hat{\mu}_{mr2}$ is given by

$$V(\hat{\mu}_{mr2}) = \frac{V(\tau_\nu^\star)}{N^2},$$

where $V(\tau_\nu^\star) = \frac{\sigma_{1w}^2}{m} P^2 + \frac{\sigma_{2w}^2}{m-1} P(1-P),$

where $\sigma_{jw}^2 = \sum_{i=1}^{N_j}(w_{ji} - w_j)^2 P_{ji}$; $j = 1, 2; w_{ji} = \frac{y_{ji} - B\, x_{ji}}{P_{ji}}$; $w_1 = \frac{\tau_{1y} - B\, \tau_{1x}}{P}$;

$w_2 = \frac{\tau_{2y} - B\, \tau_{2x}}{1-P}.$

If P is not known, it is replaced by $\hat{P}$.

13.6   Inverse Adaptive Cluster Sampling

It is a sequential sampling design in which an adaptive sampling component is added. In this method, if a sequentially sampled unit satisfies the condition C then the network to which that unit belongs is completely sampled. Thus, the final sample consists of the initial sample that is taken sequentially along with all units belonging to the networks of the k sequentially sampled units that satisfied C and the edge units of those networks. In this method, the number and size of the networks associated with a sampled edge unit are unknown. The probability of inclusion of an edge unit in the final sample is not known. Hence, any estimator that includes edge units will be biased. So, edge units are excluded from the estimation procedure.

Unbiased estimators of the population total are obtained by replacing the observed y_i with the network mean $\bar{y}_i^\star = \frac{\sum_{j \in A_i} y_j}{m_i}$,

where A_i is the index set of all units in the network to which the ith unit belongs.
(i) An unbiased estimator of the population total is given by

$$T_{IA} = \hat{M}\, \bar{y}_M^\star + (N - \hat{M})\, \bar{y}_{N-M},$$

where $\bar{y}_M^\star = \frac{\sum_{i \in S_M} \bar{y}_i^\star}{k}.$

If the sampling is with replacement, then variance of the above estimator is given by

$$V_w(T_{IA}) = \left[\frac{\sigma_M^{\star 2}}{k} + (\mu_M - \mu_{N-m})^2\right] V_n(\hat{M}) + \frac{\sigma_M^{\star 2}\, \hat{M}^2}{k} + \sigma_{N-M}^2\, E_n\left[\frac{(N-\hat{M})^2}{n-k}\right],$$

where $\sigma_M^{\star 2} = \frac{1}{M} \sum_{i \in U_M}(\bar{y}_i^\star - \mu_M)^2.$

If the sampling is without replacement, then the variance of T_{IA} is given by

$$V_{wo}(T_{IA}) = \frac{\sigma_M^{\star 2}}{k} E_n \left[\hat{M}^2 \left(1 - \frac{k}{M}\right) \right] + \sigma_{N-M}^2 E_n \left[\frac{(N-\hat{M})^2}{n-k} \left(1 - \frac{n-k}{N-M}\right) \right] + (\mu_M - \mu_{N-m})^2 V_n(\hat{M}).$$

(ii) Replacing y_i by network mean $\bar{y}_i^\star$, we can obtain the adaptive version of the estimator T_{mix} discussed in Sect. 13.2.2. While using replacement it is assumed that adaptive cluster sampling of a selected unit's network is done without considering whether the sampling is part of the first n' randomly selected units or part of the sequential sampling effort. Under the second stopping rule discussed in Sect. 13.2.2, the unbiased adaptive estimator of τ_Y is obtained by replacing y_i by network mean $\bar{y}_i^\star$ and $\bar{y}_M$ by $\bar{y}_M^\star$. So that we get the new estimator as

$$T_{mix,A} = \begin{cases} \frac{N}{n'} \sum_{i=1}^{n'} \bar{y}_i^\star & ; \text{if } n = n' \\ \\ \hat{M} \bar{y}_M^\star + (N - \hat{M}) \bar{y}_{N-M}^- & ; if \ n > n'. \end{cases}$$

Variance of the above estimator has the same formula as that for T_{mix} except that σ_M^2 is replaced by $\sigma_M^{\star 2}$.

The usual inverse adaptive cluster sampling can be improved by adding more information obtained from the final sample. In that design, the values of the edge units are used in the estimator only for the edge units that are selected in the initial sample. By using the Rao–Blackwellization technique, the estimator of the population total can be improved. Pochai (2008) proposed the following improved estimator of the population total.

$$T_{gI,A} = \begin{cases} \frac{N}{n_0} \sum_{i=1}^{n_0} \bar{y}_i^\star & ; \text{if } n_1 = n_0 \\ \\ \hat{M} \bar{y}_M^\star + \frac{N-\hat{M}}{n_1 - k} \sum_{i \in S_{N-M}} \left[y_i(1 - e_i) + \bar{y}_e \, e_i \right] & ; \text{if } n_0 < n_1 < n_2 \text{ or } n_1 = n_2 \\ & \quad \text{and } |S_M| = k \\ \\ \frac{N}{n_2} \sum_{i=1}^{n_2} \bar{y}_i^\star & ; \text{if } n_1 = n_2 \text{ and } |S_M| < k. \end{cases}$$

13.7 General Unequal Probability Inverse ACS

Sangngam (2012) applied the general unequal probability inverse sampling to adaptive cluster sampling. In this method, an initial sample is drawn by the general unequal probability inverse sampling with replacement. For the sample units belonging to class P_M, their neighborhoods are added to the sample. The procedure continues until no more units from the class P_M are observed. Thus, the final sample consists of the initial sample and all adaptively added units. For adding the units adaptively, Thompson's (1990) procedure is used. Thus, the population is divided into K mutually exclusive networks.

Let n denote the final sample size. Let ψ_k denote the set of units in the kth network and m_k denote the number of units in the kth network. The total value of the study variable in the network ψ_k is denoted by $y_k^\star = \sum_{j \in \psi_k} y_j$ and the probability of selection of that network is $z_k^\star = \sum_{j \in \psi_k} z_j$.

By using this sampling design, an unbiased estimator of the population total is given by

$$
T_{g12} = \begin{cases}
\frac{1}{n_0} \sum_{i=1}^{n_0} \frac{y_i^\star}{z_i^\star} & ; \text{if } n_1 = n_0 \\[2mm]
\hat{P}\, \bar{y}_M^\star + (1 - \hat{P})\, \bar{y}_{N-M}^\star & ; \text{if } n_0 < n_1 < n_2 \text{ or } n_1 = n_2 \text{ and } |S_M| = k \\[2mm]
\frac{1}{n_2} \sum_{i=1}^{n_2} \frac{y_i^\star}{z_i^\star} & ; \text{if } n_1 = n_2 \text{ and } |S_M| < k,
\end{cases}
$$

where $\hat{P} = \frac{m-1}{n-1}$, $\bar{y}_M^\star = \frac{1}{m} \sum_{i \in S_M} \frac{y_i^\star}{z_i^\star}$, $\bar{y}_{N-M}^\star = \frac{1}{n-m} \sum_{i \in S_{N-M}} \frac{y_i^\star}{z_i^\star}$.

An unbiased estimator of the variance of T_{g12} is given by

$$
\hat{V}(T_{g12}) = \begin{cases}
\frac{1}{n_0 (n_0-1)} \sum_{i=1}^{n_0} (\frac{y_i^\star}{z_i^\star} - T_{g1})^2 & ; \text{if } n_1 = n_0 \\[3mm]
(\bar{y}_M^\star - \bar{y}_{N-M}^\star)^2\, \hat{V}(\hat{P}) + \frac{\hat{\sigma}_M^{\star 2}}{m} \\
\left[(m-1)\,\hat{V}(\hat{P}) - \hat{P}_\star\, \hat{P}^2 \right] + \hat{\sigma}_{N-M}^{\star 2} \frac{(n-m-1)}{(n-1)(n-2)} & ; \text{if } n_0 < n_1 < n_2 \text{ or } n_1 = n_2 \\
& \quad \text{ and } |S_M| = k \\[3mm]
\frac{1}{n_2 (n_2-1)} \sum_{i=1}^{n_2} (\frac{y_i^\star}{z_i^\star} - T_{g1})^2 & ; \text{if } n_1 = n_2 \text{ and } |S_M| < k,
\end{cases}
$$

where $\hat{P}_\star = \frac{(m-1)\,(m-2)}{(n_0-1)\,(n_0-2)}$,

$\hat{\sigma}_M^{\star 2} = \frac{1}{m-1} \sum_{i \in S_M} (\frac{y_i^\star}{z_i^\star} - \bar{y}_M)^2$,

$\hat{\sigma}_{N-M}^{\star 2} = \frac{1}{n_0-m-1} \sum_{i \in S_{N-M}} (\frac{y_i^\star}{z_i^\star} - \bar{y}_{N-M})^2$,

and $\hat{V}(\hat{P}) = \frac{\hat{P}\,(1-\hat{P})}{n_0-2}$.

13.8 Exercises

1. What is inverse sampling? When it is used? Describe the two stopping rules used in inverse sampling.
2. State an unbiased estimator of the population total τ_Y under the first stopping rule when the number of units in the population satisfying the condition of interest is known.

3. State an unbiased estimator of the population total τ_Y under the first stopping rule when the number of units in the population satisfying the condition of interest is not-known.

4. Show that T_I is unbiased for τ_Y under the first stopping rule when the number of units in the population satisfying the condition of interest is not-known. Obtain its variance when the sampling is with replacement. Also obtain the variance of T_I when the sampling is without replacement.

5. Define the estimator of τ_Y based on mixed design stopping rule.

6. State the modified mixed design stopping rule to estimate τ_Y. Show that it is unbiased for τ_Y. Obtain its variance.

7. Derive an estimator of the population total in inverse sampling with replacement when the units have unequal selection probabilities.

8. Show that the estimator $(\hat{T}_Y)$ proposed by Greco and Naddeo of the population total in inverse sampling with replacement is unbiased for τ_Y. Derive variance of $\hat{T}_Y$.

9. State the unbiased estimator proposed by Greco and Naddeo of the population total in inverse sampling design in which the procedure stops if the specified number of units (k) of the first group are observed in the sample. Derive the expression of variance of the estimator. Also obtain its unbiased estimator.

10. State the unbiased estimator proposed by Greco and Naddeo of the population total in inverse sampling design in which the maximum sample size (M) is obtained without getting the specified number of units (k) of the first group.

11. Show that the estimator of τ_Y proposed by Sangngam in inverse sampling design is unbiased. Obtain variance of this estimator.

12. What is general inverse sampling? State the estimator of population total τ_Y proposed by Salehi and Seber in general inverse sampling. Obtain an unbiased estimator of the variance of this estimator.

13. State the unbiased estimator of population total τ_Y in the case of general unequal probability inverse sampling. Obtain an unbiased estimator of the variance of the estimator.

14. Describe Inverse ACS procedure. Obtain an unbiased estimator of the population total in Inverse ACS. Obtain its variance.

15. Obtain an unbiased estimator of the population total in Inverse ACS under the mixed design stopping rule. Derive variance of this unbiased estimator.

16. How can the usual Inverse ACS be improved? Discuss the improved estimator of the population total in Inverse ACS proposed by Pochai. Derive variance of this improved estimator.

17. Describe the general unequal probability Inverse ACS. Obtain an unbiased estimator of the population total by using the general inverse ACS. Derive an unbiased estimator of the variance of this estimator.

18. Obtain the modified regression estimator of the population mean in general inverse sampling. Derive an expression for the variance of this estimator.

19. State Moradi's modified regression estimator of the population mean in inverse sampling with unequal selection probabilities. Derive an expression for the variance of this estimator.

References

Christman, M.C., Lan, F.: Inverse adaptive cluster sampling. Biometrics, **57**(4), 1096–1105 (2001)

Greco, L., Naddeo, S.: Inverse sampling with unequal selection probabilities. Commun. Stat. Theory Methods **36**(5), 1039–1048 (2007)

Haldane, J.B.S.: On a method of estimating frequencies. Biometrika **33**(3), 222–225 (1945)

Lan, F.: Sequential adaptive designs to estimate abundance in rare populations. Ph. D. dissertation, American University, Washington, D.C (1999)

Mikulski, P.W., Smith, P.J.: A variance bound for unbiased estimation in inverse sampling. Biometrika, **63**, 216–217 (1976)

Moradi, M., Brown, J., Karimi, N.: An efficient and easy to carry out sampling design in environmental studies. 19th International Congress on Modelling and Simulation, Perth, Australia (pp. 12–16) (2011)

Murthy, M.N.: Ordered and unordered estimators in sampling without replacement. Sankhya **18**, 379–390 (1957)

Pathak, P.K.: On inverse sampling with unequal probabilities. Biometrika **51**(1–2), 185–193 (1964)

Pochai, N.: An improved estimator in inverse adaptive cluster sampling. Thail. Stat. **6**(1), 15–26 (2008)

Salehi, M.M., Seber, G.A.F.: A general inverse sampling scheme and its applications to adaptive cluster sampling. Aust. N. Z. J. Stat. **46**(3), 483–494 (2004)

Sampford, M.R.: An Introduction to Sampling Theory. Oliver and Boyd, London (1962a)

Sampford, M.R.: Methods of cluster sampling with and without replacement for clusters of unequal sizes. Biometrika **49**(1–2), 27–40 (1962b)

Sangngam, P.: Murthy's estimator in unequal probability inverse adaptive cluster sampling. Mod. Appl. Sci. **6**(11), 20–28 (2012)

Thompson, S.K.: Adaptive cluster sampling. J. Am. Stat. Assoc. **85**(412), 1050–1058 (1990)

Chapter 14
Two-Stage Inverse Adaptive Cluster Sampling

14.1 Introduction

It is well known that if the population under study is homogeneous then one can assign an equal probability of selection to the units in the population and can estimate the population parameters using the selected sample. But if the population has clumps in it then the equiprobable assignment may lead to poor estimates of the population parameters. In such situations, one can assign varying probabilities of selection to different units in the population. It can reduce the sampling error of the estimates that would be introduced due to the equal probability assignment. Horvitz and Thompson (1952) proposed an unbiased estimator of the total of a finite population. They also estimated the variance of their estimator when the sampling is carried out without replacement with varying probabilities of selection at each draw. Sometimes, this variance estimator takes negative values. In such cases, the estimator becomes very poor. To overcome this serious limitation of HT estimator, Des Raj (1956) proposed several unbiased estimators of the population total and derived the expression for variances of those estimators and the unbiased estimators of those variances. Des Raj's set of estimators are the ordered estimators. Corresponding to one set of estimators among them, Murthy (1957) provided an unordered estimator that is more efficient than the former estimator.

Christman and Lan (2001) considered estimators of the population total based on inverse sampling. They considered inverse sampling with stopping rules based on controlling the number of rare units sampled from the population. They named this design as inverse adaptive cluster sampling. It is a sequential sampling design in which an adaptive component is added. If a sequentially drawn unit from the population satisfies the predefined condition C then the network to which that unit belongs is completely sampled. Thus the final sample consists of the initial sample taken sequentially plus all units belonging to the networks of r sequentially sampled units that satisfy C plus the edge units of those networks. Salehi and Seber (1997a, 1997b) introduced a two-stage ACS. Salehi and Smith (2005) discussed a two-stage sequential sampling method which is a neighborhood free adaptive sampling procedure. In

© The Author(s), under exclusive license to Springer Nature Singapore Pte Ltd. 2021
R. Latpate et al., *Advanced Sampling Methods*,
https://doi.org/10.1007/978-981-16-0622-9_14

this method, a sample of primary units is selected by some conventional sampling design and then a subsample of secondary sample units within each selected primary unit is drawn. In this chapter, we consider a large geographical population that is divided into clusters. Each cluster is again divided into smaller plots. Responses can be measured on each of these plots. The variable of interest takes zero value for many population units and non-zero values for very few population units. But they are clumped. If we use the usual two-stage cluster sampling, it leads to a poor estimate of the population mean/total. Also, the clusters of the population are usually of unequal sizes. If we use SRSWR/SRSWOR then the clusters with large and small sizes are equiprobable. Hence there is a problem of over/underestimation. To overcome this problem, we use the PPS sampling at the first stage, to get the inclusion of large size clusters in the sample. At the second stage, an initial sample of a fixed number of units is selected from each of the selected PSU's. The adaptation procedure of Thompson (1990) is applied. The newly proposed method is presented in Sect. 14.3. A simulation study was undertaken using Two-Stage Inverse ACS, along with results of this study and discussion on it is presented in Sect. 14.3.

14.2   Two- Stage Inverse Adaptive Cluster Sampling

14.2.1   New Design

Let $U = \{U_1, U_2, \ldots, U_N\}$ be the finite population. Let it be divided into K non-overlapping clusters which serve as PSUs. Let N_u denote the size of the uth cluster, $(u = 1, 2, \ldots, K)$ so that we have $\sum_{u=1}^{K} N_u = N$. Note that, N_us need not be equal. At the first stage, a random sample of k clusters is drawn from the K clusters in the population by some design with inclusion probability π_u for the unit u and joint inclusion probability $\pi_{uu'}$ for primary stage units u and u'. At the second stage, a random sample of n_{ou}, $(u = 1, 2, \ldots, k)$ units is drawn by using SRSWOR from the selected uth cluster so that $n_0 = \sum_{u=1}^{k} n_{ou}$ is the total initial sample size. Each of the selected units is checked with respect to the condition C. If at least r_u $(u = 1, 2, \ldots, k)$, units satisfying condition C are found in this sample of n_{ou} units, then the sampling is stopped. Otherwise it is continued until either exactly r_u units from P_M are selected or n_{2u} (a pre-fixed number) units in total are selected from the uth cluster where $n_{ou} \leq n_{2u} \leq N_u$.

If a unit selected from the uth cluster satisfies C, then the corresponding network is completely included in the sample by using Thompson's (1990) procedure of adaptation. In this procedure, the adjacent neighboring units to the left, right, above, and below the selected unit are added to the sample. Further, if any of these units satisfy the condition C, their neighbors are also added to the sample. This process is continued till the neighboring units not satisfying condition C are observed. The resulting sample is called an adaptive sample. The set of units satisfying the condition C along with the units included in the initial sample, is called a network. These units which do not satisfy C but get added as neighbors in a network are called as the edge units. All edge units are dropped to get an unbiased estimator. Where r_u is the number of successes (that is, units satisfying the condition C) which is proportional to the sizes of selected clusters at the first stage.

$$r_u = r \frac{P_u}{\sum_{u=1}^{k} P_u},$$

where r is the total number of units satisfying the pre-decided condition C.
P_u is the draw by draw selection probability of the uth cluster.
Such kind of situations are tackled by using Des Raj estimators (Des Raj, 1956).

14.2.2 Estimators of Population Total

Murthy (1957) proposed the technique to improve the ordered estimator by unordered ones, which is a Rao–Blackwell improvement of Des Raj's estimator. Salehi and Seber (2001) used it for sequential sampling designs.

Let n_{1u} denote the number of units drawn in total from the uth cluster. This is the final sample size for the uth cluster.

HT estimator can be used to estimate the population total τ. It gives the unbiased estimator of τ as

$$\hat{\tau} = \sum_{u=1}^{k} \frac{\hat{\tau}_u}{\pi_u},$$

where π_u is the inclusion probability of the uth cluster and $hat\tau_u$ is the estimator of the population total of the uth cluster.

At the second stage of the proposed design, sample selection scheme becomes the sequential sampling. Murthy (1957) showed that corresponding to any ordered estimator of this class, an unordered estimator can be constructed and that estimator is also a minimum variance unbiased estimator.

Let $t_R(s_u')$ denote Des Raj's (1956) estimator of $\tau_u = \sum_{j=1}^{N_u} y_{uj}$, which is defined on the basis of an ordered sample s_u'. Using that technique, we can obtain an estimator of τ_u as follows:

$$\hat{\tau}_u = \frac{\sum_{s_u' \in s_u} P(s_u') \, t_R(s_u')}{P(s_u)},$$

where
$P(s_u)$ is the probability of obtaining the sample s_u.

Murthy (1957) has shown that $\hat{\tau}_u$ can be rewritten as follows:

$$\hat{\tau}_u = \sum_{j=1}^{n_{1u}} \frac{P(s_u|j)}{P(s_u)} y_{uj}$$

$$= \sum_{j=1}^{N_u} \frac{P(I_{uj}=1;s_u)}{P(s_u) \, P_{uj}} y_{uj},$$

where $P(s_u|j)$ is the conditional probability of getting sample s_u given that jth unit was selected at the first draw.

$$I_{uj} = \begin{cases} 1 & ; \text{if unit } j \text{ is selected at the first draw in the sample} \\ & \text{from } u^{th} \text{ cluster} \\ 0 & ; otherwise. \end{cases}$$

Salehi and Seber (2001) showed that Murthy's estimator given above can be obtained from a trivial estimator by using Rao–Blackwell theorem.

A trivial unbiased estimator of τ_u is given by

$$\hat{t}_u = \sum_{j=1}^{N_u} \frac{y_{uj}}{P_{uj}} I_{uj},$$

provided $P_{uj} > 0$, $j = 1, 2, \ldots, N_u$.

If s_u denotes the final sample set from the primary unit u then using Rao–Blackwell theorem we get

$$\hat{\tau}_u = E(\hat{t}_u | s_u) = \sum_{j=1}^{n_{1u}} \frac{P(s_u|j)}{P(s_u)} y_{uj}$$

$$\frac{P(s_u|j)}{P(s_u)} = \begin{cases} \frac{N_u}{n_{0u}} & ; \text{if } n_{1u} = n_{0u} \\ \frac{N_u}{n_{2u}} & ; \text{if } n_{1u} = n_{2u} \\ \frac{N_u (r_u-1)}{r_u (n_{1u}-1)} & ; \text{if } n_{0u} < n_{1u} < n_{2u} \text{ and } j \in S_{M_u} \\ \frac{N_u}{(n_{1u}-1)} & ; \text{if } n_{0u} < n_{1u} < n_{2u} \text{ and } j \in S_{N_u-M_u}. \end{cases}$$

Using these results, we get an unbiased estimator of τ_u as

$$\hat{\tau}_u = \begin{cases} \frac{N_u}{n_{0u}} \sum_{j=1}^{n_{0u}} y_{uj} & ; \text{if } n_{1u} = n_{0u} \\ \hat{M}_u \, \bar{y}_{M_u} + (N_u - \hat{M}_u) \, \bar{y}_{N_u-M_u} & ; \text{if } n_{0u} < n_{1u} < n_{2u} \\ \frac{N_u}{n_{2u}} \sum_{j=1}^{n_{2u}} y_{uj} & ; \text{if } n_{1u} = n_{2u}. \end{cases} \quad (14.1)$$

Thus, we get the estimator:

$$\hat{\tau} = \sum_{u=1}^{k} \frac{\hat{\tau}_u}{\pi_u} \text{ for the two-stage inverse sampling.}$$

In the new design, we have combined two-stage inverse cluster sampling with ACS. At the second stage, if a unit satisfies the condition C, the network to which it belongs is completely sampled. The new estimator $\hat{\tau}_u^*$ is obtained from (14.1) replacing y_{uj} by the network mean $\bar{y}_{uj}^*$.

Where, $\bar{y}_{uj}^* = \frac{\sum_{i \in A_j} y_{ui}}{n_j}$, $j = 1, 2, \cdots, n_{0u}$.

y_{ui} is the ith observed value in network A_j and uth cluster.

$\hat{M}_u = \frac{N_u(r_u-1)}{(n_{1u}-1)}$

Thus we get

$$\hat{\tau}_u^* = \begin{cases} \frac{N_u}{n_{0u}} \sum_{j=1}^{n_{0u}} \bar{y}_{uj}^* & ; \text{if } n_{1u} = n_{0u} \\ \hat{M}_u \, \bar{y}_{M_u}^* + (N_u - \hat{M}_u) \, \bar{y}_{N_u-M_u}^* & ; \text{if } n_{0u} < n_{1u} < n_{2u} \\ \frac{N_u}{n_{2u}} \sum_{j=1}^{n_{2u}} \bar{y}_{uj}^* & ; \text{if } n_{1u} = n_{2u} \end{cases} \quad (14.2)$$

Mean of the observations on the units satisfying C from the uth cluster selected in the sample is given by

$$\bar{y}_{M_u}^* = \frac{\sum_{j \in S_{M_u}} \bar{y}_{uj}^*}{r_u}$$

Mean of the observations on the units not satisfying C from the uth cluster selected in the sample is given by

$$\bar{y}_{N_u-M_u}^* = \frac{\sum_{j \in S_{N_u-M_u}} \bar{y}_{uj}^*}{(n_{1u}-r_u)}$$

The unbiased estimator for the population total is given as follows:

$\hat{\tau}^* = \sum_{u=1}^{k} \frac{\hat{\tau}_u^*}{\pi_u}.$

Below in the theorem is expressing $\hat{V}(\hat{\tau}_u^*)$.

Theorem 14.1 *An unbiased estimator of the variance of $\hat{\tau}_u^*$ is given by*

$$\hat{V}(\hat{\tau}_u^*) = \begin{cases} N_u^2 \left(1 - \frac{n_{0u}}{N_u}\right) \frac{S_{0u}^{*2}}{n_{0u}} & ; \text{if } n_{1u} = n_{0u} \\ a_u \, S_{M_u}^{*2} + \hat{V}ar(\hat{M}_u) \, (\bar{y}_{M_u}^* - \bar{y}_{N_u-M_u}^*)^2 + b_u \, S_{N_u-M_u}^{*2} & ; \text{if } n_{0u} < n_{1u} < n_{2u} \\ N_u^2 \left(1 - \frac{n_{2u}}{N_u}\right) \frac{S_{2u}^{*2}}{n_{2u}} & ; \text{if } n_{1u} = n_{2u} \end{cases}$$

where

$S_{M_u}^{*2} = \frac{\sum_{j \in S_{M_u}} (\bar{y}_{uj}^* - \bar{y}_{M_u}^*)^2}{r_u - 1},$

$S_{N_u-M_u}^{*2} = \frac{\sum_{j \in S_{N_u-M_u}} (\bar{y}_{uj}^* - \bar{y}_{N_u-M_u}^*)^2}{n_{1u} - r_u},$

$\bar{y}_{M_u}^* = \frac{\sum_{j \in S_{M_u}} \bar{y}_{uj}^*}{r_u},$

$\bar{y}_{N_u-M_u}^* = \frac{\sum_{j \in S_{N_u-M_u}} \bar{y}_{uj}^*}{n_{1u} - r_u},$

$\hat{V}(\hat{M}_u) = (1 - \frac{n_{1u}-1}{N_u}) \frac{\hat{M}_u \, (N_u - \hat{M}_u)}{(n_{1u}-2)}$

$a_u = \frac{\hat{M}_u^2}{r_u} \frac{(N_u - n_{1u} + 1) \, (n_{1u} r_u - n_{1u} - r_u) - N_u \, (n_{1u}-2)}{N_u \, (n_{1u}-2)(r_u-1)}$

$b_u = \frac{N_u \, (N_u - n_{1u}+1) \, (n_{1u}-r_u-1)}{(n_{1u}-1) \, (n_{1u}-2)},$

$S_{0u}^{*2} = \frac{\sum_{j=1}^{n_{0u}} (\bar{y}_{uj}^* - \bar{y}_{u0}^*)^2}{n_{0u}-1},$

$\bar{y}_{u0}^* = \frac{\sum_{j=1}^{n_{0u}} \bar{y}_{uj}^*}{n_{0u}},$

$S_{2u}^{*2} = \frac{\sum_{j=1}^{n_{2u}} (\bar{y}_{uj}^* - \bar{y}_{u2}^*)^2}{(n_{2u}-1)}$

$\bar{y}_{u2}^* = \frac{\sum_{j=1}^{n_{2u}} \bar{y}_{uj}^*}{n_{2u}}.$

Proof From Salehi and Seber (2001), we have

$V(\hat{\tau}_u^*) = \sum_{i=1}^{N_u} \sum_{j<i=1}^{N_u} \left(1 - \sum_{i,j \in S_u} \frac{P(S_u|i)P(S_u|j)}{P(S_u)}\right) \left(\frac{y_{u_i}}{\pi_{u_i}} - \frac{y_{u_j}}{\pi_{u_j}}\right)^2 \pi_{u_i} \pi_{u_j}.$

Unbiased estimator of $V(\hat{\tau}_u^*)$ is given by

$$\hat{V}(\hat{\tau}_u^*) = \sum_{i=1}^{n_u} \sum_{j<i=1}^{n_u} \left(\frac{P(S_u|i,j)}{P(S_u)} - \frac{P(S_u|i)P(S_u|j)}{P^2(S_u)}\right) \left(\frac{y_{u_i}}{\pi_{u_i}} - \frac{y_{u_j}}{\pi_{u_j}}\right)^2 \pi_{u_i} \pi_{u_j}.$$

(14.3)

where

$P(S_u|i)$ = Probability of getting the sample S_u from the uth cluster given that unit i is selected at the first draw.

$P(S_u|i, j)$ = Probability of getting the sample S_u from the uth cluster given that units i and j are selected at the first two draws.

$P(S_u)$ = Probability of getting the sample S_u.

In this method, we continue sampling in the uth cluster randomly without replacement until r_u units satisfying C have been selected. Then $P_i = \frac{1}{N}$ and v_u are the random variables denoting the number of units selected from the uth cluster.

To evaluate the fraction $\frac{P(S_u|i)}{P(S_u)}$, we need to obtain the number of ordered samples leading to S_u and $\{I_i = 1; S_u\}$

The last observation selected in the sample must satisfy C. Hence, after allocating one of the r_u units satisfying C as the last unit, the rest of the units can be ordered in $(v_u - 1)!$ ways. The sample S_u can be constructed in $r_u(v_u - 1)!$ ways. If unit i is in S_c or in $S_{\bar{c}}$ the event $\{I_i = 1; S_u\}$ can occur in $(r_u - 1)(v_u - 2)!$ or $r_u(v_u - 2)!$ ways. Hence, with $P_i = \frac{1}{N}$

$$
\begin{aligned}
\frac{P(S_u|i)}{P(S_u)} &= \frac{P(I_i=1; S_u)}{P(S_u)P_i} \\[2mm]
&= \frac{N_u(r_u-1)}{(v_u-1)r_u} \qquad & i = 1, 2, \ldots, r_u \\[2mm]
&= \frac{N_u}{v_u-1} \qquad & i = r_u + 1, r_u + 2, \ldots, v_u
\end{aligned}
$$

Using the above probabilities, $V(\hat{\tau}_u^*)$ can be determined.

To obtain $\hat{V}(\hat{\tau}_u^*)$, it is required to determine the fraction $\frac{P(S_u|i,j)}{P(S_u)}$.

Using the above approach for $r_u > 2$ we get

$$
\frac{P(S_u|i, j)}{P(S_u)} =
\begin{cases}
\frac{N_u(N_u-1)(r_u-2)}{(v_u-1)(v_u-2)r_u} & \text{; if } i, j \in S_c \\[2mm]
\frac{N_u(N_u-1)(r_u-1)}{(v_u-1)(v_u-2)r_u} & \text{; if } i \in S_c \text{ and } j \in S_{\bar{c}}. \\[2mm]
\frac{N_u(N_u-1)}{(v_u-1)(v_u-2)} & \text{; if } i, j \in S_{\bar{c}}
\end{cases}
$$

Substituting these values in Eq. (14.3) and simplifying the expressions proves the theorem.

Now, $\hat{\tau}^*$ is a HT type estimator of the population total. Hence, using Result 2 from Chap. 10 we get

$$
V(\hat{\tau}^*) = \sum_{u=1}^{K} \sum_{u'=1}^{K} \left(\frac{\pi_{uu'} - \pi_u \pi_{u'}}{\pi_{uu'}} \right) \tau_u^* \tau_{u'}^* + \sum_{u=1}^{K} \frac{Var(\hat{\tau}_u^*)}{\pi_u}.
$$

Note that $\pi_{uu} = \pi_u$.

An unbiased estimator of $Var(\hat{\tau}^*)$ is given by

$$
\hat{V}(\hat{\tau}^*) = \sum_{u=1}^{k} \sum_{u'=1}^{k} \left(\frac{\pi_{uu'} - \pi_u \pi_{u'}}{\pi_{uu'}} \right) \hat{\tau}_u^* \hat{\tau}_{u'}^* + \sum_{u=1}^{k} \frac{\hat{V}ar(\hat{\tau}_u^*)}{\pi_u}. \tag{14.4}
$$

Further, substituting the expression of $\hat{V}(\hat{\tau}_u^*)$ in Eq. (14.4), we get $\hat{V}(\hat{\tau}^*)$ as follows:

$$\hat{V}(\hat{\tau}^*) = \begin{cases} \sum_{u=1}^{k}\sum_{u'=1}^{k}\left(\frac{\pi_{uu'}-\pi_u\pi_{u'}}{\pi_{uu'}}\right)\hat{\tau}_u^*\hat{\tau}_{u'}^* + \sum_{u=1}^{k}\frac{N_u^2\left(1-\frac{n_{0u}}{N_u}\right)\frac{S_{0u}^{*2}}{n_{0u}}}{\pi_u} \\ \qquad\qquad \text{if } n_{1u}=n_{0u} \\[2mm] \sum_{u=1}^{k}\sum_{u'=1}^{k}\left(\frac{\pi_{uu'}-\pi_u\pi_{u'}}{\pi_{uu'}}\right)\hat{\tau}_u^*\hat{\tau}_{u'}^* + \sum_{u=1}^{k}\frac{a_u\,S_{M_u}^{*2}+\hat{V}(\hat{M}_u)\,(\bar{y}_{M_u}^*-\bar{y}_{N_u}^*-M_u)^2+b_u\,S_{N_u-M_u}^{*2}}{\pi_u} \\ \qquad\qquad \text{if } n_{0u}<n_{1u}<n_{2u} \\[2mm] \sum_{u=1}^{k}\sum_{u'=1}^{k}\left(\frac{\pi_{uu'}-\pi_u\pi_{u'}}{\pi_{uu'}}\right)\hat{\tau}_u^*\hat{\tau}_{u'}^* + \sum_{u=1}^{k}\frac{N_u^2\left(1-\frac{n_{2u}}{N_u}\right)\frac{S_{2u}^{*2}}{n_{2u}}}{\pi_u} \\ \qquad\qquad \text{if } n_{1u}=n_{2u}. \end{cases}$$

Hence the theorem. □

14.3 Simulation Study and Discussion

Area of 400 acres in the Tamhini Ghat, Maharashtra, was divided into 400 plots each of size 1 acre. A satellite image of the area showed clustering of the thorny plants. Presence of thorny plants indicates a high percentage of silica in the soil. Thorny plants adapt to such kind of nature of the soil. So, it was important to estimate the total number of thorny plants in the area. Higher value of this estimate indicates the ecological imbalance. Efforts can be taken to reestablish this balance. In view of that, a sample survey was conducted. The area of 400 acres was divided into 5 clusters C_1, C_2, C_3, C_4, and C_5 containing 130, 65, 70, 105, and 30 plots, respectively, each of size 1 acre. These 5 clusters served as PSU's. From these clusters, a random sample of 3 clusters was selected by using probability proportional to size. For instance, the selected clusters were C_1, C_2, and C_4. From these selected clusters, random samples of 6, 6, and 6 plots, respectively, were drawn using simple random sampling without a replacement method. These selected plots are shown in Fig. 14.1 by putting $\star$ in it.

Each of these plots selected from the different clusters was checked for the condition $C = \{Y > 0\}$, where Y denotes the number of thorny plants observed on a plot. The total number of units satisfying condition C is $r = 12$. The number of successes for each cluster depends on the size of the selected cluster. Networks were identified around the plots which satisfied C, by using Thompson's (1990) procedure. The units from these identified networks were added to the sample along with the corresponding edge units. This process of adding the neighbors is continued in each selected cluster so that, either at least $r_1 = 5$ plots satisfying the condition C are found or $n_{21} = 11$ plots in total are selected in the sample sequentially from C_1; either at least $r_2 = 3$ plots satisfying condition C are found or $n_{22} = 11$ plots in total are selected in the sample sequentially found from C_2 and either at least $r_4 = 4$ plots satisfying the condition C are found or $n_{24} = 11$ plots in total are selected in the sample sequentially found from C_4. The networks formed by the units selected are represented in white color in Fig. 14.1. Values of Y were recorded for all the units included in the final sample and the total number of thorny plants in that area was estimated by using the proposed estimator ($\hat{\tau}$). Further its variance was also estimated by using the formula

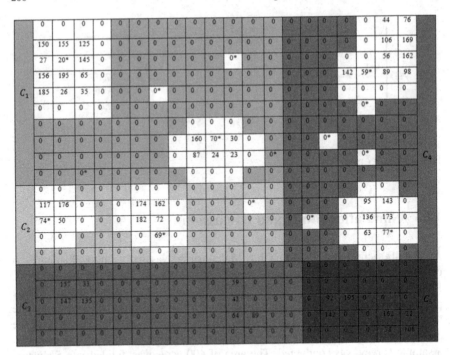

Fig. 14.1 Number of thorny plants in a plot

Table 14.1 Simulation results for $n_0 = 9$, $n_2 = 58$, number of repetitions $= 100000$

Number of successes	Two-stage inverse adaptive sampling			Inverse adaptive sampling		
r	Sample size	$\hat{\tau}^*$	$SE(\hat{\tau}^*)$	Sample size	$\hat{\tau}$	$SE(\hat{\tau})$
6	61.46	5646.48	3401.11	63.99	19049.98	15872.60
7	61.46	5650.64	3410.10	70.99	18228.51	15867.88
8	61.44	5644.61	3403.74	76.48	15905.28	14741.00
9	61.43	5650.69	3395.19	80.48	12923.12	12823.93
10	61.47	5636.36	3397.52	83.17	9957.42	10506.74
11	61.50	5619.81	3384.63	84.83	7542.00	8140.19
12	61.49	5632.03	3403.85	85.65	5921.03	6105.04

for $\hat{V}(\hat{\tau})$, given in Sect. 14.2. The functioning of the new design is demonstrated in Fig. 14.1.

To verify the efficiency of this new design Monte Carlo simulation study was made. The performance of the new design was compared with the inverse adaptive sampling design.

Table 14.2 Simulation results for $n_0 = 18$, $n_2 = 33$, number of repetitions $= 100000$

Number of successes	Two -stage inverse adaptive sampling			Inverse adaptive sampling		
r	Sample size	$\hat{\tau}^*$	$SE(\hat{\tau}^*)$	Sample size	$\hat{\tau}$	$SE(\hat{\tau})$
6	46.84	5980.36	2649.90	50.96	11414.81	12915.92
7	46.86	5968.36	2649.42	52.64	8346.97	9624.48
8	46.88	5965.62	2639.97	53.39	6335.45	6623.06
9	46.82	5959.06	2637.10	53.69	5313.16	4372.89
10	46.83	5958.59	2636.41	53.78	4894.70	3081.25
11	46.83	5950.23	2627.82	53.80	4742.19	2450.15
12	46.83	5929.91	2615.06	53.77	4705.43	2252.54

Table 14.1 gives the simulation results for $n_0 = 9$, $n_2 = 58$, and number of repetitions $= 100000$. It shows that the empirical sample size remains constant for the increase in the number of successes (r) in case of two-stage inverse adaptive sampling, but it shows an increase in the case of inverse adaptive sampling. In this setup, the bias in the estimator used in two-stage inverse adaptive sampling and the estimate of the standard error of that estimator remains more or less constant. On the contrary, the bias in the estimator used in inverse adaptive sampling gradually decreases with the increase in r. The estimate of the standard error of the estimator used in this case also gradually reduces with an increase in r. But in this setup, the estimator used in two-stage inverse adaptive sampling, though biased, appears to be more stable with a lesser variability as compared to the inverse adaptive sampling. Table 14.2 gives the simulation results for $n_0 = 18$, $n_2 = 33$, and the number of repetitions $= 100000$. It is seen that the empirical sample size remains constant for the increase in the number of successes (r) in both sampling designs. The empirical sample size is found to be consistently smaller in two-stage inverse adaptive sampling than that in inverse adaptive sampling, for the different values of r.

The estimator used in two-stage inverse adaptive sampling is observed to be more stable than that in inverse adaptive sampling. The bias introduced in this estimator gradually reduces as r increases. The estimate of the standard error of this estimator in this set up also reduced. Moreover, its value is smaller in this setup as compared to that in the first setup discussed above.

Thus, the estimator used in the new design performs better in the second setup as compared to the first. Two-stage adaptive cluster sampling has a lesser sample size. Hence, this method is cost-effective.

In the proposed method, two-stage ACS, the empirical sample size gets reduced with an increase in the initial sample size. Bias in the estimator used in this design and its standard error gets reduced with an increase in the values of the initial sample size and the number of successes. So, for larger initial sample size and a large value of the number of successes, this method performs better than the existing methods.

The expected final sample size is lesser in the new design than that in the inverse adaptive sampling design. It decreases as the values of r and n_0 increase. So, for large values of r and n_0 the new design becomes more cost-effective. The bias in the estimator used in the new design also reduces for the increase in r and n_0.

The standard error of the estimator used in the new design, also gets reduced with an increase in r and n_0. Considering this merit the new design can be considered to be more reliable as compared to the inverse adaptive sampling design.

14.4   Exercises

1. Describe two-stage inverse ACS design. Obtain an unbiased estimator of the population total under two-stage Inverse ACS design.
2. Derive an expression for the variance of the unbiased estimator obtained in Ex.1. Obtain an unbiased estimator of the variance of this estimator.
3. Describe the Inverse ACS design and its limitations. Show the advantages of two-stage Inverse ACS over Inverse ACS.

References

Christman, M.C., Lan, F.: Inverse adaptive cluster sampling. Biometrics 57(4), 1096–1105 (2001)

Des, R.: Some estimators in sampling with varying probabilities without replacement. J. Am. Stat. Assoc. 51, 269–284 (1956)

Horvitz, D.G., Thompson, D.J.: A generalization of sampling without replacement from a finite universe. J. Am. Stat. Assoc. 47, 663–685 (1952)

Murthy, M.N.: Ordered and unordered estimators in sampling without replacement. Sankhya 18, 379–390 (1957)

Salehi, M.M., Seber, G.A.F.: Adaptive cluster sampling with networks selected without replacement. Biometrika 84, 209–219 (1997a)

Salehi, M.M., Seber, G.A.F.: Two stage adaptive cluster sampling. Biometrics 53, 959–970 (1997b)

Salehi, M.M., Seber, G.A.F.: A new proof of Murthy's estimator which applies to sequential sampling. Aust. N. Z. J. Stat. 43(3), 281–286 (2001)

Salehi, M.M., Smith, D.R.: Two-stage sequential sampling: a neighbourhood-free adaptive sampling procedure. J. Agric., Biol. Environ. Stat. 10, 84–103 (2005)

Thompson, S.K.: Adaptive cluster sampling. J. Am. Stat. Assoc. 85(412), 1050–1058 (1990)

Chapter 15
Stratified Inverse Adaptive Cluster Sampling

15.1 Introduction

In forestry and environmental sciences, some species of plants and animals are rare and clustered, i.e., the abundance of zeros. The traditional sampling methods provide poor estimates of the population mean/total. In such situations, adaptive sampling is useful. In traditional stratified sampling, similar units are grouped a priori into strata, based on prior information about the population. But within a stratum, the population is rare and clumped. Hence, stratified adaptive sampling is useful for estimation of population mean/total (Thompson 1991a, b). In this design, an initial stratified random sample (units) are drawn and if selected units satisfy the condition of interest, then neighboring units are added to the sample. Again, if any of added units satisfies the condition then their neighbors are also added. This process continues till all the neighbors satisfy condition C. The process terminates if all the lastly added units do not satisfy the condition. The adaptive cluster sampling is useful for the population such as animal or plant species and mineral or fossil fuel resources.

15.2 Stratified Inverse ACS Design

In this design, there is a problem with initial sample size. In some instances, we met with all zero occurrences because of the population nature. To overcome this problem, Chirstman (2001) introduced the inverse ACS. Let the population consist of set of values $P = \{y_1, y_2, ..., y_N\}$ and is divided into two subgroups according to whether the y-values satisfy the condition C as $P_M = \{y : y_i \in C, i = 1, 2, ..., N\}$ and $P_{(N-M)} = \{y : y_i \notin C, i = 1, 2, ..., N\}$, where M is the number of units that satisfies the condition and N is population size. In general, $C = \{y; y \geq c\}$, where c is some constant specified prior to sampling. We don't know which unit belongs to which group until the unit is sampled. In this method, we select units one at a time

© The Author(s), under exclusive license to Springer Nature Singapore Pte Ltd. 2021
R. Latpate et al., *Advanced Sampling Methods*,
https://doi.org/10.1007/978-981-16-0622-9_15

until we get a predetermined number of rare units. The neighborhood of an initial unit is added to the sample whenever the units satisfy the condition of interest. Again, if added units satisfy the conditions its neighbors are added and this process continued till all the neighbors do not satisfy the condition C. If selected unit doesn't satisfy the condition, i.e., $y \in P_{N-M}$ there won't be an adaptation, and the network includes only an initial sample. In some instances, several units from the same network may be sampled initially. Hence, we consider only distinct networks. The adaptive sampling is the most powerful sampling scheme whenever units of P_M are grouped.

Let $U = \{U_{11}, U_{12}, ..., U_{1N_1}, U_{21}, U_{22}, ..., U_{2N_2}, ..., U_{L1}, U_{L2}, ..., U_{LN_L}\}$ be the finite population of N units which is first divided into subpopulations of $N_1, N_2, ...,$ N_L units. These subpopulations are non-overlapping, and together they comprise the whole populations such that $N_1 + N_2 + ..N_L = N$. These groups are called strata and size of each strata is known. An initial random sample of size $n_1, n_2, ..., n_L$ are drawn independently. Such a procedure is called stratified random sampling. If an initial random sample of size n_h is drawn from the hth strata with size N_h. The number of successes $r_h = r N_h / N$, where r is total number of successes (units satisfying the condition C). Each of the selected units is checked with respect to the condition C. If at least $r_h (h = 1, 2, ..., L)$, units satisfying the condition C are found in this sample of n_h units, then the sampling is stopped. Otherwise it is continued until either exactly r_h units from P_M are selected or $n_{h2}(h = 1, 2, ..., L)$(a pre-fixed number) units in total are selected from the hth strata where $n_{h1} \leq n_{h2} \leq N_h$. A set of neighboring units is defined for each unit by Thompson (1990). The neighborhood of an initial unit is included to the sample in hth stratum whenever the unit satisfies the condition of interest. Again if any of additional units satisfies the condition, its neighborhood is also included in the sample of hth strata. This procedure continued until all the neighborhoods are included. Terminate the procedure if lastly added units do not satisfy the condition C. If initial sample unit do not satisfy the condition, then no further units are added, and the network consist of just the initial sample unit. Those units adaptively added but do not satisfy the condition are called edge units. For estimation purpose, we drop down the edge units. For hth strata, each unit i a network A_{hi}. Some instances, several initial units belong to the same network. The distinct networks, labeled by the subscripts $z(z = 1, 2, ..., Z)$, from a partition of N_h units. Let m_{hi} denote the size of ith network of hth strata. The variable of interest related to A_{hi} is $y_{hi}^* = \sum_{j=1}^{m_{hi}} m_{hi} y_{hij}$ and is $y_{hi}^* = \sum_{j=1}^{m_{hi}} y_{hij} / m_{hi}$. Repeat the whole procedure to all the strata. The population total can estimated $\tau = \sum_{h=1}^{L} \sum_{i=1}^{N_h} y_{hi}^*$.

P_h is the draw by draw selection probability of the hth strata. Such kind of situations are tackled by using Des Raj estimators (Des Raj, 1956). Murthy (1957) proposed the technique to improve the ordered estimator by unordered ones, which is a Rao–Blackwell improvement of Des Raj's estimator. Salehi and Seber (2001) used it for sequential sampling designs. Let n_{h1} denote the number of units drawn in total from the hth strata This is the final sample size for the hth strata.

At each strata, sample selection scheme becomes the sequential sampling. Murthy(1957) showed that corresponding to any ordered estimator of this class, an unordered estimator can be constructed and that estimator is also minimum variance unbiased estimator. Let $t_R(s_h')$ denote Des Raj's (1956) estimator of $\tau_h = \sum_{i=1}^{N_h} y_{hi}^*$, which is defined on the basis of an ordered sample s_h'. Using that technique, we can obtain an estimator of τ_h as follows:

$$\hat{\tau}_h = \frac{\sum_{S_h' \in S_h} P(S_h') \, t_R(S_h')}{P(S_h)}$$

$\hat{\tau}_h = \sum_{i=1}^{n_{h1}} \frac{P(S_h|i)}{P(S_h)} \, y_{hi}^* = \sum_{i=1}^{N_h} \frac{P(I_{hi}=1;S_h)}{P(S_h) \, P_{hi}} \, y_{hi}^*$ where $P(S_h)$ is the probability of obtaining the sample S_u. Murthy (1957) has shown that $\hat{\tau}_h$ can be rewritten as follows:

Where $P(S_h|i)$ is the conditional probability of getting sample S_h given that ith unit was selected at the first draw.

$$I_{hi} = \begin{cases} 1 & ; \text{if unit } i \text{ is selected at the first draw in the sample} \\ & \quad from \ h^{th} \ stratum \\ 0 & ; otherwise. \end{cases}$$

Salehi and Seber (2001) showed that Murthy's estimator given above can be obtained from a trivial estimator by using Rao–Blackwell theorem.

A trivial unbiased estimator of τ_h is given by
$\hat{\tau}_h = \sum_{i=1}^{N_h} \frac{y_{hi}^*}{P_{hi}} I_{hi}$, provided $P_{hi} > 0, \quad for \ i = 1, 2, ..., N_h$.
If S_h denotes the final sample set from the primary unit h, then using Rao–Blackwell theorem, we get
$\hat{\tau}_h = E(\hat{t}_h|S_h) = \sum_{j=1}^{n_{h1}} \frac{P(S_h|i)}{P(S_h)} \, y_{hi}^*$

When $n_{h1} = n_h$ and $n_{h1} = n_{h2}$ then

$\frac{P(S_h|i)}{P(S_h)} = \frac{N_h}{n_h}$ and $\frac{N_h}{n_{h2}}$, respectively.

When $n_h < n_{h1} < n_{h2}$ and $i \in S_{M_h}$ then

$\frac{P(S_h|i)}{P(S_h)} = \frac{N_h(r_h-1)}{(n_{h1}-1) \, r_h}$

When $n_h < n_{h1} < n_{h2}$ and $i \in S_{N_h-M_h}$ then

$\frac{P(S_h|i)}{P(S_h)} = \frac{N_h}{n_{h1}-1}$

and $\hat{M}_h = \frac{N_h \, (r_h-1)}{n_{h1}-1}$ (Lehmann 1983).
Thus we get

$$\hat{\tau}_h = \begin{cases} \frac{N_h}{n_h} \sum_{i=1}^{n_h} \bar{y}_{hi}^* & ; \text{if } n_{h1} = n_h \\ \hat{M}_h \, \bar{y}_{M_h}^* + (N_h - \hat{M}_h) \, \bar{y}_{N_h-M_h}^* & ; \text{if } n_h < n_{h1} < n_{h2} \\ \frac{N_h}{n_{h2}} \sum_{i=1}^{n_{h2}} \bar{y}_{hi}^* & ; \text{if } n_{h1} = n_{h2}. \end{cases} \tag{15.1}$$

The unbiased estimator for the population total is given as follows:
$\hat{\tau} = \sum_{h=1}^{L} \hat{\tau}_h$
Mean of the observations on the units satisfying C from the hth strata is given by

$$\bar{y}_{M_h}^* = \frac{\sum_{i \in S_{M_h}} \bar{y}_{hi}^*}{r_h}$$

Mean of the observations on the units not satisfying C from the hth strata is given by

$$\bar{y}_{N_h-M_h}^* = \frac{\sum_{i \in S_{N_h-M_h}} \bar{y}_{hi}^*}{n_{h1}-r_h}$$

The variance of $\hat{\tau}$

$$V(\hat{\tau}) = \sum_{h=1}^{L} V(\hat{\tau}_h).$$

We get

$$\hat{V}(\hat{\tau}_h) = \begin{cases} N_h^2 \left(1 - \frac{n_h}{N_h}\right) \frac{S_{0h}^{*2}}{n_h} & ; \text{if } n_{h1} = n_h \\ a_h \, S_{M_h}^{*2} + \hat{V}(\hat{M}_h)(\bar{y}_{M_h}^* - \bar{y}_{N_h-M_h}^*)^2 + b_h \, S_{N_h-M_h}^{*2} & ; \text{if } n_h < n_{h1} < n_{h2} \\ N_h^2 \left(1 - \frac{n_{2h}}{N_h}\right) \frac{S_{2h}^{*2}}{n_{h2}} & ; \text{if } n_{h1} = n_{h2}. \end{cases}$$

$$S_{M_h}^{*2} = \frac{\sum_{i \in S_{M_h}} (\bar{y}_{hi}^* - \bar{y}_{M_h}^*)^2}{r_h - 1}$$

$$S_{N_h-M_h}^{*2} = \frac{\sum_{i \in S_{N_h-M_h}} (\bar{y}_{hi}^* - \bar{y}_{N_h-M_h}^*)^2}{n_{h1} - r_h}$$

$$\bar{y}_{M_h}^* = \frac{\sum_{i \in S_{M_h}} \bar{y}_{hi}^*}{r_h}$$

$$\bar{y}_{N_h-M_h}^* = \frac{\sum_{i \in S_{N_h-M_h}} \bar{y}_{hi}^*}{n_{h1} - r_h}$$

$$\hat{V}(\hat{M}_h) = \left(1 - \frac{n_{h1}-1}{N_h}\right) \frac{\hat{M}_h \, (N_h - \hat{M}_h)}{(n_{h1} - 2)}$$

$$a_h = \frac{\hat{M}_h^2 \, (N_h - n_{h1} + 1) \, (n_{h1}r_h - n_{h1} - r_h) - N_h \, (n_{h1} - 2)}{r_h \, N_h (n_{h1} - 2)(r_h - 1)}$$

$$b_h = \frac{N_h \, (N_h - n_{h1} + 1) \, (n_{h1} - r_h - 1)}{(n_{h1} - 1) \, (n_{h1} - 2)}$$

$$S_{0h}^{*2} = \frac{\sum_{i=1}^{n_h} (\bar{y}_{hi}^* - \bar{y}_{h0}^*)^2}{n_h - 1}$$

$$\bar{y}_{h0}^* = \frac{\sum_{i=1}^{n_h} \bar{y}_{hi}^*}{n_h}$$

$$S_{2h}^{*2} = \frac{\sum_{i=1}^{n_{h2}} (\bar{y}_{hi}^* - \bar{y}_{h2}^*)^2}{(n_{h2} - 1)}$$

$$\bar{y}_{h2}^* = \frac{\sum_{i=1}^{n_{h2}} \bar{y}_{hi}^*}{n_{h2}}.$$

15.3   Simulation Study

Area of 400 acres in the Tamhini Ghat, Maharashtra, was divided into 400 plots each of size 1 acre. A satellite image of the area showed clustering of the evergreen plants. The presence of evergreen plants indicates a high percentage of laterite in the soil. The presence of high percentage of laterite in the soil is good. So, it was important to estimate the total number of evergreen plants in the area. A higher value of this estimate indicates the ecological balance. In view of that, a sample survey was conducted. The area of 400 acres was divided into 5 strata $C_1, C_2, C_3, C_4,$ and C_5 containing 130,65,70,105, and 30 plots, respectively, each of size 1 acre. From these strata random sample of size $n_h (h = 1, 2, ..., L)$ was selected without replacement. These selected plots are shown in Fig. 15.1 by putting * in it. Each of these plots

0	0	0	0	0	0	0	0	0	0	0	0	0	0	0	0	0	0		0	244	85	
170	135	120	0	0	0	0*	0	0	0	0	0	0	0*	0	0		0	111	150		
227	200*	165	0	0	0	0	0	0	0	0*	0	0	0	0	0		0	0	56*	152	
157	190	35	0	0	0	0	0	0	0	0	0	0	0*	0	145	65*	99	85			
180	46	95	0	0	0	0*	0	0	0	0	0	0	0	0	0	0	0				
0	0	0	0	0	0	0	0	0	0	0	0	0	0*	0	0	0*	0	0			
0	0	0	0	0	0	0	0	0	0	0	0	0	0	0	0	0	0	0			
0	0*	0	0	0*	0	0	0*	265	50*	55	0	0	0	0*	0	0	0	0			
0	0	0	0	0	0	0	0	67	34	33	0	0*	0	0	0	0*	0	0			
0	0	0*	0	0	0	0	0	0	0	0	0	0	0	0	0	0	0				
0	0	0	0	0	0	0	0	0	0	0	0	0	0	0	0	0	0	0			
217	140	0	0	0	124	164	0	0	0*	0	0*	0	0	0	0	120	150	0			
64*	54	0	0	0	50*	64	0	0	0	0	0	0*	0	0	0	155	188	0			
0	0	0	0*	0	0	98*	0	0	0	0*	0	0	0	0	0	68	37*	0			
0	0	0	0	0*	0	0	0	0*	0	0*	0	0	0	0	0	0	0	0*			
0	0	0	0	0	0	0	0*	0	0	0	0*	0*	0	0	0	0*					
0	165*	35	0	0	0*	0	0	0	83*	0	0	0	0	0	0*	0					
0	148	139	0	0	0	0	0	74	0	0	0*	189*	165	0	0	0					
0	0	0	0*	0	0	0*	56	79*	0	0	132	0	0*	66*	212						
0	0	0*	0	0*	0	0	0	0	0	0*	0*	0	58	208*							

Fig. 15.1 Number of evergreen plants (Y) on a plot

selected from the different clusters was checked for the condition $C = \{Y > 0\}$, where Y denotes the number of evergreen plants observed on a plot. The total number of units satisfying the condition C is $r = 24$. The number of successes for each stratum depends on the size of the strata. Networks were identified around the plots which satisfy C, by using Thompson's (1990) procedure. The units from these identified networks were added to the sample along with the corresponding edge units. This process of adding the neighbors is continued in each stratum so that, either at least $r_1 = 8$ plots satisfying the condition C are found or $n_{12} = 20$ plots in total are selected in the sample sequentially from C_1; either at least $r_2 = 4$ plots satisfying the condition C are found or $n_{22} = 20$ plots in total are selected in the sample sequentially found from C_2; either at least $r_3 = 4$ plots satisfying the condition C are found or $n_{32} = 20$ plots in total are selected in the sample sequentially found from C_3. Either at least $r_4 = 6$ plots satisfying the condition C are found or $n_{42} = 20$ plots in total are selected in the sample sequentially found from C_4 and either at least $r_5 = 2$ plots satisfying the condition C are found or $n_{52} = 20$ plots in total are selected in the sample sequentially found from C_5. The networks formed by the units selected are represented in white color in Fig. 15.1. Values of Y were recorded for all the units included in the final sample and the total number of evergreen plants in that area was estimated by using the proposed estimator $\hat{\tau}_h$. Further, its variance was also estimated by using the formula for $\widehat{V(\hat{\tau}_h)}$, given in Sect. 15.2. The functioning of the new design is demonstrated in Fig. 15.1. To verify the efficiency of this new design, Monte Carlo simulation study was performed for 100,000 repetitions.

The results are given in Table 15.1.

Table 15.1 Simulation results for $n_0 = 10$, $n_2 = 20$, number of repetitions $= 100000$

Number of successes	Inverse adaptive stratified random sampling	
r	$\hat{\tau}$	$SE(\hat{\tau})$
12	6530.77	3656.11
11	6512.01	3649.48
10	6541.33	3660.93
9	6797.64	3954.36
8	6924.74	4086.44

The result showed that as the number of successes (r) decreases the bias reduces and finally it converges with the true population total ($=6{,}923$), but standard error increases.

15.4   Exercises

1. Describe stratified ACS design.
2. Describe stratified Inverse ACS design.
3. Obtain an unbiased estimator of the population total in stratified Inverse ACS design.
4. Obtain an unbiased estimator of the variance of the estimator obtained in Exercise 3.

References

Des, R.: Some estimators in sampling with varying probabilities without replacement. J. Am. Stat. Assoc. **51**, 269–284 (1956)

Lehmann, E.L.: Theory of Point Estimation. Chapman and Hall, New York (1983)

Murthy, M.N.: Ordered and unordered estimators in sampling without replacement. Sankhya **18**, 379–390 (1957)

Salehi, M.M., Seber, G.A.F.: A new proof of Murthy's estimator which applies to sequential sampling. Australian and New Zealand J. Stat. **43**(3), 281–286 (2001)

Thompson, S.K.: Adaptive cluster sampling. J. Am. Stat. Assoc. **85**(412), 1050–1058 (1990)

Thompson, S.K.: Adaptive cluster sampling: designs with primary and secondary units. Biometrics **47**(3), 1103–1115 (1991a)

Thompson, S.K.: Stratified adaptive cluster sampling. Biometrika **78**(2), 389–397 (1991b)

Chapter 16
Negative Adaptive Cluster Sampling

16.1 Introduction

The adaptive cluster sampling (ACS) design has been discussed in Chap. 10. Although the ACS design is found appropriate for sampling from a rare and clustered population, it suffers from the drawback of losing control of the final sample size. Several suggestions have been made by the different researchers for limiting this final sample size of the adaptive cluster sample. For instance, Salehi and Seber (1997a, 1997b) suggested a two-stage sampling design in which primary units are selected using a conventional design and secondary units within the selected primary units are sub-sampled using ACS. Brown (1994) proposed a design in which initial sample is selected sequentially until the final sample size reaches a specified value. Lee (1998) developed a two-phase design, in which the first-phase sample is an ACS sample based on an auxiliary variable and the second-phase sample is selected from the first phase using probability proportional to size (PPS) with replacement sampling design. This design controls the number of measurements of the study variable but it cannot control that of the auxiliary variable. Salehi and Seber (2002) proposed an estimator of the population mean. Bahl and Tuteja (1991) proposed ratio and product type exponential estimators for estimating the mean of a finite population. In Chap. 12, we have described the procedure of adaptive cluster double sampling (ACDS) proposed by Medina and Thompson (2004). It is a method based on combining the idea of double sampling and ACS. It requires the availability of an inexpensive and easy to measure auxiliary variable. ACDS controls the final sample size to some extent. But it requires the second-phase sample.

We were interested in finding a sampling design that will consider the type of relationship between the two variables, reduce the cost of sampling, and will be more precise than the earlier developed sampling designs. In Sect. 16.2, we have proposed a new method, negative adaptive cluster sampling (NACS). In this method, the process of adding the units to the initial sample is different than that of ACS. In this sampling design, we consider two highly negatively correlated variables X and

Y. X is highly abundant in the population whereas Y is rare. Taking observations on X is rather economical and easy as compared to Y. In Sect. 16.3, we have proposed the different estimators of population total and have derived variances of these estimators. A sample survey is presented in Sect. 16.4, along with results of this survey and discussion on it.

16.2 Negative Adaptive Cluster Sampling

If X and Y are positively correlated then we do not get abundant auxiliary information. Because the population under study is rare and clustered. Hence ACDS cannot be used in such a situation. Medina and Thomson have completely ignored this point. Irrespective of the type of relation between the auxiliary and interest variable, they have suggested the use of ACDS. If we have a rare and cluster population with respect to the variable of interest, then to have an abundant auxiliary variable there must be a strong negative correlation between the two variables. In that situation, a new sampling design is proposed. Let us see some practical situations, where such type of negative relationship is observed.

The plateaus in Western Ghats of Sahyadri from Goa to Varandha Ghat (Bhor, Maharashtra, India) are rich in aluminum ore—Laterite. Due to which the thorny plants are rarely observed. They are abundantly available on basalt-kind plateaus. But there are some rare patches of thorny plants. It indicates the absence of aluminum in that part. If the interest is to estimate the total number of thorny plants in that area, NACS can be effectively used. Here aluminum content of the soil is the auxiliary variable. Its presence can be detected easily. The estimate of the total number of these thorny plants can be obtained by using NACS.

The plateaus in the Western Ghats (Maharashtra, India) of Sahyadri from Tamhini Ghat to Mumbai are dominated by the presence of Basalt rocks. In this area, Neem, Ziziphus, and thorny plants are highly abundant. But there are intermediate patches of semi-evergreen plants. The estimate of the total number of these evergreen plants can be obtained by using NACS.

In another example, suppose we want to estimate the total population of fish in a specified region under the sea. Let this region be subdivided to form a grid of locations. There are ample bushes of a specific variety of plants under the seawater. The fish are detracted by that specific variety of bushes present in that region. But there are some rare sea plants in that region which provide food for the fish and hence they get attracted toward these plants. So, the fish present in these locations can be counted. Here, we consider the number of bushes as the auxiliary variable and the number of fish as the variable of interest. Further, it is observed that if the number of bushes of a specific variety of plants is greater than C, then no fish will be found at that particular location.

The situations presented above show the negative correlation between the two variables. In such a situation, we propose a new sampling design. In ACS the units in the initial sample are identified whether they satisfy the desired condition C with respect to the variable of interest or not. Further, the networks are expanded around the units in the initial sample that satisfies the condition C. Here, we propose different adaptive procedures. The variables are negatively correlated and the adaptive procedure involves the auxiliary variable instead of the variable of interest. We get the clusters of units during the adaptation. Hence, this method is called the negative adaptive cluster sampling.

In ACDS, by using the adaptation technique the first-phase units are decided by using an auxiliary variable. Then by using some traditional methods such as SRSWOR, the second-phase units are selected. In NACS, the adaptation is used to discover the networks in the population with reference to the auxiliary variable. Further, the networks corresponding to the variable of interest are identified. There is no second phase in NACS. That is how NACS is different than ACDS. So in general, NACS is not ACDS. But NACS can be looked upon as a particular case of ACDS where the entire networks identified in the first phase, corresponding to the variable of interest are considered as the second-phase units.

Secondly, ACDS does not consider the type of relationship between the auxiliary variable and the variable of interest. In contrast to this, NACS requires a negative relationship between the auxiliary variable and the variable of interest. The networks corresponding to the auxiliary variable and the variable of interest are discovered by using exactly the opposite conditions on the two variables. The use of auxiliary variables is justified by ACDS in the first-phase sampling. Hence we use the auxiliary information in NACS for adaptation purpose. We assume that the population information of the auxiliary variable is known. In NACS, the networks are formed by using ACS with auxiliary information. The corresponding Y is observed only for those units which satisfy the condition C_X. Here, the population is rare and clustered and we observe Y only for the units that satisfy the condition C_X. So, there is a substantial reduction in sample size with respect to Y. This reduced sample size is called the effective sample size. Hence, the design is called NACS. Consider a population of N units which can be observed and measured with respect to variables X and Y which are negatively correlated. Suppose the population is rare with respect to the variable of interest (Y); equivalently we can say that it is highly abundant with respect to the auxiliary variable X. Taking observations on X is easy and inexpensive.

16.2.1 Procedure of NACS

Form a grid of population containing N grid points of equal size and shape. Draw an initial sample of size n grid points from this grid using SRSWOR or SRSWR method. Check whether each of the selected units satisfies the condition C_X or does not satisfy the condition C_X. Add the unit to the left, right above, and below to each unit included in the initial sample that satisfies the condition C_X. These units are called neighbors of that unit. If any of these neighbors satisfy the condition C_X, add their neighbors also to the sample. Continue this way till the neighbors that do not satisfy the condition C_X are found. The set of neighbor units satisfying the condition

C_X along with the corresponding unit selected in the initial sample that satisfies the condition C_X constitutes a network. Thus, the networks are formed around the units selected in the initial sample that satisfy C_X. Note that a unit selected in the initial sample which does not satisfy the condition C_X forms a network of size one. Suppose K distinct clusters are formed with respect to X population. A cluster includes the units in a network and the corresponding edge units. Edge units do not satisfy the condition C_X. If all edge units in a cluster are dropped, we get a network. From the K clusters, we get the K networks.

Observe the values of the variable of interest corresponding to all the units in these K networks. Further using the following estimators, the population total of Y can be estimated. Estimates of the standard error of these estimators can be obtained. If we drop the auxiliary information to get modified Hansen–Hurwitz and Horvitz–Thompson estimators then NACS reduces to ACS.

16.2.2 Difference in ACS, ACDS, and NACS

In ACS, the units in the initial sample are identified whether they satisfy the desired condition C with respect to the variable of interest or not. Further, the networks are expanded around the units in the initial sample that satisfies condition C.

In NACS, we consider an auxiliary variable along with the variable of interest. These two variables are assumed to be highly negatively correlated. The adaptive procedure involves the auxiliary variable instead of the variable of interest. We get the clusters of units during the adaptation. Hence this method is called the negative adaptive cluster sampling.

In ACDS, by using the adaptation technique the first-phase units are decided by using an auxiliary variable. Then by using some traditional methods such as SRSWOR, the second phase units are selected. In NACS, the adaptation is used to discover the networks in the population with reference to the auxiliary variable. Further, the networks corresponding to the variable of interest are identified. There is no second phase in NACS. That is how NACS is different than ACDS. So in general, NACS is not ACDS. But NACS can be looked upon as a particular case of ACDS where the entire networks identified in the first phase, corresponding to the variable of interest are considered as the second-phase units.

Secondly, ACDS does not bother about the type of relationship between the auxiliary variable and the variable of interest. In contrast to this NACS requires a negative relationship between the auxiliary variable and the variable of interest. The networks corresponding to the auxiliary variable and the variable of interest are discovered by using exactly the opposite conditions on the two variables. The use of auxiliary variables is justified by ACDS in the first-phase sampling. In NACS, we assume that the population information of the auxiliary variable is known. The networks are formed by using ACS with auxiliary information. The corresponding Y is observed only for those units which satisfy the condition C_X which is based on auxiliary information.

Here, the population is rare and clustered and we observe Y only for the units that satisfy the condition C_X. So, there is a substantial reduction in sample size with respect to Y. This reduced sample size is called the effective sample size.

16.3 Different Estimators in NACS

16.3.1 Modified Ratio Type Estimator

Chutiman and Chiangpradit (2014) proposed a ratio estimator of the population total of the variable of interest. It is based on the Des Raj estimators of the population totals of the auxiliary and the variable of interest. Des Raj estimator itself is an ordered estimator.

We consider a ratio estimator which is based on the HT estimators of the population totals of the two variables. HT estimator is an unbiased and unordered estimator. So, the computational difficulty involved in our estimator is much lesser than that in the ratio estimator proposed by Chutiman and Chiangpradit (2014).

Consider a population $U = \{1, 2, \ldots, i, \cdots, N\}$ which is partitioned into K networks, denoted as $\psi_1, \psi_2, \ldots, \psi_k$. Let the values of the auxiliary variable X be known for all the population units.

Suppose a survey is conducted by using NACS. The information on the variable of interest Y is collected for all the units selected in the final sample.

Define $y_{k.} = \sum_{k \in \psi_k} y_k$ and $x_{k.} = \sum_{k \in \psi_k} x_k$. The modified HT estimator of the population total of the variable Y can be obtained by using the inclusion probabilities of networks.

We can estimate the population total $\tau_y = \sum_U y_k = \sum_{\psi_k} y_{k.}$ by defining the estimator:

$$(\hat{\tau}_y)_{HT} = \sum_s \check{y}_k$$
$$= \sum_{s_k} \sum_{\psi_k} \frac{y_k}{\pi_k}$$
$$= \sum_{s_k} \frac{(\sum_{\psi_k} y_k)}{\pi_k}$$
$$= \sum_{s_k} \frac{y_{k.}}{\pi_k^*}$$
$$= \sum_{s_k} \check{y}_{k.}.$$

Note: (i) s_k is the set of units selected in the final adaptive sample from the kth network.

(ii) $\pi_k = \pi_k^*$, where π_k denotes the inclusion probability of kth unit and π_k^* denotes the inclusion probability of kth cluster which includes kth unit.

The population total of X is $\tau_x = \sum_U x_k = \sum_{\psi_k} x_{k.}$.

The estimator of τ_x is

$$(\hat{\tau}_x)_{HT} = \sum_s \check{x}_k$$
$$= \sum_{s_k} \sum_{\psi_k} \frac{x_k}{\pi_k}$$
$$= \sum_{s_k} \left(\frac{\sum_{\psi_k} x_k}{\pi_k} \right)$$
$$= \sum_{s_k} \check{x}_{k.}.$$

The generalized population ratio for total is
$$\tau_{RAD} = \frac{\tau_y}{\tau_x} \tau_x$$
$$= \hat{R} \tau_x$$
Note that $E(\hat{\tau}_x)_{HT} = \tau_x$ and $E(\hat{\tau}_y)_{HT} = \tau_y$.

Estimator of τ_{RAD} is
$$\hat{\tau}_{RAD} = \frac{(\hat{\tau}_y)_{HT}}{(\hat{\tau}_x)_{HT}} \tau_x.$$
$$= \hat{R} \tau_x$$
Using the Taylor linearization technique about the point (τ_x, τ_y) gives the approximation:
$$\hat{\tau}_{RAD} = (\hat{\tau}_y)_{HT} + \frac{\tau_y}{\tau_x}(\tau_x - (\hat{\tau}_x)_{HT}).$$

Theorem 16.1 *The approximate variance of $\hat{\tau}_{RAD}$ is given by*
$$AV(\hat{\tau}_{RAD}) = \sum_S \sum_{\psi_k} \Delta_{kl}^* \check{E}_{k.} \check{E}_{l.}$$
Where
$$\Delta_{kl}^* = \pi_{kl}^*$$
$$\check{E}_{k.} = \frac{E_{k.}}{\pi_k^*}$$
and
$$E_{k.} = y_{k.} - R\,x_{k.}$$
The variance estimator of the modified ratio estimator is
$$\hat{V}(\hat{\tau}_{RAD}) = \sum_{k,l \in \psi_k} \sum_{\psi_k \in S} \check{\Delta}_{kl}^* \check{e}_{k.} \check{e}_{l.}$$
Where
$$\check{\Delta}_{kl}^* = \frac{\Delta_{kl}^*}{\pi_{kl}^*}$$
$$\check{e}_{k.} = \frac{y_{k.} - \hat{R}\,x_{k.}}{\pi_k^*}.$$

Proof Let
$$\hat{\tau}_{RAD} = \hat{R}\,\tau_x.$$
Hence, $V(\hat{\tau}_{RAD}) = \tau_x^2\, V(\hat{R})$.

Approximate variance of $\hat{\tau}_{RAD} = AV(\hat{\tau}_{RAD}) = \tau_x^2 AV(\hat{R})$

Using Taylor's linearization technique (described in Särndal et al. 1992), $AV(\hat{R})$ can be determined.

Here, $\hat{R}$ is a function of the two random variables $(\hat{\tau}_y)_{HT}$ and $(\hat{\tau}_x)_{HT}$.

Thus, $\hat{R} = f((\hat{\tau}_y)_{HT}, (\hat{\tau}_x)_{HT}) = \frac{(\hat{\tau}_y)_{HT}}{(\hat{\tau}_x)_{HT}}$

Hence, $\frac{\partial \hat{R}}{\partial (\hat{\tau}_y)_{HT}} = \frac{1}{(\hat{\tau}_x)_{HT}}$

and
$$\frac{\partial \hat{R}}{\partial (\hat{\tau}_x)_{HT}} = -\frac{(\hat{\tau}_y)_{HT}}{(\hat{\tau}_x^2)_{HT}}.$$
Hence, we get
$$a_1 = \frac{\partial \hat{R}}{\partial (\hat{\tau}_y)_{HT}} \text{ evaluated at the expected value point } (\tau_y, \tau_x)$$
$$= \frac{1}{\tau_x}.$$
$$a_2 = \frac{\partial \hat{R}}{\partial (\hat{\tau}_x)_{HT}} \text{ evaluated at the expected value point } (\tau_y, \tau_x)$$
$$= -\frac{R}{\tau_x}.$$

Hence, $E_{k.} = a_1 y_{k.} + a_2 x_{k.} = \frac{(y_{k.} - R\, x_{k.})}{\tau_x}$.

Using Taylor's linearization $\hat{R}$ can be written as follows:

$\hat{R} = R + \frac{1}{\tau_x} \sum_S \frac{E_{k.}}{\pi_k^*}$.

Hence, $AV(\hat{R}) = \frac{1}{\tau_x^2} V \left[\sum_S \frac{E_{k.}}{\pi_k^*} \right]$

$$= \frac{1}{\tau_x^2} V \left[\sum_S \breve{E}_{k.} \right]$$

$$= \frac{1}{\tau_x^2} \sum \sum_U \Delta_{kl}^* \breve{E}_{k.} \breve{E}_{l.}$$

$$AV(\hat{\tau}_{RAD}) = \sum \sum_U \Delta_{kl}^* \breve{E}_{k.} \breve{E}_{l.}$$

$$\hat{V}(\hat{\tau}_{RAD}) = \sum_{k,l \in \psi_k} \sum_{\psi_k \in S} \breve{\Delta}_{kl}^* \breve{e}_{k.} \breve{e}_{l.}$$

Hence the theorem.   □

16.3.2 Modified Regression Estimator

The modified regression estimator $\hat{\tau}_{RADD}$ is a function of HT estimators $(\hat{\tau}_x)_{HT}$ and $(\hat{\tau}_y)_{HT}$.

$\hat{\tau}_{RADD} = (\hat{\tau}_y)_{HT} + (\sum_{\psi_k} x_{k.} - (\hat{\tau}_x)_{HT}) \hat{\beta}_1$.

By using weighted least square method, we get

$\hat{\beta}_1 = (\sum_s \frac{x_k^2}{\sigma_k^2 \pi_k^*})^{-1} (\sum_s \frac{x_k. y_k.}{\sigma_k^2 \pi_k^*})$.

By using Taylor linearization technique about the points (τ_x, τ_y), $\hat{\beta}_1$ is approximated by

$\hat{\beta}_1^0 = \beta_1 + T^{-1}(t - \hat{T}\beta_1)$.

Where

$\hat{T} = \sum_s \frac{x_k^2}{\sigma_k^2 \pi_k^*} \; ; T = \sum_{\psi_k} x_{k.}^2 \; ; t = \sum_{\psi_k} x_{k.} y_{k.} \; ; \hat{t} = \sum_s \frac{x_k. y_k.}{\sigma_k^2 \pi_k^*}$.

The approximate variance of $\hat{\beta}$ given by

$AV(\hat{\beta}) = T^{-1} V T^{-1}$.

Let $E_{k.}$ is the residual of population fit.

$E_{k.} = y_{k.} - x_{k.} \beta_1$.

$V = \sum \sum_U \Delta_{kl}^* (\frac{x_k. E_k.}{\pi_k^*})(\frac{x_l. E_l.}{\pi_l^*})$.

The estimator of the variance of $\hat{\beta}$ is

$\hat{V}(\hat{\beta}) = (\sum_s \frac{x_k^2}{\sigma_k^2 \pi_k^*})^{-1} \hat{V} (\sum_s \frac{x_k^2}{\sigma_k^2 \pi_k^*})^{-1}$,

where

$\hat{V} = \sum \sum_s \breve{\Delta}_{kl}^* (\frac{x_k. e_k.}{\pi_k^*})(\frac{x_l. e_l.}{\pi_l^*})$,

where $e_{k.}$ is the sample residual fit.

$e_{k.} = y_{k.} - x_{k.} \hat{\beta}_1$

$\hat{\tau}_{RADD} = (\hat{\tau}_y)_{HT} + (\sum_{\psi_k} x_{k.} - (\hat{\tau}_x)_{HT}) \hat{T}^{-1} \sum_s \frac{x_k. y_k.}{\sigma_k^2 \pi_k^*}$

$$= \sum_s (1 + (\sum_{\psi_k} x_{k.} - (\hat{\tau}_x)_{HT}) \hat{T}^{-1} \frac{x_k.}{\sigma_k^2}) \breve{y}_{k.}$$

It shows that the regression estimator can be expressed as a linear function of the π expanded values $\breve{y}_{k.}$ for $k \epsilon S$.

Thus,

$\hat{\tau}_{RADD} = \sum_s g_{ks}^* \check{y}_k.$

With the sample dependent weights

$g_{ks}^* = 1 + (\sum_{\psi_k} x_k. - (\hat{\tau}_x)_{HT}) \hat{T}^{-1} \frac{x_k.}{\sigma_k^2.}$

The regression estimator relates to the hypothetical population fit of the model ζ, which produces $\hat{\beta}_1$ given by equation of fitted values.

$y_k^0. = x_k.\beta_1$ and the population fit of residuals $E_k. = y_k. - y_k^0..$

The regression estimator becomes

$\hat{\tau}_{RADD} = \sum_s g_{ks}^* (\check{y}_k^0. + \check{E}_k.)$

$\qquad\quad = \sum_{\psi_k} y_k^0. + \sum_s g_{ks}^* \check{E}_k.$

Theorem 16.2 *The approximate variance of* $\hat{\tau}_{RADD}$ *is*

$AV(\hat{\tau}_{RADD}) = \sum\sum_U \Delta_{kl}^* \check{E}_k.\check{E}_l.$

The estimator of $AV(\hat{\tau}_{RADD})$ *is*

$\widehat{AV}(\hat{\tau}_{RADD}) = \sum\sum_s \check{\Delta}_{kl}^* (g_{ks}^* \check{e}_{ks}^*)(g_{ls}^* \check{e}_{ls}^*).$

Where

$\check{e}_{ks}^* = \frac{e_{ks}^*}{\pi_k^*}$ *since* $e_{ks}^* = \check{y}_k. - x_k.\hat{\beta}_1$

Proof $\hat{\tau}_{RADD} = \sum_{\psi_k} y_k^0. + \sum_s g_{ks}^* \check{E}_k.$

Hence, $AV(\hat{\tau}_{RADD}) = V(\sum_s g_{ks}^* \check{E}_k.)$

By using Result 5.10.1 from Särndal et al. (1992), we get

$AV(\hat{\tau}_{RADD}) = \sum\sum_U \Delta_{kl}^* \check{E}_k.\check{E}_l.$ Estimator of this $AV(\hat{\tau}_{RADD})$ is given by

$\widehat{AV}(\hat{\tau}_{RADD}) = \sum\sum_s \check{\Delta}_{kl}^* (g_{ks}^* \check{e}_{ks}^*)(g_{ls}^* \check{e}_{ls}^*)$

Hence the theorem.   □

16.3.3 Product Estimator

Since the two variables are negatively correlated, it is of interest to define a product estimator of the population total based on the HT estimators of the population totals of the two variables. The product estimator of the population total τ_y is defined as

$\hat{\tau}_{RADE} = (\hat{\tau}_y)_{HT} \; exp \left(\frac{\tau_x - (\hat{\tau}_x)_{HT}}{\tau_x + (\hat{\tau}_x)_{HT}} \right).$

Theorem 16.3 *Bias in the product estimator is given by*

$B(\hat{\tau}_{RADE}) = \frac{3}{16} \frac{\tau_y}{\tau_x^2} \sum\sum_{i \neq j \in \psi_k} (\pi_i \pi_j - \pi_{ij})(\frac{x_i}{\pi_i} - \frac{x_j}{\pi_j})^2 - \frac{1}{4\tau_x} \sum\sum_{i \neq j \in \psi_k} (\pi_i \pi_j - \pi_{ij})(\frac{x_i}{\pi_i} - \frac{y_j}{\pi_j})^2.$

Variance of $\hat{\tau}_{RADE}$ *is given by*

$V(\hat{\tau}_{RADE}) = \frac{1}{2} \sum\sum_{i \neq j \in \psi_k} (\pi_i \pi_j - \pi_{ij})(\frac{y_i}{\pi_i} - \frac{y_j}{\pi_j})^2 + \frac{1}{8} \frac{\tau_y^2}{\tau_x^2} \sum\sum_{i \neq j \in \psi_k} (\pi_i \pi_j - \pi_{ij})(\frac{x_i}{\pi_i} - \frac{x_j}{\pi_j})^2 - \frac{1}{2} \frac{\tau_y}{\tau_x} \sum\sum_{i \neq j \in \psi_k} (\pi_i \pi_j - \pi_{ij})(\frac{x_i}{\pi_i} - \frac{y_j}{\pi_j})^2.$

Proof Let $e_y = \frac{(\hat{\tau}_y)_{HT} - \tau_y}{\tau_y}$

and

$e_x = \frac{(\hat{\tau}_x)_{HT} - \tau_x}{\tau_x}$.

Hence, $E(e_y) = E(e_x) = 0$.

$E(e_y^2) = \frac{1}{2\,\tau_y^2} \sum \sum_{i \neq j \in \psi_k} (\pi_i\,\pi_j - \pi_{ij})\,(\frac{y_i}{\pi_i} - \frac{y_j}{\pi_j})^2$.

$E(e_x^2) = \frac{1}{2\,\tau_x^2} \sum \sum_{i \neq j \in \psi_k} (\pi_i\,\pi_j - \pi_{ij})\,(\frac{x_i}{\pi_i} - \frac{x_j}{\pi_j})^2$.

$E(e_x e_y) = \frac{1}{2\,\tau_x\,\tau_y} \sum \sum_{i \neq j \in \psi_k} (\pi_i\,\pi_j - \pi_{ij})\,(\frac{x_i}{\pi_i} - \frac{y_j}{\pi_j})^2$.

Hence,

$\hat{\tau}_{RADE} = \tau_y\,(1 + e_y)\,exp\left[\frac{\tau_x - \tau_x\,(1+e_x)}{\tau_x + \tau_x\,(1+e_x)}\right]$

$= \tau_y\,(1 + e_y)\,exp\left[\frac{-e_x}{2+e_x}\right]$

$= \tau_y\,(1 + e_y)\,exp\left[\frac{-1}{2}\,e_x(1 + \frac{e_x}{2})^{-1}\right]$

$\approx \tau_y(1 + e_y)exp\left[\frac{-1}{2}\,e_x(1 - \frac{1}{2}\,e_x + \frac{1}{4}\,e_x^2)\right]$.

By neglecting the terms involving e_x with power three and above we get

$\hat{\tau}_{RADE} \approx \tau_y\,(1 + e_y)exp\left[\frac{-1}{2}(e_x - \frac{1}{2}\,e_x^2)\right]$

$\approx \tau_y\,(1 + e_y)\left[1 - \frac{1}{2}(e_x - \frac{1}{2}\,e_x^2) + \frac{1}{8}\,(e_x - \frac{1}{2}\,e_x^2)^2\right]$

$\approx \tau_y\,(1 + e_y)\left[1 - \frac{1}{2}(e_x - \frac{1}{2}\,e_x^2) + \frac{1}{8}\,e_x^2\right]$

$= \tau_y\,(1 + e_y)\left[1 - \frac{1}{2}\,e_x + \frac{3}{8}\,e_x^2\right]$

$= \tau_y\left[1 + e_y - \frac{1}{2}\,e_x + \frac{3}{8}\,e_x^2 - \frac{1}{2}\,e_x\,e_y\right]$.

Bias in $\hat{\tau}_{RADE} = E(\hat{\tau}_{RADE}) - \tau_y$

$= \tau_y\,E\left[e_y - \frac{1}{2}\,e_x + \frac{3}{8}\,e_x^2 - \frac{1}{2}\,e_x\,e_y\right]$

$= \frac{3\,\tau_y}{16\,\tau_x^2} \sum \sum_{i \neq j \in \psi_k} (\pi_i\,\pi_j - \pi_{ij})(\frac{x_i}{\pi_i} - \frac{x_j}{\pi_j})^2 - \frac{1}{4\,\tau_x} \sum \sum_{i \neq j \in \psi_k} (\pi_i\,\pi_j - \pi_{ij})(\frac{x_i}{\pi_i} - \frac{y_j}{\pi_j})^2$.

Now,

Variance of $\hat{\tau}_{RADE} = V(\hat{\tau}_{RADE}) = E(\hat{\tau}_{RADE} - \tau_y)^2$

$= E\left(\tau_y\,(1 + e_y - \frac{1}{2}\,e_x + \frac{3}{8}\,e_x^2 - \frac{1}{2}\,e_x\,e_y) - \tau_y\right)^2$

$\approx \tau_y^2\,E(e_y - \frac{1}{2}\,e_x)^2$

$= \tau_y^2\,E(e_y^2 + \frac{1}{4}\,e_x^2 - e_x\,e_y)$.

Hence,

$V(\hat{\tau}_{RADE}) = \frac{1}{2} \sum \sum_{i \neq j \in \psi_k} (\pi_i\,\pi_j - \pi_{ij})(\frac{y_i}{\pi_i} - \frac{y_j}{\pi_j})^2 + \frac{1}{8}\,\frac{\tau_y^2}{\tau_x^2} \sum \sum_{i \neq j \in \psi_k} (\pi_i\,\pi_j - \pi_{ij})(\frac{x_i}{\pi_i} - \frac{x_j}{\pi_j})^2 - \frac{1}{2}\,\frac{\tau_y}{\tau_x} \sum \sum_{i \neq j \in \psi_k} (\pi_i\,\pi_j - \pi_{ij})(\frac{x_i}{\pi_i} - \frac{y_j}{\pi_j})^2$.

Hence the theorem. $\qquad\qquad\square$

16.4 Sample Survey

A survey was conducted by using NACS. The interest was to estimate the total number of evergreen plants which are rare in that region due to the presence of Basalt rocks.

The area of 100 acres in the Tamhini Ghat, Maharashtra, was divided into 100 plots each of size 1 acre and the percentage of silica observed on each of these plots

24 *	25	86 *	60	52	35	65	50	60	1
40	30	30	75	18	19	55	30	4	14
45	48 *	56	23	15 *	17	53	30	13	12
47	47	23	25	80	60	45	45	35	70
48	50	25 *	35	57	68	40	23 *	80	40
49	43	36	65	58	58	90	45	90	30
45	35 *	56	85	19	30	18	18	40 *	50
48	53	65	55	13	16 *	15	18	30	60
70	30	17	18*	15	48	44	44	35	50
30	30	18	17	15	43	36	50	80	36

* in a square indicates selection in initial sample.

Fig. 16.1 Percentage of Silica (SiO_2) observed on the different plots of the square region

was measured. The time required to measure the percentage of silica in a sample from one-acre plot is fairly lesser than the time required to measure the number of evergreen plants in one acre. Secondly, testing a soil sample for the percentage of silica is much cheaper than the cost incurred in counting the number of evergreen plants in one acre. The cost of testing a soil sample was two dollars and that of counting the number of evergreen plants in one acre was twenty dollars. So, we considered the percentage of silica in one acre as the auxiliary variable.

The nature of the soil in the Western Ghats is of two types: Basalt rocks and Laterite. After studying the nature of the soil, we had observed the abundance of evergreen plants whenever the silica content of the soil is less than or equal to 20%. Hence, we considered $C_x = \{X \leq 20\}$ as the condition for adaptation.

A random sample of 10 plots was drawn from this area by using SRSWOR. The plots selected in the initial sample from this population related to the auxiliary variable X (percentage of silica in a plot) are shown by putting $\star$ in that plot (Fig. 16.1).

Then the procedure, NACS was used. The networks are formed around the plots selected in the initial sample which satisfied the condition C_x. Each plot with $C_x = \{X > 20\}$ and selected in the initial sample formed a network of size 1 (shown in orange color). There were such 6 networks of size 1 selected in the initial sample from the above population of X variable. There was 1 network of size 13 (shown in blue color) and another network of size 4 (shown in purple color). Thus the total number of distinct networks in the sample was 8.

Here, the clusters are formed by using auxiliary information and domain knowledge of silica content and evergreen plants. These two variables are negatively correlated. It means that the abundance of silica in soil leads to the rare evergreen plants. A cluster involves the network units and edge units. The edge units of clusters of size more than 1 are dropped to get the networks. Select only those networks which satisfy the condition $C_x = \{X \leq 20\}$ and are measured for the survey variable (number of evergreen plants) as shown in Fig. 16.2.

*		*							405
				35	20			306	130
	*			100*	65			107	108
			*						
			*				*		
				15		40	40	*	
				120	65*	95	30		
		75	32*	91					
		36	35	81					

Fig. 16.2 Plot-wise number of evergreen plants observed plots in the population

Results and Discussion:

For computational efficiency in estimation of each estimator, r number of repetitions were performed where r varied as 5000, 10000, 20000, and 100000. It is very difficult to take all possible samples. In our study, the population size was 100 and the initial sample size was 10. Thus the number of possible samples is 1.731×10^{13}. This is a very large number. Hence we took r repetitions for NACS. We required an initial sample size (say n). It was varied as 10, 15, 20, 25, 35, and 45.

For establishing the condition under which the estimators used in NACS are more efficient than the traditional estimators, we have repeated the simulations for different replications.

The estimated population total over r possible samples is given by
$$\hat{\tau}_y = \frac{\sum_{i=1}^{r} (\hat{\tau}_y)_i}{r}.$$
The estimated variance of the estimator of total is given by
$$\widehat{MSE}(\hat{\tau}_y) = \frac{\sum_{i=1}^{r} ((\hat{\tau}_y)_i - \tau_y)^2}{r},$$
where $(\hat{\tau}_y)_i$ is the value of the relevant estimator for the ith sample.

The estimates of τ_y along with the corresponding estimates of standard error (SE) were obtained by using the HT- and HH-type estimators $((\hat{\tau}_y)_{HT}$ and $(\hat{\tau}_y)_{HH})$ as well as the regression estimator $\hat{\tau}_{RADD}$ under ACS. The results are shown in Table 16.1.

The estimates of τ_y along with the corresponding estimates of standard errors (SE) were obtained by using the modified ratio and regression estimators ($\hat{\tau}_{RAD}$ and $\hat{\tau}_{RADD}$) under NACS, two-phase estimator ($\hat{\tau}$) (Särndal and Swensson 1987) under ACDS. The results are presented in Table 16.2.

For presenting the cost–benefit analysis of the new design, we calculated the expected sampling costs in ACS and expected effective sampling costs in NACS.

Expected sampling cost in ACS is based on the final sample size (n_s) and the expected effective sampling cost in NACS is based on the expected effective sample size (n_e) and expected final sample size (n_s) in NACS.

Final sample size (n_s) $= n + \sum_{k=1}^{K} (n_k - X_k)\delta_k$,
where

n_k: Size of the kth network in the population, $k = 1, 2, \ldots, K$.

X_k: Number of units included in the initial sample from the kth network, $k =$

Table 16.1 Estimated values and SE of different estimators in ACS for the different values of r and n

Number of repetitions (r)	Initial sample size (n)	$(\hat{\tau}_y)_{HT}$	$\hat{SE}((\hat{\tau}_y)_{HT})$	$(\hat{\tau}_y)_{HH}$	$\hat{SE}((\hat{\tau}_y)_{HH})$	$(\hat{\tau}_y)_{RADD}$	$\hat{SE}((\hat{\tau}_y)_{RADD})$
5000	10	2031.62	1316.18	2031.03	1561.31	2030.03	1326.19
	15	2033.26	973.95	2030.81	1266.50	2030.05	969.91
	20	2032.51	756.07	2031.37	1092.98	2033.93	754.95
	25	2031.08	599.86	2029.82	976.26	2030.24	599.23
	35	2032.12	383.25	2031.88	829.61	2031.98	384.17
	45	2030.19	244.15	2031.92	729.06	2030.18	241.85
10000	10	2029.31	1315.63	2038.90	1552.43	2031.27	1315.07
	15	2033.75	969.66	2032.07	1259.94	2033.64	969.54
	20	2032.51	755.62	2039.92	1091.23	2032.63	752.70
	25	2029.71	595.55	2031.45	970.16	2034.34	595.94
	35	2030.47	382.99	2030.45	828.79	2034.92	383.94
	45	2031.99	241.71	2035.60	723.35	2033.48	241.11
20000	10	2033.84	1314.15	2031.48	1537.71	2030.40	1308.67
	15	2030.97	967.08	2030.63	1258.89	2031.22	969.27
	20	2032.02	678.40	2033.18	1086.86	2031.85	752.64
	25	2032.05	598.03	2031.33	968.46	2032.84	594.92
	35	2031.53	383.40	2031.13	824.78	2032.79	381.93
	45	2031.71	237.25	2030.99	720.96	2033.05	237.62
100000	10	2029.16	1309.28	2031.04	1536.88	2031.43	1307.76
	15	2031.48	970.77	2034.84	1255.06	2032.35	968.57
	20	2029.15	754.17	2031.81	1078.86	2030.34	752.49
	25	2029.00	592.94	2031.10	965.18	2031.72	592.12
	35	2031.42	382.12	2033.61	810.07	2030.42	381.46
	45	2031.08	236.11	2031.18	720.00	2030.91	237.15

Table 16.2 Estimated values and SE of different estimators in NACS and ACDS for the different values of r and n

Number of repetitions	Initial sample size	NACS				ACDS	
r	n	$(\hat{\tau}_y)_{RAD}$	$\hat{SE}((\hat{\tau}_y)_{RAD})$	$(\hat{\tau}_y)_{RADD}$	$\hat{SE}((\hat{\tau}_y)_{RADD})$	$\hat{\tau}$	$\hat{SE}(\hat{\tau})$
5000	10	2174.91	1598.58	2033.39	1319.39	2023.72	1463.54
	15	2091.30	1132.63	2031.59	971.84	2025.94	1125.09
	20	2063.51	863.58	2032.81	756.87	2041.87	910.41
	25	2068.98	683.87	2033.67	597.45	2031.20	770.94
	35	2062.73	442.43	2030.12	384.63	2029.86	602.93
	45	2042.49	298.33	2031.42	241.82	2034.65	511.70
10000	10	2154.28	1597.53	2030.73	1316.04	2037.79	1460.22
	15	2114.63	1131.08	2032.16	970.07	2032.13	1115.55
	20	2080.71	856.13	2031.54	754.93	2030.93	909.49
	25	2071.95	677.41	2031.32	596.46	2025.96	769.57
	35	2051.42	440.30	2033.42	382.25	2035.08	598.60
	45	2042.54	293.42	2031.70	238.85	2033.21	511.19
20000	10	2178.62	1596.31	2033.10	1310.00	2027.63	1457.46
	15	2117.47	1129.47	2033.74	970.02	2035.04	1107.99
	20	2083.95	855.61	2033.67	753.12	2027.68	908.82
	25	2074.24	679.72	2033.52	596.23	2031.29	768.96
	35	2051.13	439.57	2032.67	381.35	2030.20	597.88
	45	2047.81	290.23	2031.25	238.84	2028.44	507.91
100000	10	2183.83	1592.43	2032.20	1309.67	2033.26	1451.42
	15	2114.00	1128.52	2031.48	963.04	2028.92	1107.41
	20	2085.86	854.92	2031.62	751.92	2030.62	894.39
	25	2066.96	673.99	2031.29	596.11	2031.44	757.87
	35	2050.03	424.25	2031.90	381.20	2032.70	594.83
	45	2041.78	284.18	2031.19	235.91	2033.18	506.43

$1.2, \ldots, K.$

$$\delta_k = \begin{cases} 1 & ; \text{if the initial sample includes a sampling unit from} \\ & k^{th} \ network \\ 0 & ; otherwise. \end{cases}$$

Effective sample size $(n_e) = \sum_{j=1}^{n_s} \delta_{C_x}(j)$
    Where

$$\delta_{C_x}(j) = \begin{cases} 1 ; \text{if } U_j \text{ satisfies the condition } C_x \\ 0 ; otherwise. \end{cases}$$

Since the expected sample size under ACS is the total size of the included clusters for the variable of interest and expected effective sample size under NACS is the total size of the included networks for the variable of interest. Hence, we get $n_e < n_s$. In ACS, we consider only a variable of interest and adaptation is made to get clusters. They include the network units and edge units. Thus, the expected sampling cost in ACS $= 20E(n_s)$.

In NACS, we consider two variables, auxiliary and interest variable. Here, using auxiliary information we determine the clusters that give us the expected sample size. Edge units are dropped from these clusters to get networks. Only, networks of interest variables are observed to get the expected effective sample size.

In ACS, we consider only the variable of interest and adaptation is made to get clusters. Clusters include the network units and edge units.

Expected sampling cost in NACS $= 2E(n_s) + 20E(n_e)$.

Values of $E(n_s)$ and $E(n_e)$ are obtained by averaging the values of n_s and n_e over the r repetitions. The results are presented in Table 16.5.

To evaluate the performance of NACS, we compared the performance of the proposed modified regression estimator with that of the other estimators. The results are shown in Table 16.3.

ACDS controls the final sample size to some extent. But it requires the second-phase sample. We were interested in finding a sampling design that will consider the

Table 16.3 Estimated values of the different estimators in NACS and their standard errors for initial sample of size 45 and number of repetitions equal to 100000

Estimator	Design	Estimate	Estimate of SE	Relative efficiency of NACS
$\hat{\tau}_{RADD}$	ACS	2030.91	237.15	1.011
$\hat{\tau}_{RAD}$	NACS	2041.78	284.18	1.451
$\hat{\tau}_{RADD}$	NACS	2031.19	235.91	1.000
$\hat{\tau}$	ACDS	2033.18	506.43	4.608
$(\hat{\tau}_y)_{Reg}$	SRSWOR	1981.99	558.94	5.614

type of relationship between the two variables, reduce the cost of sampling, and will be more precise than the earlier developed sampling designs.

NACS differs from ACDS. In ACDS the second-phase units are selected by using some conventional sampling technique and hence the sampling variations are introduced in the second phase as well. Due to this the standard error (SE) of the estimator is increased. On the contrary, in NACS, there are no second-phase units. We take an observation of all the units included in the final adaptive sample. So, sampling variations introduced in ACDS at the second phase are completely wiped out in NACS. Hence NACS performs better than ACDS under the specified conditions.

To understand the working of NACS, let us consider the following hypothetical situation.

Suppose we have a population of 10 units. Let X be the auxiliary variable and Y be the variable of interest. The values of X for these 10 units be given as $(1, 2, 3, 8, 7, 10, 15, 14, 4, 3)$. The corresponding values of Y be given as $(20, 15, 12, 0, 0, 0, 0, 0, 10, 11)$. Here, X and Y are highly negatively correlated. We use the auxiliary information to select the sample with condition $C_x = \{X \leq 5\}$ (say). The random sample of size 3 is drawn by using SRSWOR from X, say $(2, 7, 15)$. The corresponding Y values are $(15, 0, 0)$. By using the adaptation condition C_x we get three networks with X values $(1, 2, 3), (7)$ and (15). Since values 7 and 15 do not satisfy the condition C_x, the corresponding Y values are not observed. We observe Y values corresponding to X values included in the first network. These values are $(20, 15, 12)$. These values will be used for further estimation part.

In NACS, we do not take observations related to the variable of interest on the edge units in the discovered networks. So, we have introduced the term effective sample size (n_e). We have already mentioned that $n_e < n_s$. Thus, along with the merit of controlling the final sample size, NACS also has the merit of reducing the cost of sampling. Hence, NACS is superior as compared to ACS under this setup.

If we use the same estimator in ACS and NACS, the two designs are equally efficient because the two designs differ only at the design stage. The two designs differ in costs. As said earlier, NACS is more cost-effective than ACS. The degree of correlation affects the performance of NACS. The two variables must be highly negatively correlated. The significance of the correlation can be tested by using the t-test. In case of a weak correlation, the method is not recommended.

If the two variables are positively correlated then NACS reduces to ACS. There is no problem with losing control of the auxiliary variables. In ACDS and NACS as well, we consider only one auxiliary variable. ACDS completely ignores the type of correlation between the two variables. This drawback is covered in NACS. The effective sample size in NACS is smaller than the final sample size in the corresponding ACS. It finally leads to a reduction in the cost of sampling in NACS.

NACS is not simply ACS for variable X but it is different. There are many more things that can be studied related to NACS. In NACS, we observe the values of the variable of interest corresponding to only the units included in networks of that variable. There is no double sampling. Hence, sampling efforts are reduced in NACS as compared to ACDS.

If we drop the auxiliary information then NACS reduces to ACS. In that case we have used the condition $C_y = \{Y > 0\}$ for adaptation. The modified HH and HT estimators were used for ACS. The expected final sample size and the expected effective sample size are computed. The expected final sample size is greater than the expected effective sample size.

The modified ratio and regression estimators were used for NACS. For estimation of parameters, the auxiliary information is utilized. The modified regression estimator is more efficient than the modified ratio estimator. Even though, both the estimators are biased, they gave more stable estimates. As the initial sample size increased the standard error of the estimate decreased.

The modified regression estimator in NACS gave us better results as compared to its use in ACDS and ACS. This estimator is found to be more efficient than the conventional regression estimator in SRSWOR as shown in Table 16.3. In classical cluster sampling, comparisons are often made on the basis of cost. It is often less expensive (in terms of time and money) to sample units within a cluster than to select a new cluster. The same is true for NACS.

The relative bias in the ratio estimator $\hat{\tau}_{RAD}$ used in ACS showed a consistent reduction with the increase in the initial sample size. This estimator was observed to be positively biased. The regression estimator $\hat{\tau}_{RADD}$ used in SRSWOR also showed a reduction in the relative bias with an increase in the initial sample size. This estimator was found to be negatively biased. The other newly proposed product estimator $\hat{\tau}_{RADE}$ was found to be positively biased. The relative bias in this estimator also showed a reduction with an increase in the initial sample size.

Among the above three estimators in Table 16.4, the product estimator has the least values of the relative bias for the different number of repetitions and the initial sample sizes. This estimator was observed to be superior to the other two estimators. In NACS, the product estimator is superior to the modified ratio estimator and inferior to the modified regression estimator as shown in Tables 16.1 and 16.4.

The total cost involved in NACS is expected to be much lesser as compared to ACS. Since it is assumed that the auxiliary variable is abundant and the interest variable is rare, the cost involved in selection, acquisition, and measurement of units with respect to the auxiliary variable is expected to be much smaller than that involved in ACS. ACS involves the measurement cost of edge units and network units.

Further, the cost involved in making observations on the edge units in NACS is definitely lesser than that in ACS. There is no cost involved to measure edge units in NACS because, at the stage of formation of networks, edge units are dropped without inspecting the variable of interest. NACS assumes that the values of the auxiliary variable corresponding to all units in the population are known. For large geographical areas, it is very difficult to get such information. It limits the applicability of NACS. So, further research is required in that direction. The next chapter deals with the idea of double sampling in NACS is given.

The proposed NACS methodology is studied on a small pilot study. If the population size is large we will get a more precise idea about the proposed estimators. We had presented a number of samples ranging from 5000 to 100000 of sizes varying from 15 to 45 each from PPS with replacement and without replacement.

Table 16.4 Estimated values, relative bias, and SE of different estimators using various sampling designs

Number of repetitions	Initial sample size	ACS			SRSWOR			NACS		
r	n	$\hat{\tau}_{RAD}$	$\hat{S}E(\hat{\tau}_{RAD})$	R.Bias $\hat{\tau}_{RAD}$	$\hat{\tau}_{RADD}$	$\hat{S}E(\hat{\tau}_{RADD})$	R.Bias $\hat{\tau}_{RADD}$	$\hat{\tau}_{RADE}$	$\hat{S}E(\hat{\tau}_{RADE})$	R.Bias $\hat{\tau}_{RADE}$
5000	10	2210.39	1617.20	0.088	1644.51	1573.59	-0.190	2112.10	1426.17	0.039
	15	2102.83	1116.68	0.035	1800.16	1251.84	-0.113	2072.56	1037.91	0.024
	20	2081.72	855.22	0.024	1861.89	1026.63	-0.083	2073.19	1031.48	0.027
	25	2055.09	689.35	0.011	1907.20	880.70	-0.060	2033.73	630.42	0.001
	35	2047.36	450.05	0.008	1960.42	677.89	-0.034	2040.69	404.21	0.004
	45	2036.71	298.54	0.002	1975.47	560.23	-0.027	2035.80	257.11	0.002
10000	10	2198.03	1613.36	0.082	1614.02	1566.68	-0.205	2106.95	1428.10	0.037
	15	2119.59	1122.55	0.043	1757.91	1198.66	-0.134	2044.44	1034.71	0.006
	20	2086.72	864.86	0.027	1868.74	1023.86	-0.079	2065.56	794.98	0.017
	25	2065.10	677.29	0.016	1909.50	881.30	-0.059	2058.69	621.74	0.013
	35	2048.99	439.73	0.008	1937.47	697.98	-0.046	2049.77	393.83	0.009
	45	2041.08	287.84	0.004	1980.39	556.09	-0.024	2030.59	267.07	-0.0001
20000	10	2176.74	1592.60	0.071	1644.80	1573.61	-0.190	2094.24	1422.64	0.031
	15	2117.40	1134.99	0.042	1797.84	1221.34	-0.114	2070.86	1029.37	0.019
	20	2082.63	861.30	0.025	1854.13	1024.60	-0.087	2049.24	796.45	0.008
	25	2057.96	683.46	0.013	1914.27	881.05	-0.051	2042.47	628.31	0.005
	35	2051.99	445.81	0.010	1964.50	694.98	-0.032	2037.87	403.77	0.003
	45	2039.77	295.42	0.004	1982.50	558.42	-0.023	2033.73	261.34	0.001
100000	10	2185.95	1598.78	0.076	1632.23	1575.20	-0.196	2100.97	1428.57	0.034
	15	2115.53	1130.26	0.041	1784.16	1217.45	-0.121	2069.654	1031.73	0.019
	20	2082.77	859.85	0.025	1864.19	1016.43	-0.082	2055.78	793.03	0.012
	25	2066.74	680.64	0.017	1903.56	878.48	-0.062	2045.53	626.45	0.007
	35	2051.39	440.21	0.010	1952.62	691.26	-0.038	2038.69	403.88	0.003
	45	2044.35	288.84	0.006	1977.69	561.55	-0.026	2034.82	257.82	0.001

Table 16.5 Comparison of sample sizes and costs of sampling using ACS and NACS

		ACS			NACS		
Number of repetitions	Initial sample size	Expected final sample size	Expected sampling cost		Expected sample size	Expected effective sample size	Expected sampling cost
r	n	$E(n_s)$	(in dollars)		$E(n_s)$	$E(n_e)$	(in dollars)
5000	10	21.32	426.40		21.32	13.40	310.64
	15	28.22	564.40		28.22	16.43	385.04
	20	33.84	676.80		33.84	18.17	431.08
	25	38.90	778.00		38.90	19.35	464.80
	35	48.10	962.00		48.10	20.75	511.20
	45	56.53	1130.60		56.53	21.42	541.46
10000	10	21.38	427.60		21.38	13.45	311.76
	15	28.21	564.20		28.21	16.44	385.22
	20	33.87	677.40		33.87	18.19	431.54
	25	38.90	778.00		38.90	19.38	465.40
	35	48.05	961.00		48.05	20.72	510.50
	45	56.58	1131.60		56.58	21.44	541.96
20000	10	21.40	428.00		21.40	13.49	312.60
	15	28.18	563.60		28.18	16.40	384.36
	20	33.92	678.40		33.92	18.25	432.84
	25	38.93	778.60		38.93	19.38	465.46
	35	48.08	961.60		48.08	20.72	510.56
	45	56.52	1130.40		56.52	21.42	541.44
100000	10	21.39	427.80		21.39	13.47	312.18
	15	28.22	564.40		28.22	16.44	385.24
	20	33.88	677.60		33.88	18.23	432.36
	25	38.92	778.40		38.92	19.36	465.04
	35	48.06	961.20		48.06	20.73	510.72
	45	56.53	1130.60		56.53	21.41	541.26

Table 16.1 showed that as the initial sample size increased the standard error of HH and HT estimator decreased. Hence, NACS is consistent with the statistical regularity principle. The modified regression estimator is more efficient than the modified ratio estimator as shown in Table 16.1.

16.5 Exercises

1. Describe NACS design.
2. Discuss the practical situations where NACS is used.

3. Bring out the difference between ACDS and NACS.
4. Compare ACS, ACDS, and NACS.
5. Derive modified ratio type estimator of the population total used in NACS.
6. State and prove the theorem that gives an approximate variance of the estimator derived in Exercise 5. Obtain the estimate of this approximate variance.
7. Derive modified regression type estimator of the population total used in NACS.
8. State and prove the theorem that gives an approximate variance of the estimator derived in Exercise 7.
9. Define product type estimator of the population total used in NACS.
10. State and prove the theorem that gives the bias in the product estimator and its variance.
11. Define expected effective sample size and expected final sample size.
12. Derive expressions for the expected effective sample size and expected final sample size in NACS.

References

Bahl, S., Tuteja, R.K.: Ratio and product type exponential estimators. J. Inf. Optim. Sci. **12**, 159–163 (1991)

Brown, J.A.: The applications of adaptive cluster sampling to ecological studies. In: Fletcher, Manley, B.F.J. (eds.), Statistics in Ecological and Environmental Monitoring, Otago Conference Series, vol. 2, pp. 86–97. University of Otago Press, New Zealand (1994)

Chutiman, N., Chiangpradit, M.: Ratio estimator in adaptive cluster sampling without replacement of networks. J. Prob. Stat. **2014**(2), 1–6, Article ID 726398 (2014)

Lee, K.: Two phase adaptive cluster sampling with unequal probabilities selection. J. Korean Stat. Soc. **27**, 265–278 (1998)

Medina, M.H.F., Thompson, S.K.: Adaptive cluster double sampling. Biometrika **91**, 877–891 (2004)

Salehi, M.M., Seber, G.A.F.: Adaptive cluster sampling with networks selected without replacement. Biometrika **84**, 209–219 (1997a)

Salehi, M.M., Seber, G.A.F.: Two stage adaptive cluster sampling. Biometrics **53**, 959–970 (1997b)

Salehi, M.M., Seber, G.A.F.: Unbiased estimators for restricted adaptive cluster sampling. Australian and New Zealand J. Stat. **44**(1), 63–74 (2002)

Särndal, C.E., Swensson, B.: A general view of estimation for two phases of selection with applications to two phase sampling and nonresponse. Int. Stat. Rev. **55**(3), 279–294 (1987)

Särndal, C.E., Swensson, B., Wretman, J.: Model Assisted Survey Sampling. Springer Inc, New York (1992)

Chapter 17
Negative Adaptive Cluster Double Sampling

17.1 Introduction

We have discussed the methods of ACDS and NACS in detail in the previous chapter. In this chapter, we have proposed another new method for estimating the mean/total of the variable of interest. This method is a two-phase variant of the NACS obtained by combining the idea of double sampling and NACS. We have named this method as negative adaptive cluster double sampling (NACDS). Here, we assume that the auxiliary information is easily available and is less expensive. It can be seen as a variant of ACDS also. When the two variables have a positive correlation, the selection of units for observing the variable of interest based on the condition related to the auxiliary variable is justified. But, the nature of the underlined population is rare and patchy. The auxiliary information is also rare and patchy. Hence, we cannot exploit the auxiliary information at the design and estimation stage. To exploit the auxiliary information at the design and estimation stage; auxiliary information must be abundantly available, easy to measure, and less costly. In such cases, we advocate our new method, NACDS. This design covers the merits of double sampling and NACS. In NACDS, we assume that the complete auxiliary information is available and we have knowledge about the negative relationship between the variable of interest and the auxiliary variable. According to this relationship, the condition of adaptation related to the auxiliary variable is reversed. The NACDS in detail is given in Sect. 17.2.

In Sect. 17.3, we have proposed regression and ratio estimators given by Särndal and Swensson (1987), of population total of the variable of interest, using NACDS. The estimates of the variances of these estimators are also given. Section 17.4 discusses a sample survey and the results and discussion related to this method and survey.

© The Author(s), under exclusive license to Springer Nature Singapore Pte Ltd. 2021
R. Latpate et al., *Advanced Sampling Methods*,
https://doi.org/10.1007/978-981-16-0622-9_17

17.2   Negative Adaptive Cluster Double Sampling (NACDS)

Medina and Thompson (2004) introduced ACDS. It considers the auxiliary variable which is easy to measure and inexpensive. But it does not take into account the type of relationship between the two variables. If all information on auxiliary variable is available and the two variables are negatively correlated then we propose the new method, negative adaptive cluster double sampling (NACDS). It is a combination of NACS and double sampling. It is a cost-effective method. To exploit the auxiliary information at design and estimation stage; auxiliary information must be available, easy to measure, and less costly. In such cases, we advocate our new method NACDS. This design covers the merits of double sampling and NACS. According to the relationship between the two variables, the condition of adaptation related to the auxiliary variable is reversed.

17.2.1   Sampling Design and Notations

Let U=$\{U_1, U_2, \cdots, U_N\}$ be a finite population of N units. Let Y and X be the interest and auxiliary variable, respectively. They are known to be highly negatively correlated. Let X_i and Y_i, $i = 1, 2, \cdots, N$ be the values of X and Y, respectively, associated with the unit U_i. It is assumed that the information on auxiliary variable can be obtained from all the units selected in the sample. The goal is to estimate the population total of Y given by $\tau_Y = \sum_{i=1}^{N} Y_i$.

An initial sample of size n units is drawn from the population by using SRSWOR. We denote this initial sample drawn as S_0. From S_0, obtain an adaptive cluster sample S_1 by using the following procedure:

Denote the condition of interest with respect to X values by C_X. According to the negative correlation the condition is reversed for adaptation. Now following the procedure given by Thompson (1990), we add the neighbors of the units in S_0 that satisfy the condition C_X. The units to the right, left, above, and below a unit are called the neighbors of that unit. If any of these neighbors satisfy C_X then their neighbors are also added to the sample. This is continued till the neighbors not satisfying C_X are obtained. The units added to the sample S_0 adaptively which satisfy the condition C_X constitute a network. The units added to the sample S_0 adaptively which do not satisfy the condition C_X are called the edge units. The set of units in a network along with its edge units is called a cluster. The set of units included in all such clusters is called an adaptive cluster sample. We denote it by S_1. Thus, indirectly we are assuming that the condition C_X for the additional sampling and a set of neighboring units for each $U_i \in U$ have been defined.

Let K denote the number of distinct clusters formed by S_0. Mark the corresponding K clusters in the Y population and drop down the edge units to get K networks. This completes the first phase of the design.

From each of these selected networks, draw a sample by using SRSWOR. The sizes of these samples may be different. Suppose m_i denotes the number of units selected from the ith selected network. Collection of all these units selected be denoted by S_2. This completes the second phase of sampling design. Now, note the values of X and Y for all the units included in S_2. This data is used to estimate the population parameter. In this design, the X value associated with every unit in the adaptive cluster sample S_1 has to be measured. Hence, the procedure does not control the number of observations on the auxiliary variable, but only the number of observations on the survey variable.

The first-phase sample $S_1(S_1 \subset U)$ of size n_{s_1} is drawn by a design denoted by P (.) such that P (S_1) is the probability of choosing S_1. The inclusion probabilities are defined as follows:

π_k: Probability that unit k is included in $S_1 = \sum_{k \in S_1} P(S_1)$,

π_{kl}: Probability that unit k and l are included in $S_1 = \sum_{k,l \in S_1} P(S_1)$

with $\pi_{kk} = \pi_k$.

Let $\Delta_{kl} = \pi_{kl} - \pi_k \pi_l$.

We assume that $\pi_k > 0$ for all k and $\pi_{kl} > 0$ for all $k \neq l$.

Given S_1, the second phase sample $S_2(S_2 \subset S_1)$ of size n_{s_2} is drawn according to a sampling design $P(.|S_1)$ such that $P(S_2|S_1)$ is the conditional probability of choosing S_2.

The inclusion probabilities given S_1 are defined by

$\pi_{k|S_1}$: Probability that kth unit is included in S_2 given $S_1 = \sum_{k \in S_2} P(S_2|S_1)$.

$\pi_{kl|S_1}$: Probability that kth and lth units are included in S_2 given S_1

$= \sum_{k,l \in S_2} P(S_2|S_1)$.

With $\pi_{kk|S_1} = \pi_{k|S_1}$.

Let $\Delta_{kl|S_1} = \pi_{kl|S_1} - \pi_{k|S_1} \pi_{l|S_1}$.

We assume that for any S_1, $\pi_{k|S_1} > 0$, for all $k \in S_1$ and $\pi_{kl|S_1} > 0$ for all $k \neq l \in S_1$.

For all $k, l \in S_1$ and any S_1,

$\pi_k^* = \pi_k \pi_{k|S_1}$.

$\pi_{kl}^* = \pi_{kl} \pi_{kl|S_1}$.

With $\pi_{kk}^* = \pi_k^*$.

Let $\Delta_{kl}^* = \pi_{kl}^* - \pi_k^* \pi_l^*$

$\check{y}_k = \frac{y_k}{\pi_k}$

$\check{y}_k^* = \frac{y_k}{\pi_k^*}$

and

$\check{\Delta}_{kl} = \frac{\Delta_{kl}}{\pi_{kl}}$.

17.3 Different Estimators

17.3.1 Regression Estimator

Using the ideas from Särndal and Swensson (1987), we propose the regression type estimator of the population total of Y. We assume that the relationship between Y and X can be modeled through a regression model ζ such that $E_\zeta[Y_j] = X'_j\beta$ and
$$V_\zeta[Y_j] = \sigma_j^2 , j = 1, 2, \cdots, m_i; \quad i = 1, 2, \cdots, n.$$
Where Y_j's are independent
and
$$\beta = \begin{bmatrix} \beta_0 \\ \beta_1 \end{bmatrix}.$$
The model ζ is just a tool to express the relationship between Y and X in the finite population.

If all the N points (Y_i, X_i) are available, then the weighted least square estimator of β and the associated residuals are
$$\beta = \left(\frac{\sum_U X_j X'_j}{\sigma_k^2} \right)^{-1} \left(\frac{\sum_U X_j Y_j}{\sigma_k^2} \right)$$
and
$$E_j = Y_j - X'_j \beta$$
The estimators of β and E_j are given by
$$b = \left(\frac{\sum_{S_2} x_k x'_k}{\sigma_k^2 \pi_k^*} \right)^{-1} \left(\frac{\sum_{S_2} x_k y_k}{\sigma_k^2 \pi_k^*} \right)$$
and
$$e_j = y_j - \hat{y}_j = y_j - x'_j b, \quad j \in S_2.$$
Where $\pi_k^* = \pi_k \, \pi_{k|S_1}$.

The first-phase regression estimator of τ_Y is given by
$$\hat{\tau}_{Y_1 R} = \left[\sum_{S_1} \frac{\hat{y}_k}{\pi_k} + \sum_{S_2} \frac{(y_k - \hat{y}_k)}{\pi_k^*} \right]$$

Theorem 17.1 *Approximate variance of the estimator $\hat{\tau}_{Y_{1}R}$ is given by*
$$AV(\hat{\tau}_{Y_{1}R}) = \left[\sum \sum_U \Delta_{kl} \, \check{y}_k \, \check{y}_l + E_1 (\sum \sum_{S_1} \Delta_{kl|S_1} \check{E}_k \, \check{E}_l) \right].$$
Where $\check{E}_k = \frac{E_k}{\pi_k^*}$.
Unbiased estimator of $AV(\hat{\tau}_{Y_1R})$ is given by
$$\widehat{AV}(\hat{\tau}_{Y_{1}R}) = \left[\sum \sum_{S_2} \frac{\check{\Delta}_{kl} \, \check{y}_k \, \check{y}_l}{\pi_{kl|S_1}} + \sum \sum_{S_2} \frac{\Delta_{kl|S_2} \, g_{kS_2} \, g_{lS_2} \check{e}_{kS_2} \, \check{e}_{lS_2}}{\pi_{kl|S_1}} \right].$$
Where $\check{e}_{kS_2} = \frac{e_k}{\pi_k^*}$
$$g_{kS_2} = 1 + (\sum_{S_1} \frac{x_k}{\pi_k} - \sum_{S_2} \frac{x_k}{\pi_k^*})' (\sum_{S_2} \frac{x_k x'_k}{\sigma_k^2 \pi_k^*})^{-1} \frac{x_k}{\sigma_k^2}.$$

Proof The first-phase regression estimator of τ_y is given by
$$\hat{\tau}_{Y_{1}R} = \left[\sum_{S_1} \frac{\hat{y}_k}{\pi_k} + \sum_{S_2} \frac{(y_k - \hat{y}_k)}{\pi_k^*} \right].$$
The estimation error of τ_y can be written as
$$\hat{\tau}_{Y_{1}R} - \tau_y = (\sum_{S_1} \frac{y_k}{\pi_k} - \sum_U y_k) + (\sum_S \frac{E_k}{\pi_k^*} - \sum_{S_1} \frac{E_k}{\pi_k}).$$

Using the operators E_I, V_I, E_{II} and V_{II} as in Sect. 4.3, Särndal et al. (1992), we obtain

$$AV(\hat{\tau}_{Y_{1R}}) = V_I \, E_{II}(\hat{\tau}_{Y_{1R}} - \tau_y) + E_I \, V_{II}(\hat{\tau}_{Y_{1R}} - \tau_y)$$
$$= V_{FPU} + V_{SPU}.$$

Where

$$V_{FPU} = \sum \sum_U \Delta_{kl} \, \breve{y}_k \, \breve{y}_l$$
$$V_{SPU} = E_1(\sum \sum_{S_1} \Delta_{kl|S_1} \, \breve{\breve{E}}_k \, \breve{\breve{E}}_l)$$
$$\breve{\breve{E}}_k = \frac{E_k}{\pi_k^*}$$

An unbiased estimator of $AV(\hat{\tau}_{Y_{1R}})$ is given by

$$\widehat{AV}(\hat{\tau}_{Y_{1R}}) = \hat{V}_{FPU} + \hat{V}_{SPU}$$
$$= \left[\sum \sum_{S_2} \frac{\Delta_{kl} \, \breve{y}_k \, \breve{y}_l}{\pi_{kl|S_1}} + \sum \sum_{S_2} \frac{\Delta_{kl|S_2} \, g_{k_{S_2}} \, g_{l_{S_2}} \, \breve{\breve{e}}_{k_{S_2}} \, \breve{\breve{e}}_{l_{S_2}}}{\pi_{kl|S_1}} \right].$$

Hence the theorem. $\qquad\square$

17.3.2 Ratio Estimator

If we assume the above regression model without intercept then the ratio estimator of the population total of Y can be defined as

$$\hat{\tau}_{YRE} = \left(\sum_{S_1} \frac{\tau_{x_k}}{\pi_k} \right) \hat{B}.$$

Where $\hat{B} = \frac{\hat{\tau}_{yht}}{\hat{\tau}_{xht}}$.

τ_{x_k} : Total of X observations in the kth network selected in the sample by using design s_1.

$\tau_{x_k} = \sum_{S_1} x_k$

π_{I_k} = Inclusion probability of the kth network in the sample selected by using design s_I.

$\hat{\tau}_{yht}$: HT estimator of the population total of Y = $\sum_{S_2} \frac{y_k}{\pi_{i|k} \, \pi_k}$.

$\hat{\tau}_{xht}$: HT estimator of the population total of X = $\sum_{S_2} \frac{x_k}{\pi_{i|k} \, \pi_k}$.

The prediction equation is

$$\hat{y}_k = \hat{B} \, x_k \, , k \in S_1.$$

The residuals are

$$e_{k_{S_2}} = y_k - \hat{B} \, x_k.$$

Its g weights are

$$g_{k_{S_2}} = (\sum_{S_1} \breve{x}_k) / (\sum_{S_2} \breve{x}_k).$$

Where

$$\breve{x}_k = \frac{x_k}{\pi_k}$$

and

$$\breve{\breve{x}}_k = \frac{x_k}{\pi_k^*}.$$

Theorem 17.2 *The approximate variance of $\hat{\tau}_{YRE}$ can be written as*

$$AV(\hat{\tau}_{YRE}) = \left[\sum \sum_U \Delta_{kl} \, \breve{y}_k \, \breve{y}_l + E_1(\sum \sum_{S_1} \Delta_{kl|S_1} \, \breve{\breve{E}}_k \, \breve{\breve{E}}_l) \right].$$

Unbiased estimator of $AV(\hat{\tau}_{YRE})$ is given by

$$\widehat{AV}(\hat{\tau}_{YRE}) = \left[\sum \sum_{S_2} \frac{\breve{\Delta}_{kl} \, \breve{y}_k \, \breve{y}_l}{\pi_{kl|S_1}} + \sum \sum_{S_2} \frac{\Delta_{kl|S_2} \, g_{k_{S_2}} \, g_{l_{S_2}} \breve{e}_{k_{S_2}} \breve{e}_{l_{S_2}}}{\pi_{kl|S_1}} \right].$$

Proof The proof of this theorem can be given easily by substituting

$g_{k_{S_2}} = (\sum_{S_1} \breve{x}_k)/(\sum_{S_2} \breve{x}_k)$

in the proof of Theorem 17.1.

Hence the theorem.   □

17.4 Sample Survey

A sample survey was conducted by using NACDS. The area of 400 acres in the Tamhini Ghat, Maharashtra, India, was divided into 400 plots each of size 1 acre. A random sample of 12 plots was drawn from this area by using SRSWOR. The percentage of silica content of the soil (X) was measured on these selected plots. Silica is abundant in the soil from Tamhini Ghat to Mumbai. But, there are intermediate patches of laterite where the occurrence of evergreen plants is more. We considered the condition $C_X = \{X \leq 20\} = \{Percentage of silica \leq 20\}$. Further, the plots in the sample satisfying C_X were located. Then the clusters were formed around these plots by using the procedure given by Thompson (1990). Each plot with $\{X > 20\}$ and selected in the initial sample formed a cluster of size one. Here the clusters were formed by using auxiliary information and the domain knowledge of silica content and evergreen plants. The two variables, percentage of silica content X and number of evergreen plants Y are negatively correlated. After forming such clusters in the X population, the edge units of clusters of size more than one were dropped to get networks. The networks are formed by using percent of silica content. The corresponding networks of the number of evergreen plants are located. Figures 17.1 and 17.2 illustrate this methodology.

These plots formed the first-phase sample S_1. Let K denote the number of distinct networks represented in this sample. A random sample of m_i (say), $(i = 1, 2, \cdots , K)$ units was drawn from the ith network among these K networks by using SRSWOR. The collection of all so selected units formed the second phase sample S_2. In our study, there were 12 networks formed in S_1. We took $m_1 = m_2 = m_3 = 2, m_4 = 3, m_5 = 4, m_6 = 2, m_7 = 2, m_8 = 2, m_9 = m_{10} = 4, m_{11} = 3$ and $m_{12} = 0$.

This set of units formed S_2. Values of the variables X and Y corresponding to the plots included in the second phase sample were recorded together to form bivariate data. Using this data, the total number of evergreen plants in that area was estimated by using the given estimators.

Results and Discussion:

For the computational efficiency in estimation, r number of repetitions were performed, where r varied as 5000, 10000, 20000, and 100000. We considered the initial sample sizes as 5, 10, 15, 20, and 25 for each repetition.

24	25	86	60	52	35	65	50	60	1	22	23	83	48	30	56	43	52	1	4
40	30	30	75	18	19*	55	30	4	14	38	27	27	14	14	49	23	6	6	10+
45	48	56	23	15	17	53	30	13	12*	43	45	53	11	12*	47	23	7	8	7
47	47	23	25	80	60	45	45	35	70	45	44	20	76	55	39	38	27	61	34
48	50	25	35	57	68	40	23	80	40	46	47	22	53	63	34	26	72	31	37
49	43	36	65	58	58	90	45	90	30	47	40	33	54	53	84	38	82	21	30
45	35	56	85	19	30	18	18	40	50	43	32	53	25	25	42	41	32	41	22
48	53	65	55	13	16*	15	18	30	60	46	50	62	29	17	18	51	22	51	40
70	30	17	18	15	48	44	44	35	50	68	27	24	29	43	19*	12	28	41	27
30	30	18	17	15	43	36	50	80	36	28	27	25	24	22	14	43	72	27	27
29	31	93	68	61	45	66	52	63	25	27	29	93	42	32	59	47	57	27	42
45	36	37	83	27	29	56	32	37	48	43	36	37	88	26	52	27	41	43	58
50	54	63	31	20	27	54	32	76	77	16	18	63	85	24	50	27	7	14	20
52	53	30	15	18	20	46	47	38	75	18*	21	20	70	57	42	42	12	11*	9
53	57	32	19	18*	70	41	25	83	45	51	20	32	47	65	37	20	10	18	15
54	50	43	73	16	78	91	47	93	24	26	42	43	59	55*	87	42	87	48	38
50	42	63	93	67	68	29	28	20	19	12	24	63	33	27	35	35	37	34	30
53	60	72	63	28	40	26	20	19*	13	11	23	70	35	29	65	85	5	8	10
75	37	24	26	22	26	45	46	12	10	9	28	24	45	45	22	26	14	9*	5
35	37	25	25	24	58	37	52	18	16	14	48	25	52	24	77	47	77	9	11

* in a square indicates selection in initial sample.

Fig. 17.1 Silica (SiO_2) percentage on the different plots of Tamhini Ghat

The estimated population total over r repetitions is given by

$$\hat{T}_Y = \frac{\sum_{i=1}^r \hat{T}_{Y_i}}{r}.$$

Where $\hat{T}_{Y_i}$ denotes the estimated value of an estimator of the population total of the variable Y for the ith repetition.

The estimated mean square error of the estimator of population total of the variable Y is given by:

$$\widehat{MSE}(\hat{T}_Y) = \frac{\sum_{i=1}^r (\hat{T}_{Y_i} - T_Y)^2}{r}.$$

Table 17.1 gives the values of $\hat{T}_Y$ and $\widehat{SE}(\hat{T}_Y)$ using regression estimator in NACS and Ratio and Regression estimators in NACDS.

The final adaptive sample size in NACS is denoted by n_s and it is given as $n_s = n+$ number of units added by adaptation procedure.

After dropping down the edge units, the left over sample size is called as the effective sample size (n_e).

Effective sample size $(n_e) = \sum_{j=1}^{n_s} \delta_{C_x}(j)$, where

$$\delta_{C_x}(j) = \begin{cases} 1 & \text{; if } U_j \text{ satisfies the condition } C_x \\ 0 & \text{; } otherwise \end{cases}$$

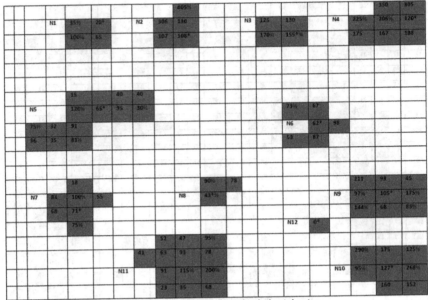

N1 to N12 denote the network numbers. % in a plot indicates the selection at phase two.

Fig. 17.2 Number of evergreen plants observed on the plots in the population

Effective sample size at the second phase is denoted by n_e^* and it is given as

$$n_e^* = \sum_{i=1}^{K} \delta_i \, m_i,$$

where

$$\delta_i = \begin{cases} 1 & ; \text{if } i^{th} \text{ cluster is selected in the sample in the} \\ & \quad first \; phase \\ 0 & ; otherwise \end{cases}$$

$$i = 1, 2, \cdots, K.$$

The total expected sample size for the NACDS is the sum of expected sample size at first stage and second stage.

Expected sample size for NACDS $= n_s + \sum_{i=1}^{K} \delta_i \, m_i$

The results are given in Table 17.2.

It was observed that the expected final sample size and the expected effective sample size increase as the initial sample size increases. But a remarkable reduction in the expected effective sample size at the second phase of the design was observed.

In general, the relation $E(n_s) > E(n_e) > E(n_e^*)$ was observed.

If the costs of sampling per unit with respect to X and Y variables are C_1 and C_2, respectively, then the total cost of sampling in ACS design $= (C_1 + C_2) \, E(n_s)$. The total cost of sampling in NACS design $= C_1 \, E(n_s) + C_2 \, E(n_e)$ On the other hand, the total cost of sampling in NACDS $= C_1 \, E(n_s) + C_2 \, E(\sum_{i=1}^{K} \delta_i \, m_i)$. It can be easily seen that $\sum_{i=1}^{k} m_i \leq n_e$.

Table 17.1 Comparison of NACS with NACDS

Number of repetitions	Initial sample size	Regression estimator in NACS		Regression estimator and ratio estimator in NACDS			
r	n	$\hat{\tau}_y$	$\hat{SE}(\hat{\tau}_y)$	$(\hat{\tau}_y)_{Reg}$	$\hat{SE}(\hat{\tau}_y)_{Reg}$	$(\hat{\tau}_y)_{Ratio}$	$\hat{SE}(\hat{\tau}_y)_{Ratio}$
5000	5	9092.12	9813.04	9139.53	9295.75	9555.85	10461.68
	10	9073.51	6291.11	9095.31	6493.03	9206.55	6778.47
	15	9067.73	4982.09	9073.74	5154.54	9230.12	5373.81
	20	9057.88	4200.33	9057.41	4378.46	9106.03	4595.43
	25	9041.63	3588.37	9043.17	3809.32	9144.12	3992.87
10000	5	9091.96	9448.93	9117.21	9239.01	9458.85	10383.60
	10	9081.00	6252.73	8971.90	6440.05	9223.40	6738.29
	15	9055.30	4937.97	8959.31	5147.22	9114.49	5328.22
	20	9028.43	4142.43	9098.10	4357.54	9173.93	4577.09
	25	9020.88	3621.87	9062.58	3808.90	9110.27	3964.64
20000	5	9114.37	9619.88	9095.66	9226.77	9421.81	10328.04
	10	9090.99	6226.61	9086.71	6418.04	9224.23	6724.55
	15	9035.02	4981.76	9071.33	5134.35	9097.89	5313.16
	20	9022.01	4158.73	9056.96	4346.72	9047.11	4529.84
	25	9039.70	3627.15	9042.25	3788.86	9107.88	3947.24
100000	5	9196.61	9470.26	9069.59	9203.23	9431.05	10295.4
	10	9138.80	6097.93	9067.15	6350.09	9222.82	6676.93
	15	9077.27	4816.89	9052.93	5101.60	9115.33	5266.15
	20	9056.61	4079.37	9049.11	4332.31	9118.94	4496.88
	25	9047.59	3566.08	9032.29	3748.15	9136.57	3902.81

Hence, the cost of sampling in NACDS is usually lesser than that in NACS.

We have calculated the expected costs of sampling in ACS, NACS, and NACDS and the results are given in Table 17.3.

It was observed that, Expected Sampling Cost in ACS > Expected Sampling Cost in NACS > Expected Sampling Cost in NACDS. Thus the new design is cost-effective as compared to NACS and ACS. Also, the NACS estimator is more precise as compared to NACDS. The regression estimator is more precise as compared to the ratio estimator for NACDS. Similar to ACDS, in our method also we can start selecting the second phase units before completing the selection of the first-phase units. Due to this, the survey does not require us to visit the same plots again. It saves the traveling cost, time required to collect the information. In the first phase, we consider only auxiliary information. Hence, there is considerable cost saving as compared to ACS.

In NACS, it is required to take observations on interest variable related to all units included in the networks. In NACDS, we take a subsample from each of the networks

Table 17.2 Estimated values of final sample sizes, effective sample sizes, and effective sample sizes for second phase of NACDS

Number of repetitions	Initial sample size	Expected final sample size	Expected effective sample size	Expected effective sample size at second phase
r	n	n_s	n_e	n_e^*
5000	5	12.99	8.84	3.07
	10	24.32	16.19	5.65
	15	35.33	23.26	8.12
	20	45.31	29,28	10.26
	25	55.14	35.18	12.31
10000	5	12.81	8.65	3.01
	10	24.38	16.27	5.68
	15	35.51	23.48	8.20
	20	45.66	29.67	10.37
	25	54.99	35.04	12.28
20000	5	12.97	8.82	3.06
	10	24.43	16.31	5.69
	15	35.53	23.50	8.21
	20	45.55	29.57	10.34
	25	54.99	35.04	12.29
100000	5	12.84	8.67	3.02
	10	24.57	16.47	5.74
	15	35.35	23.29	8.14
	20	45.45	29.45	10.31
	25	54.99	35.05	12.29

discovered at the first phase. Usually, taking observations on the interest variable is costly hence NACDS is definitely cost-effective.

Remark: The estimators used in NACDS and ACDS are the same. Hence, in terms of precision, both the methods are equally efficient.

The domain knowledge of the population is utilized to conduct the survey of rare population. If the study variable is rare and we have abundant auxiliary information then there is a negative correlation. The auxiliary information is utilized at the design and estimation stages in NACDS. The regression estimator is more precise as compared to the ratio estimator for NACDS. But, the authors presented only the simulation study. Also, they haven't presented the real-world example and implementation. ACDS and NACDS are equally efficient. The expected effective sample size for NACDS is very small as compared to ACS and NACS. Hence, NACDS is cost-effective. There is only one condition for adaptation on the auxiliary variable.

Table 17.3 Expected sampling costs in ACS, NACS and NACDS

Number of repetitions (r)	Initial sample size (n)	Expected sampling cost in ACS	Expected sampling cost in NACS	Expected sampling cost in NACDS
5000	5	324.75	259.97	115.72
	10	608.00	477.71	214.21
	15	883.25	687.49	308.99
	20	1132.75	867.93	392.43
	25	1378.50	1044.92	473.17
10000	5	320.25	254.68	113.68
	10	609.50	479.89	215.14
	15	887.75	693.53	311.53
	20	1141.50	878.73	396.23
	25	1374.75	1040.97	471.97
20000	5	324.25	259.41	115.41
	10	610.75	481.04	215.54
	15	888.25	694.09	311.84
	20	1138.75	875.90	395.15
	25	1374.75	1040.97	472.22
100000	5	321.00	255.27	114.02
	10	614.25	485.46	217.21
	15	883.75	688.30	309.55
	20	1136.25	872.60	394.10
	25	1374.75	1041.22	472.22

If the type of correlation between the auxiliary variable and study variable is known to be negative, it is more suited methodology for environmental, ecological, forestry, social science, and medical surveys.

17.5   Exercises

1. Write the situations in which NACDS is used.
2. Describe NACDS design in detail.
3. Obtain regression estimator of the population total under NACDS design.
4. State and prove the theorem that gives the approximate variance of the regression estimator of the population total under NACDS design. Also, obtain its unbiased estimator.
5. Obtain ratio estimator of the population total under NACDS design.

6. State and prove the theorem that gives the approximate variance of the ratio estimator of the population total under NACDS design. Also, obtain its unbiased estimator.
7. Derive an expression for the total cost of sampling in NACDS design.
8. Compare NACDS with NACS in terms of efficiency and cost.

References

Medina, M.H.F., Thompson, S.K.: Adaptive cluster double sampling. Biometrika **91**, 877–891 (2004)

Särndal, C.E., Swensson, B.: A general view of estimation for two phases of selection with applications to two phase sampling and nonresponse. Int. Stat. Rev. **55**(3), 279–294 (1987)

Särndal, C.E., Swensson, B., Wretman, J.: Model Assisted Survey Sampling. Springer Inc, New York (1992)

Thompson, S.K.: Adaptive cluster sampling. J. Am. Stat. Assoc. **85**(412), 1050–1058 (1990)

Chapter 18
Two-Stage Negative Adaptive Cluster Sampling

18.1 Introduction

An environmentalist or ecologist often comes across a population that is rare and clustered. The information is completely available on some auxiliary variable which is inexpensive and easy to obtain. On the basis of the domain knowledge about the situation, it is known that the auxiliary variable and the variable of interest have a high negative correlation. In such a situation, if one tries to use the traditional sampling design to estimate the population parameters then very poor estimates may be obtained. We have discussed the methods of NACS and NACDS for such situations in Chap. 16 and 17.

Salehi and Seber (1997a, b) used two-stage sampling procedure in adaptive sampling to overcome the problems associated with ACS as discussed in Chap. 11. But they divided the population under study into hypothetical clusters before the two-stage sampling procedure starts. This is hardly achieved in practice. Muttlak and Khan (2002) proposed two-stage ACS design. In this method, first, the population is divided into two types of clusters: large and small, by using ACS, a two-stage sample is drawn from the large networks and a single-stage sample is drawn from the rest. But it may become expensive, if taking observations on interest variable is costly. However the method of Muttlak and Khan (2002) overcomes the shortcoming of the Salehi and Seber (1997a, b) method. In such a situation, the use of an easy to measure auxiliary variable is expected to reduce the cost of sampling.

In keeping the view above, the two-stage NACS design has been introduced and described in Sect. 18.2. In this design, an initial sample is drawn from the population using SRSWOR at the first stage. Using the auxiliary information, adaptive clusters are discovered with the help of the initial sample. At the second stage, a pre-fixed number of units are drawn from each of the clusters formed. The values of the auxiliary variable and the variable of interest of the selected units at the second stage are recorded. Section 18.3 discusses the different proposed estimators of population total of the variable of interest by using two-stage NACS. Variances of these estimators along with their estimators are also given. Section 18.4 discusses an application of the proposed method and compares the performances of the different proposed estimators and the results and discussion on this application.

18.2 Two-Stage NACS

Consider a rare and clustered population with respect to a variable of interest Y. Let X be an auxiliary variable that is highly negatively correlated with Y. Assume that the values of X for all population units are easily available. The interest is to estimate the population total τ_y of the variable Y. We propose the following sampling design and termed it the "Two- Stage Negative Adaptive Cluster Sampling" design.

Divide the population under study into N units (plots) of equal size and shape. Draw an initial random sample of size n_0 by using SRSWOR from this population. Record the values of the X variable corresponding to these units. Let C_x denote the condition related to the X variable. Check whether each of the units selected in the above sample satisfies C_x. If a unit satisfies C_x then add the units to the left, right, above, and below it, to the sample. These four units together constitute the neighborhood of that unit. We assume that the neighborhoods are symmetric. Further, if any of these neighborhoods satisfy the condition C_x, then add the neighbors of those units also to the sample. Continue this way adding the neighbors to the sample till the neighbors not satisfying C_x are observed. These units are called the edge units of a network. The units connected to each other satisfying the condition C_x constitute a network. The set of units in a network along with the edge units constitute a cluster. Thus due to the adaptive process, networks of units satisfying C_x are formed. Note that the units not satisfying C_x but included in the initial sample are treated as networks of size one. The number of units included in the initial sample along with those adaptively added (including the edge units) is called the final sample size n_S. Record the values of the X variable corresponding to all network units adaptively added to the sample. Assume that due to adaptive procedure, K distinct networks are formed in the final sample. Treat this set of K networks as a sample of K clusters drawn from the K^* clusters in the population. The corresponding networks of the variable of interest (Y) are called the (primary stage units) PSUs. This completes the first stage of the design.

In the second stage, draw a random sample of unequal size n_k from the kth network included in the final sample, ($k = 1, 2, \cdots, K$), by using SRSWOR. Note that, for the networks of size one that gets included in the sample, $n_k = 1$. The values of n_k for the other networks can be fixed by the surveyor as per his judgment. Now record the values of the Y variable for all these units selected at the second stage. Thus, bivariate data on (X, Y) is obtained for all the units included at the second stage.

Two-stage NACS design differs from two-stage ACS design. In a two-stage ACS design, it is assumed that the population is already divided into known different groups. At the first stage in that design, a random sample is drawn from these predefined groups. At the second stage, an initial sample is drawn randomly from each of these selected groups and then adaptive procedure by Thompson (1990) is used. The values of the variable of interest are recorded for the units selected in the final sample. No auxiliary information is used in that design.

In two-stage NACS design, at the first stage, using the auxiliary information and Thompson's adaptive procedure, a set of clusters in the population is discovered. These clusters serve as PSUs. At the second stage, from each of these clusters a random sample of units is selected. The set of these selected units serve as secondary -stage units (SSU's). The number of units selected from the PSUs may be different.

A two-stage ACS design may lead to a poor estimate of the population total if most of the clusters selected at the first stage are empty. This risk has been covered in two-stage NACS design.

By using ACS at the first stage, we discover the majority of non-empty clusters. At the second stage, the samples are drawn from the majority of non-empty clusters. Thus the risk of getting a poor estimate is reduced in our design.

Apparently, the proposed design looks like adaptive cluster double sampling as discussed in Chap. 12. But this design does not have complete auxiliary information. In two-stage NACS design, it is assumed that the information related to the auxiliary variable is available for all units in the population. If complete auxiliary information is known then we get a more precise estimate of the population. It is assumed that the two variables are negatively correlated. Hence, it is called as two-stage NACS design. Also, we can control the final sample size like ACDS. There is a substantial reduction in sample size for the variable of interest Y.

18.3   Different Estimators

The following estimators are used to estimate the population total.

18.3.1   Composite HT Estimator

Consider a population $U = \{1, 2, \cdots, i, \cdots, N\}$ which is partitioned into N_K networks, denoted as $\psi_1, \psi_2, \cdots, \psi_k, \cdots, \psi_{N_K}$. The set of networks is symbolically represented as $\psi_K = \{1, 2, \cdots, k, \cdots, N_K\}$. An initial sample of size n_0 is drawn from U by using SRSWOR. For the units satisfying the condition C_x, we apply the adaptive procedure to get the networks. The corresponding networks of the study variable (Y) are called as the PSUs. We denote by the notation S_I, the set of networks formed in the sample.

For every $k \in S_I$, a sample S_k of elements is drawn from $\psi_K (S_k \subset \psi_K)$ by using SRSWOR to get the SSUs. The size of S_k for networks of sizes more than one are unequal. The resulting sample of elements is denoted by $S = \cup_{k \in S_I} S_k$. The total number of elements in S is given by $n_S = \sum_{k \in S_I} n_{S_k}$.

Inclusion probability of kth PSU in the population at the first stage is

$$\pi_{Ik} = 1 - \left[\frac{\binom{N-z_k}{n_0}}{\binom{N}{n_0}} \right].$$

Where z_k is the number of units from the kth network selected at the first stage and n_0 is the number of units selected in the initial sample.

Inclusion probability of kth and lth PSU in the population at the first stage is

$$\pi_{Ikl} = 1 - \left[\frac{\binom{N-z_k}{n_0} + \binom{N-z_l}{n_0} - \binom{N-z_k-z_l}{n_0}}{\binom{N}{n_0}} \right]$$

Let $\Delta_{Ikl} = \pi_{Ikl} - \pi_{Ik}\pi_{Il}$.

With $\Delta_{Ikk} = \pi_{Ik}(1 - \pi_{Ik})$ and $\check{\Delta}_{Ikl} = \frac{\Delta_{Ikl}}{\pi_{Ikl}}$.

The inclusion probabilities of the units available at the second stage of the design (SSUs) are given as

$\pi_{i|k}$- Inclusion probability of ith unit from the kth PSU selected at the first stage.

$\pi_{ij|k}$- Inclusion probability of ith and jth unit from the kth PSU selected at the first stage.

$\Delta_{ij|k} = \pi_{ij|k} - \pi_{i|k}\pi_{j|k}$.

With $\Delta_{ii|k} = \pi_{i|k}(1 - \pi_{i|k})$ and $\Delta_{ij|k} = \frac{\Delta_{ij|k}}{\pi_{ij|k}}$

π_i - The inclusion probability of the ith unit in the sample is $\pi_{Ik}\pi_{i|k}$, if $i \in \psi_k$.

and

$$\pi_{ij} = \begin{cases} \pi_{Ik}\pi_{i|k} & ; \text{if } i = j \in \psi_k \\ \pi_{Ik}\pi_{ij|k} & ; \text{if } i \text{ and } j \in \psi_k, \ i \neq j \\ \pi_{Ikl}\pi_{i|k}\pi_{j|l} & ; \text{if } i \in \psi_k, \ j \in \psi_l, \ k \neq l \end{cases}$$

We propose the Composite Horvitz–Thompson (HT) estimator as

$$\hat{\tau}_{CHT} = \sum_{S_I} \frac{\hat{\tau}_{htk}}{\pi_{Ik}} \tag{18.1}$$

Where,

$\hat{\tau}_{htk} = \sum_{S_k} \check{y}_{i|k}$ is the HT estimator of τ_k with respect to second stage of the design.

τ_k is the sum of Y values in the kth network in the population.

Where

$\check{y}_{i|k} = \frac{y_i}{\pi_{i|k}}$

Theorem 18.1 *An unbiased estimator of variance of* $\hat{\tau}_{CHT}$ *is given by*

$$\hat{V}(\hat{\tau}_{CHT}) = \sum_{S_I}\sum \check{\Delta}_{Ikl} \frac{\hat{\tau}_{htk}}{\pi_{Ik}} \frac{\hat{\tau}_{htl}}{\pi_{Il}} + \sum_{S_I} \frac{\hat{V}_k}{\pi_{1k}^2}. \tag{18.2}$$

Proof In order to facilitate the proof, we use the following notations:

$$E_I E_{II}(\hat{\tau}_{CHT}) = E_{P_I}\left[E(\hat{\tau}_{CHT}|S_1)\right].$$
$$V_I E_{II}(\hat{\tau}_{CHT}) = V_{P_I}\left[E(\hat{\tau}_{CHT}|S_1)\right].$$
$$E_I V_{II}(\hat{\tau}_{CHT}) = E_{P_I}\left[V(\hat{\tau}_{CHT}|S_1)\right].$$

Here the subscript I indicates the expected value or variance with respect to the design $P_I(.)$ used in stage I and II indicates the conditional expected value or conditional variance with respect to the design used in stage two given S_I.

With these notations, we have

$$E_{II}(\hat{\tau}_{CHT}) = E(\hat{\tau}_{CHT}|S_1)$$
$$= \sum_{S_1} E_{P_I}\left(\tfrac{\hat{\tau}_{htk}}{\pi_{Ik}}\right)$$
$$= \sum_{S_1} \tfrac{\tau_{htk}}{\pi_{Ik}}$$
$$= \sum_{S_1} \check{\tau}_{htk}.$$

$$V_{II}(\hat{\tau}_{CHT}) = V(\hat{\tau}_{CHT}|S_1)$$
$$= \sum_{S_1} V\left(\tfrac{\hat{\tau}_{htk}}{\pi_{Ik}}|s_1\right)$$
$$= \sum_{S_1} V_{P_I}\left(\tfrac{\hat{\tau}_{htk}}{\pi_{Ik}}\right)$$
$$= \sum_{S_1} \tfrac{V_k}{\pi_{Ik}^2}.$$

$$V(\hat{\tau}_{CHT}) = V_I E_{II}(\hat{\tau}_{CHT}) + E_I V_{II}(\hat{\tau}_{CHT})$$
$$= V_I\left(\sum_{S_1}\check{\tau}_{htk}\right) + E_I\left(\sum_{S_1}\tfrac{V_k}{\pi_{Ik}^2}\right)$$
$$= \sum\sum_{U_1} \Delta_{Ikl}\check{\tau}_{htk}\check{\tau}_{htl} + \sum_{U_1}\tfrac{V_k}{\pi_{Ik}}$$
$$= V_{PSU} + V_{SSU}.$$

Further, to obtain an unbiased estimator of $V(\hat{\tau}_{CHT})$ we proceed as follows:

$$E_{II}(\hat{\tau}_{htk}.\hat{\tau}_{htl}) = \tau_{htk} + V_k \qquad for \ \ k = l$$
$$= \tau_{htk}.\tau_{htl} \qquad for \ \ k \neq l.$$

Consider

$$a = E\left(\sum\sum_{S_1}\check{\Delta}_{Ikl}.\tfrac{\hat{\tau}_{htk}}{\pi_{Ik}}.\tfrac{\hat{\tau}_{htl}}{\pi_{Il}}\right)$$
$$= E_I\left[\sum\sum_{S_1}\check{\Delta}_{Ikl}.\tfrac{E_{II}(\hat{\tau}_{htk}.\hat{\tau}_{htl})}{\pi_{Ik}.\pi_{Il}}\right]$$
$$= E_I\left[\sum\sum_{S_1}\check{\Delta}_{Ikl}.\tfrac{\tau_{htk}.\tau_{htl}}{\pi_{Ik}.\pi_{Il}}\right] + E_I\left[\sum_{S_1}\check{\Delta}_{Ikk}.\tfrac{V_k}{\pi_{Ik}^2}\right]$$
$$= \sum\sum_{U_1}\Delta_{Ikl}\check{\tau}_{htk}\check{\tau}_{htl} + \sum_{U_1}\left(\tfrac{1}{\pi_{Ik}} - 1\right)V_k.$$

Thus,

$$a = V_{PSU} + \sum_{U_1}\left(\tfrac{1}{\pi_{Ik}} - 1\right)V_k$$

Let $b = E\left[-\sum_{S_1}\tfrac{1}{\pi_{Ik}}.\left(\tfrac{1}{\pi_{Ik}} - 1\right)\hat{V}_k\right]$

$$= -E_I\left[\sum_{S_1}\tfrac{1}{\pi_{Ik}}.\left(\tfrac{1}{\pi_{Ik}} - 1\right)E_{II}(\hat{V}_k)\right]$$
$$= -E_I\left[\sum_{S_1}\tfrac{1}{\pi_{Ik}}.\left(\tfrac{1}{\pi_{Ik}} - 1\right)V_k\right]$$
$$= -\sum_{U_1}\left(\tfrac{1}{\pi_{Ik}} - 1\right)V_k$$

a and b together imply that $E(\hat{V}_{PSU}) = V_{PSU}$.

Now,

$$E(\hat{V}_{SSU}) = E_I\left[\sum_{S_1}\tfrac{E_{II}(\hat{V}_k)}{\pi_{Ik}^2}\right]$$
$$= E_I\left(\sum_{S_1}\tfrac{V_k}{\pi_{Ik}^2}\right)$$

$$= \sum_{U_1} \frac{V_k}{\pi_{Ik}^2}$$
$$= V_{SSU}.$$

Thus,

$$E\left[\hat{V}(\hat{\tau}_{CHT})\right] = E\left[\hat{V}_{PSU} + \hat{V}_{SSU}\right]$$
$$= V_{PSU} + V_{SSU}$$
$$= V(\hat{\tau}_{CHT}).$$

Hence, an unbiased estimator of $V(\hat{\tau}_{CHT})$ is given by

$$\hat{V}(\hat{\tau}_{CHT}) = \sum\sum_{S_1} \check{\Delta}_{Ikl} \cdot \frac{\hat{\tau}_{hik}}{\pi_{Ik}} \cdot \frac{\hat{\tau}_{hil}}{\pi_{Il}} + \sum_{S_1} \frac{V_k}{\pi_{Ik}^2}.$$

Hence the theorem. ☐

18.3.2 Two-Stage Regression Estimator

We assume that the auxiliary information x_i is available on all units in the population and the subsampling is carried out from the selected clusters. The two-stage samples are drawn by using NACS and model ζ is fitted on the basis of data points (x_i, y_i) corresponding to units $i \in S$, where S is the total two-stage sample.

Thus, $S = \cup_{k \in S_I} S_k$.

The final sample size is $n_S = \sum_{S_I} n_{S_k}$.

The population point scatter (y_i, x_i), $i = 1, 2, \cdots, N$ can be described by the general linear model ζ such that $E_\zeta(y_i) = x_i' \beta$ and $V_\zeta(y_i) = \sigma_i^2$.

y_i is the value of the characteristic of interest variable for the ith unit in the sample and y_i's are uncorrelated under the model.

In a hypothetical population fit of the above model to all N points (y_i, x_i), $i = 1, 2, \cdots, N$ the weighted least square estimator,

$$B = (\sum_U \frac{x_i x_i'}{\sigma_i^2})^{-1} (\sum_U \frac{x_i y_i}{\sigma_i^2}).$$

In terms of more familiar regression analysis notations, we can express

$$B = (X \Sigma^{-1} X')^{-1} (X \Sigma^{-1} Y).$$

Where $X = (x_1, x_2, \cdots, x_N)'$, $Y = (y_1, y_2, \cdots, y_N)'$ and

$$\Sigma = \begin{bmatrix} \sigma_1^2 & 0 & 0 & \dots & 0 \\ 0 & \sigma_2^2 & 0 & \dots & 0 \\ \vdots & \vdots & \vdots & \ddots & \vdots \\ 0 & 0 & 0 & \dots & \sigma_N^2 \end{bmatrix}$$

From the regression theory, it is known that B is the best linear unbiased estimator of β under this model. Note that in the above model, the model weight $\frac{1}{\sigma_i^2}$ is attached to the ith observation.

We say that B corresponds to the hypothetical population fit of the model ζ to the population data points (x_i, y_i), $i = 1, 2, \cdots, N$.

The population quantity B is unknown. It can be estimated by using sample data. By applying the estimation principle, we can write B as

$B = T^{-1} t.$

Where

$T = \sum_{\psi K} \frac{x_i x_i'}{\sigma_i^2}$

and

$t = \sum_{\psi K} \frac{x_i y_i}{\sigma_i^2}$

The ω estimators of T and t are

$\hat{T} = \sum_S \frac{x_i x_i'}{\sigma_i^2 \pi_i}$

and

$\hat{t} = \sum_S \frac{x_i y_i}{\sigma_i^2 \pi_i}$

$\hat{B} = T^{-1} \hat{t} = \left(\sum_S \frac{x_i x_i'}{\sigma_i^2 \pi_i} \right)^{-1} \left(\sum_S \frac{x_i y_i}{\sigma_i^2 \pi_i} \right).$

$\hat{B}$ estimates B with certain bias. B in turn estimates the model parameter β under a hypothetical complete enumeration. In the above expression of $\hat{B}$ the sampling weights are $\frac{1}{\pi_i} = \frac{1}{\pi_{Ii} \pi_{k|i}}$.

The fitted values in the above model are given by

$y_i = x_i' \hat{B}.$

The residuals are given by $e_{is} = y_i - \hat{y}_i$.

The values $\hat{y}_i$ can be calculated for all units in the population. The residuals can be calculated only for the units included in the sample.

The two-stage regression estimator of the population total τ_y is

$$(\hat{\tau}_y)_{Reg} = \sum_{\psi K} \hat{y}_i + \sum_S \frac{e_{is}}{\pi_i}$$

$$= \sum_{\psi K} \sum_{\psi_k} \hat{y}_i + \sum_S \frac{(y_i - \hat{y}_i)}{\pi_i}$$

$$= \sum_{\psi K} \sum_{\psi_k} x_i' \hat{B} + \sum_S \frac{y_i}{\pi_i} - \sum_S \frac{\hat{y}_i}{\pi_i} \qquad (18.3)$$

$$= \sum_{\psi K} \sum_{\psi_k} x_i' \hat{B} + \sum_S \frac{y_i}{\pi_i} - \sum_S \frac{x_i' \hat{B}}{\pi_i}.$$

Consider the first term in (18.3).

$\sum_{\psi K} \sum_{\psi_k} x_i' \hat{B} = \sum_{\psi K} \tau_{xk} \hat{B}.$
Where $\tau_{xk} = \sum_{\psi_k} x_i.$

Consider the third term in (18.3)

$\sum_S \frac{x_i' \hat{B}}{\pi_i} = \sum_{S_I} \frac{1}{\pi_{Ik}} \sum_{S_k} \frac{x_i'}{\pi_{i|k}} \hat{B}$

$\quad = \sum_{S_I} \frac{\hat{\tau}_{htk}}{\pi_{Ik}} \hat{B}.$

Where $\hat{\tau}_{htk} = \sum_{S_k} \frac{x_i}{\pi_{i|k}}.$

Substituting these values in (18.3) we get

$$(\hat{\tau}_y)_{Reg} = \sum_S \frac{y_i}{\pi_i} + \left[\sum_{\psi_K} \tau_{xk} - \sum_{S_I} \frac{\hat{\tau}_{htk}}{\pi_{1k}} \right] \hat{B}. \tag{18.4}$$

We have $\hat{B} = (\sum_S \frac{x_i x_i'}{\sigma_i^2 \pi_i})^{-1} (\sum_S \frac{x_i y_i}{\sigma_i^2 \pi_i})$.

Hence, $\hat{B} = T^{-1} (\sum_S \frac{x_i y_i}{\sigma_i^2 \pi_i})$.

Substituting this value in (18.4) we get

$$\begin{aligned}
(\hat{\tau}_y)_{Reg} &= \sum_S \frac{y_i}{\pi_i} + \left[\sum_{\psi_K} \tau_{xk} - \sum_{S_I} \frac{\hat{\tau}_{htk}}{\pi_{Ik}} \right] \hat{T}^{-1} (\sum_S \frac{x_i y_i}{\sigma_i^2 \pi_i}) \\
&= \sum_S (1 + [\sum_{\psi_K} \tau_{xk} - \sum_{S_I} \frac{\hat{\tau}_{htk}}{\pi_{1k}}]) T^{-1} \frac{x_i}{\sigma_i^2} \frac{y_i}{\pi_i} \\
&= \sum_S g_{is} \frac{y_i}{\pi_i}.
\end{aligned} \tag{18.5}$$

Where g weights are
$$g_{is} = (1 + [\sum_{\psi_K} \tau_{xk} - \sum_{S_I} \frac{\hat{\tau}_{htk}}{\pi_{1k}}]) T^{-1} \frac{x_i}{\sigma_i^2}.$$

Theorem 18.2 *An unbiased estimator of $V(\hat{\tau}_y)_{Reg}$ is given by*
$$\hat{V}(\hat{\tau}_y)_{Reg} = \sum\sum_{S_I} \check{\Delta}_{Ikl} \frac{\hat{\tau}_{Ek}}{\pi_{Ik}} \frac{\hat{\tau}_{El}}{\pi_{Il}} + \sum_{S_I} \frac{\hat{V}_{Ek}}{\pi_{Ik}}.$$

Proof The estimation procedure of variance of regression estimator is given by Särndal et al. (1992).
$$\begin{aligned}
V(\hat{\tau}_y)_{Reg} &= V((\hat{\tau}_y)_{Reg} - \tau_y) \\
&= V_I E_{II}[(\hat{\tau}_y)_{Reg} - \tau_y] + E_I V_{II}[(\hat{\tau}_y)_{Reg} - \tau_y] \\
&= V_I(Q_{S_I}) + E_I V_{II}(R_S) \\
&= V_{PSU} + V_{SSU}.
\end{aligned}$$

Subscript I indicates expected value or variance with respect to the design $P_I(.)$ used in stage I and II indicates the conditional expected value or conditional variance with respect to the set of designs $P_i(.), i \in S_I$, used in stage two given S_I.

Now consider
$$\begin{aligned}
(\hat{\tau}_y)_{Reg} - \tau_y &= \sum_S g_{is} \frac{y_i}{\pi_i} - \sum_{\psi_K} y_i \\
&= (\sum_{S_I} \frac{\tau_{Ek}}{\pi_{Ik}} - \sum_{\psi_K} \tau_{Ek}) + (\sum_{S_I} \frac{1}{\pi_{Ik}} \sum_{S_k} g_{is} \frac{E_i}{\pi_{i|k}} - \sum_{\psi_K} E_i) \\
&= Q_{S_I} + R_S.
\end{aligned}$$

Where
E_i is the population fit residual $= y_i - x_i' B$ and $\tau_{Ek} = \sum_{\psi_k} E_i$.

To derive the approximate variance we set $g_{is} = 1, \forall i$. This simplifies the term R_S in the above expression. We get
$$\begin{aligned}
AV(\hat{\tau}_y)_{Reg} &= AV_{PSU} + AV_{SSU} \\
&= V_I(\sum_{S_I} \check{\tau}_{Ek}) + E_I(\sum_{S_I} \frac{V_{Ek}}{\pi_{Ik}}) \\
&= \sum\sum_{\psi_K} \Delta_{Ikl} \check{\tau}_{Ek} \check{\tau}_{El} + \sum_{S_I} \frac{V_{Ek}}{\pi_{Ik}} \\
&= \sum\sum_{\psi_K} \Delta_{Ikl} \frac{\tau_{Ek}}{\pi_{Ik}} \frac{\tau_{El}}{\pi_{Il}} + \sum_{\psi_K} \frac{V_{Ek}}{\pi_{Ik}}.
\end{aligned}$$

Where
$$V_{Ek} = \sum \sum_{\psi_k} \Delta_{ij|k} \frac{E_i}{\pi_{i|k}} \frac{E_j}{\pi_{j|k}}$$
An unbiased estimator of $V(\hat{\tau}_y)_{Reg}$ is given by:
$$\hat{V}(\hat{\tau}_y)_{Reg} = \hat{V}_{PSU} + \hat{V}_{SSU}.$$
Where
$$\hat{V}_{PSU} = \sum \sum_{S_I} \check{\Delta}_{Ikl} \frac{\hat{\tau}_{Ek}}{\pi_{Ik}} \frac{\hat{\tau}_{El}}{\pi_{Il}} - \sum_{S_I} \frac{1}{\pi_{Ik}} \left(\frac{1}{\pi_{Ik}} - 1 \right) \hat{V}_{Ek}$$
and
$$\hat{V}_{SSU} = \sum_{S_I} \frac{\hat{V}_{Ek}}{\pi_{Ik}}$$
$$\hat{\tau}_{Ek} = \sum_{S_k} \frac{g_{is} e_{is}}{\pi_{i|k}}$$
$$\hat{V}_{Ek} = \sum \sum_{S_k} \check{\Delta}_{ij|k} \frac{g_{is} e_{is}}{\pi_{i|k}} \frac{g_{js} e_{js}}{\pi_{j|k}}.$$
Hence,

$$\hat{V}(\hat{\tau}_y)_{Reg} = \sum_{S_I} \sum \check{\Delta}_{Ikl} \frac{\hat{\tau}_{Ek}}{\pi_{Ik}} \frac{\hat{\tau}_{El}}{\pi_{Il}} + \sum_{S_I} \frac{\hat{V}_{Ek}}{\pi_{Ik}} \qquad (18.6)$$

Hence the theorem.   □

Robinson and Särndal (1983) has shown that the modified regression estimators are asymptotically unbiased and consistent.

18.4 Sample Survey

The plateaus in Western Ghats of Sahyadri from Goa to Varandha Ghat (Bhor, Maharashtra, India) are rich in aluminum ore—Laterite. Due to which the thorny plants are rarely observed. They are abundantly available on basalt-kind plateaus. But there are some rare patches of thorny plants. It indicates the absence of aluminum in that part. A sample survey was conducted in an area of 400 acres of land in Varandha Ghat. The interest was to estimate the total number of thorny plants in that area. Here, the aluminum content of the soil was the auxiliary variable. Its presence could be detected easily. Two-stage NACS was used to estimate the total number of thorny plants. The area of 400 acres was divided into 400 square plots of area one acre each. The percentage of aluminum content in the soil of each plot X was recorded. For instance, a random sample of 10 plots was selected from the study area without replacement. The adaptive procedure was used for the plots having $\{X \leq 20\}$. The networks formed in the X population were marked. This completed stage I of the sampling design. The observations on variable X in the population along with the networks are shown in Fig. 18.1.

In Fig. 18.1, each square represents a plot of one-acre area. In a square ⋆ indicates the selection of the corresponding plot in the initial sample. The networks formed by the adaptive procedure are shown in blue color. The edge unit of the networks is shown in violet color.

Figure 18.2 displays the corresponding different networks selected in the sample in blue color. The number in each square gives the number of the thorny plants (Y)

22.0	29.9	58.9	39.6	58.3	57.0	24.5	37.6	26.2	40.5	29.0	52.7	34.1	22.7	46.6	31.9	10.2	7.94	16.6	13.7
8.0	6.0	10.9	6.2	13.3	17.6	26.3	28.8	33.4	40.4	24.1	56.2	49.0	22.8	52.8	26.8	7.2	3.1	9.9	5.6
18.3	19.0	8.1	12.8	18.6	17.4	54.4	47.9	48.9	29.2	43.3*	54.1	55.6	41.1	40.2	55.0	13.5	11.7*	16.4	4.6
7.4	1.7	14.7	11.9*	7.3	18.3	42.0	39.2	37.4	58.5	39.8	21.0	52.7	44.6	23.4	49.4	5.3	16.7	12.2	9.8
2.7	18.0	17.9	6.8	17.6	4.4	25.0	52.5	34.9	45.0	49.9	48.2	36.9	30.1	38.9	53.0	26.8	28.7	38.4	22.2
36.9	24.8	21.3	29.4	59.1	28.8	51.0	46.9	38.9	39.4	25.9	26.7	30.6	56.8	55.7	36.3	21.2	24.8	49.5	38.1
44.7	26.0	39.8	26.7	27.1	53.3	50.6	30.6	30.5	52.4	53.9	39.7	28.6	32.5	56.9	52.2	44.9	46.8	51.1	50.9
29.4	29.0	41.3	32.5	59.1	22.5	41.4	49.1	28.6	23.5	52.0	51.2	33.3	57.8	46.0	23.3	31.6	32.6	36.3	32.0
52.5	23.8	48.2	53.7	34.1	54.8	46.2	29.5	24.7	31.8	22.7	32.4	57.9*	21.2	43.1	41.4	30.7	31.9*	48.5	49.6
30.4	48.3	42.0*	32.8	25.0	49.1	39.2	33.4	52.2	51.1	33.1	37.9	24.4	27.2	39.9	28.6	43.3	47.6	34.5	27.0
40.8	57.6	27.7	44.0	46.2	50.9	32.8	59.8	39.7	48.1	51.0	26.2	27.2	46.5	58.6	52.0	48.5	58.0	28.8	27.7
41.2	51.3	32.7	51.3	23.9	3.5	3.6	8.8	22.6	55.7	23.9	52.7	45.1	40.2	42.6	47.3	57.9	46.5	36.4	51.9
32.5	53.9	56.1	54.9	54.7	3.8	12.6	12.5	33.8	45.9	40.7	53.5	48.2	54.2	21.9	41.6	44.2	38.7	25.9	41.3
55.9	32.8	59.1	45.2	38.9	27.9	12.9*	57.2	50.8	51.2	44.1	25.2	32.9	21.3	34.5	41.0	35.9	39.5	22.2	33.7
30.9	45.7	59.1	53.0	33.0	38.7	44.7	39.2	25.3	37.5	31.5	55.7	50.8	39.1	42.0	27.3	52.3	38.8	56.3	27.9
48.2	58.9	41.9	56.9	51.4	47.5	41.6	47.1	26.7	55.7	53.8	49.1	27.1	45.3	34.5	21.7	52.9	56.6*	52.9	43.6
24.2	38.2	25.2	52.3	41.2*	47.6	24.7	46.1	21.0	37.9	13.3	6.4	17.2	25.7	26.9	37.5	47.0	21.9	33.7	36.1
33.3	50.4	22.2	28.5	30.2	26.6	39.3	45.4	30.7	36.2	16.1	6.5*	7.95	43.9	46.8	52.5	21.2	40.2	50.4	45.1
29.9	43.9	39.1	47.2	39.4	53.5	39.9	23.6	59.6	44.5	12.6	8.5	45.4	33.2	58.8	21.2	49.8	50.6	54.5	52.8
38.2	34.9	36.3	37.2	35.8	53.9	48.1	28.2	22.4	41.6	35.1	23.4	36.3	39.1	42.5	25.6	55.3	38.2	47.6	53.7

* **In a plot indicates its selection in the initial sample at the first stage**

Fig. 18.1 Percentage of aluminum content in the soil of plots in the population

actually observed on the corresponding plot. In the second stage of the design, random samples of unequal sizes were drawn from the networks selected in the sample by using SRSWOR. The set of plots in total so selected constituted the sample at the second stage. Values of the variables X and Y corresponding to the plots included in this second stage sample were recorded together to form bivariate data.

Using this data, the total number of thorny plants in the area was estimated by using the proposed estimators. To avoid information loss, the selected networks of size 1 are merged with some other selected network whose size is more than 1.

Results and Discussion

For computational efficiency in estimation of each estimator, r number of repetitions were performed, where r varied as 5000, 10000, 20000, and 100000 and the initial sample size was varied as 25, 30, 35, and 40.

The estimated population total over r repetitions is given by

$$\hat{\tau} = \frac{\sum_{i=1}^{r} (\hat{\tau}_y)_i}{r}$$

The estimated variance of the estimator of total is given by

$$\widehat{MSE}(\hat{\tau}_y) = \frac{\sum_{i=1}^{r} ((\hat{\tau}_y)_i - \tau_y)^2}{r}.$$

Where

$(\hat{\tau}_y)_i$ is the value of the relevant estimator for the ith sample.

0	0	0	0	0	0	0	0	0	0	0	0	0	0	0	0	95	143	44	76
150	155	125	160	70	30	0	0	0	0	0	0	0	0	0	0	136	173	106	169
27	20	145	87	24	23	0	0	0	0	0*	0	0	0	0	0	63	77*	56	162
156	195	65	92*	162	22	0	0	0	0	0	0	0	0	0	0	142	59	89	98
185	26	35	142	38	108	0	0	0	0	0	0	0	0	0	0	0	0	0	0
0	0	0	0	0	0	0	0	0	0	0	0	0	0	0	0	0	0	0	0
0	0	0	0	0	0	0	0	0	0	0	0	0	0	0	0	0	0	0	0
0	0	0	0	0	0	0	0	0	0	0	0	0*	0	0	0	0	0*	0	0
0	0	0*	0	0	0	0	0	0	0	0	0	0	0	0	0	0	0	0	0
0	0	0	0	0	0	0	0	0	0	0	0	0	0	0	0	0	0	0	0
0	0	0	0	0	174	162	117	0	0	0	0	0	0	0	0	0	0	0	0
0	0	0	0	0	182	72	74	0	0	0	0	0	0	0	0	0	0	0	0
0	0	0	0	0	0	69*	0	0	0	0	0	0	0	0	0	0	0	0	0
0	0	0	0	0	0	0	0	0	0	0	0	0	0	0	0	0	0	0	0
0	0	0	0	0	0	0	0	0	0	0	0	0	0	0	0	0	0*	0	0
0	0	0	0	0*	0	0	0	0	0	59	157	33	0	0	0	0	0	0	0
0	0	0	0	0	0	0	0	0	0	43	147*	135	0	0	0	0	0	0	0
0	0	0	0	0	0	0	0	0	0	64	89	0	0	0	0	0	0	0	0
0	0	0	0	0	0	0	0	0	0	0	0	0	0	0	0	0	0	0	0

Fig. 18.2 Number of thorny plants in plots of the population

The expected final sample size in ACS (n_s), expected effective sample size in NACS (n_e) and expected effective sample in two-stage NACS (n_e^*)were calculated for the different initial sample sizes and a number of repetitions. The results are given in Table 18.1.

In general, the relation $E(n_e^*) < E(n_e) < E(n_s)$ was observed. The difference between them showed a consistent increase for the increase in the initial sample size and increase in the number of repetitions.

Table 18.2 gives a comparison between the performance of the HT estimator in ACS and the regression estimator in NACS. The regression estimator in NACS has shown consistently a better performance. A remarkable reduction in the standard error of the estimator used in NACS was observed as compared to that in ACS. The estimates of the population total obtained in NACS were closer to the true value as compared to that in ACS.

Table 18.3 gives a comparison of the performance of the Composite HT estimator used in two-stage ACS, modified regression estimator in two stages NACS and the regression estimator used in ACDS.

In general, the standard error of the modified regression estimator in two-stage NACS was found to be the smallest among the other two estimators. Though this

Table 18.1 Expected sample sizes in two-stage negative adaptive cluster sampling

Number of repetitions	Initial sample size	Expected sample size	Expected effective sample size	Expected effective sample size at stage II
r	n_0	n_s	n_e	n_e^*
5000	25	57.13	35.51	20.17
	30	64.61	38.70	22.02
	35	71.17	40.92	23.33
	40	78.35	43.80	25.01
10000	25	57.34	35.77	20.32
	30	64.72	38.82	22.08
	35	71.72	41.50	23.66
	40	78.28	43.76	24.99
20000	25	57.06	35.46	20.14
	30	64.68	38.76	22.06
	35	71.67	41.43	23.62
	40	78.18	43.63	24.92
100000	25	57.00	35.39	20.10
	30	64.63	38.70	22.02
	35	71.68	41.46	23.63
	40	78.11	43.56	24.88

Table 18.2 The performance of the estimators of population total in ACS and NACS

Number of repetitions	Initial sample size	Estimators in ACS		Estimators in NACS	
r	n_0	$(\hat{\tau}_y)_{HT}$	$\hat{SE}((\hat{\tau}_y)_{HT})$	$(\hat{\tau}_y)_{Reg}$	$\hat{SE}(\hat{\tau}_y)_{Reg}$
5000	25	5607.55	2740.63	5541.74	2152.5
	30	5628.43	2386.57	5537.37	1840.40
	35	5634.98	2116.44	5488.27	1612.08
	40	5604.27	1889.93	5494.96	1428.91
10000	25	5615.08	2735.85	5498.47	2133.38
	30	5630.50	2386.11	5511.75	1839.13
	35	5587.48	2136.66	5484.61	1601.84
	40	5642.95	1889.39	5514.59	1411.25
20000	25	5600.14	2745.98	5507.78	2132.94
	30	5611.66	2395.80	5496.40	1833.40
	35	5644.97	2118.25	5502.41	1595.60
	40	5641.07	1888.96	5526.75	1409.64
100000	25	5621.66	2752.54	5504.73	2120.70
	30	5632.83	2401.75	5510.37	1826.97
	35	5637.05	2118.11	5504.40	1584.22
	40	5631.41	1890.98	5499.82	1403.57

Table 18.3 The performance of the estimators of population total in two-stage ACS, two-stage NACS, and ACDS

Number of repetitions	Initial sample size	Estimators in two-stage ACS		Estimators in two-stage NACS		Estimators in ACDS	
r	n_0	$(\hat{\tau}_y)_{CHT}$	$\hat{SE}((\hat{\tau}_y)_{CHT})$	$(\hat{\tau}_y)_{Reg}$	$\hat{SE}(\hat{\tau}_y)_{Reg}$	$(\hat{\tau}_y)_{Reg}$	$\hat{SE}(\hat{\tau}_y)_{Reg}$
5000	25	5630.28	2037.09	4246.87	1442.40	5499.80	2162.89
	30	5534.72	1778.70	4726.78	1281.96	5521.87	1911.49
	35	5545.52	1519.06	5197.80	789.42	5519.04	1644.74
	40	5515.10	1355.41	5428.99	519.29	5474.57	1470.12
10000	25	5607.22	2020.43	4258.36	1220.33	5508.80	2161.75
	30	5565.78	1774.00	4819.58	1185.57	5504.97	1864.42
	35	5534.03	1537.03	5244.47	769.67	5483.98	1641.66
	40	5512.25	1342.90	5492.78	518.50	5493.10	1465.47
20000	25	5618.80	2039.70	4347.34	1650.59	5518.32	2165.35
	30	5551.93	1757.35	4795.36	1048.28	5524.50	1886.80
	35	5532.36	1527.24	5199.69	427.28	5523.68	1633.60
	40	5504.43	1357.03	5497.29	294.91	5485.18	1453.53
100000	25	5619.88	2038.00	4273.07	1751.98	5518.31	2169.74
	30	5548.22	1748.21	4809.55	827.11	5505.67	1875.64
	35	5520.31	1521.62	5193.84	716.27	5509.64	1646.47
	40	5509.31	1350.04	5506.94	91.82	5498.02	1461.08

estimator used in the proposed design is biased, the bias reduced nearly to zero for the sample of size 40. By using Table 18.3, the modified regression estimator is consistent and asymptotically unbiased. Thus, the design proposed is more precise as compared to the other existing designs. The two-stage ACS is discussed in Chap. 11. In this design, clusters are formed by using usual cluster formation criteria. The adaptation procedure is applied at the second stage. Hence, there is no control on the total sample size. This method leads to either over or underestimate. The adjusted two-stage ACS was proposed by Muttlak and Khan (2002). In this design, the first-stage cluster is formed by using auxiliary information. But at the second stage auxiliary information is dropped. Also, the surveyor has to enumerate each and every unit of a large network and then draw a sample for stage two. Hence, this method is not cost-effective.

But in the two-stage NACS design, networks are formed naturally and units are selected according to auxiliary information. It is less costly. By using adaptation procedures, networks are formed on the basis of auxiliary information. The corresponding network of study variable is called the primary stage unit. The second stage units are drawn from PSUs. Here, PSUs are not measured with respect to the variable of interest. Only SSUs are measured with respect to the variable of interest. Hence, our methodology is cost-effective. If c_1 and c_2 denote the cost of sampling per unit with respect to X and Y variables, respectively then

Expected Cost of sampling in ACS = $c_2\, n_s$.

Expected Cost of sampling in NACS = $c_1\, N + c_2\, n_e$.

Expected Cost of sampling in two-stage NACS = $c_1\, N + c_2 \sum_{i=1}^{k} \delta_i\, m_i$

Table 18.4 Expected sampling costs under different designs

Number of repetitions	Initial sample size	Expected final sample size	Expected effective sample size	Expected effective sample size at stage II	Expected sampling costs under different designs		
r	n_0	n_s	n_e	n_e^*	ACS	$NACS$	$Two stage NACS$
5000	25	57.13	35.51	20.17	3427.80	2930.60	2010.20
	30	64.61	38.70	22.02	3876.60	3122.00	2121.20
	35	71.17	40.92	23.33	4270.20	3255.20	2199.80
	40	78.35	43.80	25.01	4701.00	3428.00	2300.60
10000	25	57.34	35.77	20.32	3440.40	2946.20	2019.20
	30	64.72	38.82	22.08	3883.20	3129.20	2124.80
	35	71.72	41.50	23.66	4303.20	3290.00	2219.60
	40	78.28	43.76	24.99	4696.80	3425.60	2299.40
20000	25	57.06	35.46	20.14	3423.60	2927.60	2008.40
	30	64.68	38.76	22.06	3880.80	3125.60	2123.60
	35	71.67	41.43	23.62	4300.20	3285.80	2217.20
	40	78.18	43.63	24.92	4690.80	3417.80	2295.20
100000	25	57.00	35.39	20.10	3420.00	2923.40	2006.00
	30	64.63	38.70	22.02	3877.80	3122.00	2121.20
	35	71.68	41.46	23.63	4300.80	3287.60	2217.80
	40	78.11	43.56	24.88	4686.6	3413.6	2292.80

$$= c_1 N + c_2 n_e^*.$$

Where, m_i : Number of units selected from the ith network selected at the first stage.

$$\delta_i = \begin{cases} 1 & ; \text{if } i^{th} \text{ network in the population is selected at the first stage} \\ 0 & ; otherwise. \qquad (i = 1, 2, \cdots, K) \end{cases}$$

By considering $c_1 = 2$ and $c_2 = 60$, we have calculated the costs of sampling in the above mentioned designs for the different values of n_0 which varied as 25, 30, 35, and 40 for r repetitions. It varied as 5000, 10000, 20000, and 100000. These costs are represented in Table 18.4.

It was observed that, the cost of sampling in two-stage NACS < cost of sampling in NACS < cost of sampling in ACS. Thus the newly two-stage NACS design is cost-effective as compared to the other existing designs applicable for the rare and clustered populations.

We utilized the auxiliary information at design and estimation stage. Using the domain knowledge, we construct the two-stage NACS design. NACS is more efficient than ACS. The proposed two-stage NACS is more efficient than NACS, ACDS, and two-stage ACS. The estimator proposed is biased for the population total of the variable of interest but for the large initial sample, the bias gets reduced to nearly zero. Since collecting observations on the auxiliary variable is inexpensive and here the adaptation is used at the first stage, the cost of sampling involved in the proposed design is much smaller than the other competitive designs.

The sample survey was conducted by using this proposed design at Western Ghat of Maharashtra, India. The performance of the estimators was compared by using the survey data. The results showed that the proposed design is more efficient as compared to other considered designs and it is cost-effective as well. Thus, when the gathering of information on the auxiliary variable is possible and inexpensive and there is a negative correlation between the two variables; two-stage NACS can be recommended. Such type of situation is frequently observed in environmental, ecological, forestry, mining, and medical fields. Hence, the proposed method can be used in these areas effectively.

18.5 Exercises

1. Describe two-stage NACS.
2. Distinguish between two-stage ACS and two-stage NACS.
3. Discuss the composite Horvitz–Thompson estimator of the population total used in two-stage NACS.
4. State and prove the theorem that gives an unbiased estimator of the variance of the composite Horvitz–Thompson estimator of the population total used in two-stage NACS.
5. Derive two-stage regression estimator of the population total used in two-stage NACS.
6. State and prove the theorem that gives an unbiased estimator of the variance of the two-stage regression estimator of the population total used in two-stage NACS.
7. Obtain an expression for the expected cost of sampling in two-stage NACS and compare this cost with ACS and NACS.

References

Muttlak, H.A., Khan, A.: Adjusted two stage adaptive cluster sampling. Environ. Ecol. Stat. 9(1), 111–120 (2002)

Robinson, P.M., Särndal, C.E.: Asymptotic properties of the generalized regression estimator in probability sampling. Sankhya B45, 240–248 (1983)

Salehi, M.M., Seber, G.A.F.: Adaptive cluster sampling with networks selected without replacement. Biometrika 84, 209–219 (1997a)

Salehi, M.M., Seber, G.A.F.: Two stage adaptive cluster sampling. Biometrics 53, 959–970 (1997b)

Särndal, C.E., Swensson, B., Wretman, J.: Model Assisted Survey Sampling. Springer Inc, New York (1992)

Thompson, S.K.: Adaptive cluster sampling. J. Am. Stat. Assoc. 85(412), 1050–1058 (1990)

Chapter 19
Balanced and Unbalanced Ranked Set Sampling

19.1 Introduction

In many studies, it is observed that the measurement of the units depends upon a host of extraneous factors like difficulty of reaching the sampling unit, costs involved, destruction of the sampling unit (especially when repeated measurements are required in the sample), etc. McIntyre (1952) proposed a sampling method to provide the precise estimates for the grass and pasture yield in Australia. The measurement of units in this process was to be obtained by the process of harvesting the forage, then moving and clipping the browse, dry, and weighting. This process is laborious, expensive, time-consuming, and a difficult task. Ranking a large number of observational units (quadrats) is rather easy than the actual measurements. This ranking method becomes popular with ranked set sampling (RSS). RSS, an environmental sampling scheme, is commonly utilized to reduce costs and potentially increase precision when actual measurements of units are costly and or time-consuming but the ranking of the set of units according to the concomitant variable or some other rough gauging methods can be rather easy than actual measurements. Let us explain by the following different examples:

Example 19.1 The infestation problem in poplar trees caused by the bark-eating caterpillar *Inderbela quadrinotato* is commonly found in many forest areas in India. The larva of the caterpillar makes tunnel into the trunk of the tree and feeds bark tissues covering them with fecal ribbons. The estimation of the number or density of caterpillars in a forest stand by some conventional methods like SRS requires the measurement of the number of caterpillars in each sampled tree. These methods are not only difficult but very costly as well as it is compulsory to remove the ribbon to confirm the presence of active insects inside. In this case, the ranking of the trees based upon the length of ribbon made by this insect is easy in comparison with the actual measurements.

© The Author(s), under exclusive license to Springer Nature Singapore Pte Ltd. 2021
R. Latpate et al., *Advanced Sampling Methods*,
https://doi.org/10.1007/978-981-16-0622-9_19

Example 19.2 Martin et al. (1980) evaluated RSS for estimating shrub phytomass in Appellation Oak forest. In this study, counting the number of each vegetation type in randomly selected blocks of forest stands is essentially required which is rather costly and time-consuming. However, the ranking of a small number of blocks of forest stand can be done easily by visual inspection.

Example 19.3 Samawi (1999) took the study of the determination of normal ranges of bilirubin level of blood for newborn babies using RSS. While using other methods of sampling, the blood samples are necessary from each sampled baby for testing in the laboratory. However, by the use of RSS, the ranking of bilirubin levels of a small number of babies can be done by observing their physical parts like chest, face, etc., to see the yellowish color of the body. As the yellowish color increases the bilirubin in the blood goes higher.

Example 19.4 The estimation study of the mercury contamination (mg/kg) in fish is a time-consuming and difficult task besides resulting in the destruction of the fishes by methods other than RSS. It requires selecting some live or freshly dead fishes from the catch, measures the length, remove a fillet from each fish, homogenized with blender, and analyzed for mercury concentration with a gas chromatography. RSS methodology was suggested by Murff and Sager (2006) for such studies. It was seen that much gain in the efficiency can be actually made by the RSS over the SRS methodology.

With the classical designs, the entire selection of sample units in the sample is made by the actual measurement of each of the selected units. For example, in SRS, the selected units require measurement cost and time toward each and every unit of the population. This does not guarantee to have an optimum sample due to the equal probabilities of selection of units of the population. Similar situations also prevail in other classical designs. With RSS, however, the researcher has a greater advantage of saving time and money on the selection of units. Despite this advantage, the objective is the same, that is, to select the units (based on actual measurement or ranking method) and then estimate some function $Z(y)$ (equivalently the population total, $Z(y) = \hat{Y} = \sum_{i=1}^{N} y_i$, the population mean $\mu = \frac{1}{N} \sum_{i=1}^{N} y_i$, or finite population variance $\frac{1}{N-1} \sum_{i=1}^{N} (y_i - \mu)^2$, etc.). Moreover, RSS also gains the advantages of the stratified random sampling, in which, the sampling unit from various order statistics is for measurement. The classical estimators used in conjunction with RSS are unbiased and the variance of RSS estimators is smaller for an equivalent amount of sampling effort (Takahasi and Wakimoto 1968).

For a long time, RSS did not attract the attention it deserved, but there has been a surge in interest in it over the past thirty years or so. After the mathematical foundation of RSS laid by Takahasi and Wakimoto (1968), many theoretical developments have been made. Dell (1969) and Dell and Clutter (1972) provided various expressions of the variance of RSS estimator. Estimating the population variance based on judgment ordered ranked set samples was considered by Stokes (1980). On estimation of population variance, one may refer to MacEachern et al. (2002), Chen and Lim (2011), Perron and Sinha (2004), and Sengupta and Mukhuti (2006). The kernel estimators of probability density functions using RSS were given by Barabesi and Fattorini (2002) and Lim et al. (2014). Frey (2014) considered bootstrap confidence bands for the CDF using RSS. Other works on the estimation of distribution function is found in Wolfe (2004). Lam et al. (2002) given non-parametric estimators for the cumulative distribution function and the mean when auxiliary information is available. Using RSS, the population proportions and quantiles were also estimated (Chen et al. 2006; Chen 2001; Zhu and Wang 2005). Other different and useful versions of RSS are available, these are included in Chap. 21. The successful application by the use of RSS and its other modifications have been implemented in the different areas, namely, agriculture, forestry, ecology, epidemiology, etc.

The basic steps to obtain a ranked set sample of size $n = km$, where k is the number of sampling units selected for measurements in each cycle (set size) and m is the total number of cycles of the process are given below.

1. The k units are randomly selected from the population. These k sampling units are ranked based on a perception of relative values for the variable under interest, say Y. This may be done based on personal judgment, expert judgment including visual inspection or measurement of a covariate or concomitant variable correlated with the study variable, whose actual measurement is relatively easier and inexpensive. Then the unit with rank 1 is identified and taken for the measurement of Y and the remaining units are discarded.
2. Other k units are randomly selected from the population and ranked by the same way for the variable Y. The unit with rank 2 is identified and taken for the measurement of Y and the remaining units are discarded.
3. The above process is continued until one identified the unit with rank k for the measurement of Y. This whole process of step 1 to step 3 is referred to as a cycle.
4. By repeating the process for m cycles, one gets the desired sample of size n.

Note that (i) there are k^2m sampling units taken from the population, but only km units (m units for each rank order) are retained for the actual measurements. So for each fully measured unit, there are $k - 1$ additional units for ranking in each cycle. Ranking k^2m units with using km of them for measurement is thus a sampling method of boosting the efficiency (ii) all measured units are independent, (iii) under a stable ranking condition, observations from the same judgment class are identically distributed, (iv) It is expected that ranking is cheap and measuring units are expensive. In order to retain the reliability of the rankings, it is usually suggested that it be carried out in small set sizes, say $k = 3, 4, 5$, etc. to reduce ranking errors, but $m \geq 2$ may be adjusted in obtaining the requisite sample of size n.

In the above procedure, each rank order statistic is measured the same number of times, i.e., m, termed as balanced RSS. This was the original concept given by McIntyre (1952). Since RSS contains more information than SRS of the same sample size, RSS is always more efficient than SRS. Further improvement in RSS may be done by appropriately chosen the unbalance RSS.

In the following sections, details about balanced and unbalanced RSS along with the allocation procedures are given. Some important properties are also given.

19.2 Balanced RSS

In balanced RSS(BRSS), the number of measurements made on each ranked unit is the same for all the ranks (i.e., m). This maintains the unbiasedness of SRS and, in addition, increases its representativeness of the true underlying population. An elucidation of BRSS for the cycle j is presented (Fig. 19.1).

$Y_{(1:1)_j}, Y_{(1:2)_j}, \ldots Y_{(1:k)_j}$

$Y_{(2:1)_j}, Y_{(2:2)_j}, \ldots Y_{(2:k)_j}$

$\vdots$

$Y_{(k:1)_j}, Y_{(k:2)_j}, \ldots Y_{(k:k)_j},$

where, $Y_{(i:1)_j}, Y_{(i:2)_j}, \ldots Y_{(i:k)_j}$ denotes the y-value of 1st, 2nd, $\ldots$, kth observations for measuring ith order statistics under jth cycle for set size k of BRSS. It is to be noted that the diagonal units of Fig. 19.1, viz., $Y_{(1:1)_j}, Y_{(2:2)_j}, \ldots, Y_{(k:k)_j}$, a collection of independent order statistics from k disjoint collections of k simple random samples, are actually measured under jth cycle. The pdf of ith order statistic is given by

$$f_{(i)}(y) = \frac{k!}{(i-1)!(k-i)!} [F(y)]^{i-1} [1-F(y)]^{k-i} f(y). \tag{19.1}$$

The following equality has an important role in RSS.

$f(y) = \frac{1}{k} \sum_{i=1}^{k} f_{(i)}(y_{(i:i)}).$

Theorem 19.1 *The unbiased estimator under BRSS is given by*

$$\bar{Y}_{(k)BRSS} = \frac{1}{km} \sum_{i=1}^{k} \sum_{j=1}^{m} Y_{(i:k)_j} \tag{19.2}$$

With its variance by
$V(\bar{Y}_{(k)BRSS}) = \frac{1}{k^2 m} \sum_{i=1}^{k} \sigma_{(i:k)}^2$
where, $Y_{(i:k)_j}$ represents the measured value of the ith rank order statistic under jth cycle when set size is k.

Proof Let μ and σ^2 be the mean and variance of the population. Taking expectations both side of (19.2), we get
$E(\bar{Y}_{(k)BRSS}) = \frac{1}{km} \sum_{i=1}^{k} \sum_{j=1}^{m} E(Y_{(i:k)_j}).$

Fig. 19.1 Elucidation of k^2 units in k sets each of size k for cycle j

$Y_{(1:1)_j}, Y_{(1:2)_j}, \ldots Y_{(1:k)_j}$

$Y_{(2:1)_j}, Y_{(2:2)_j}, \ldots Y_{(2:k)_j}$

$\vdots$

$Y_{(k:1)_j}, Y_{(k:2)_j}, \ldots Y_{(k:k)_j}$

Since, for fixed i, $Y_{(i:k)_j}$, $j = 1, 2, \ldots, m$, are i.i.d. with mean $\mu_{(i:k)}$, (say) and variance $\sigma^2_{(i:k)}$ (say), therefore, using $f(y) = \frac{1}{k} \sum_{i=1}^{k} f_{(i)}(y)$, we get

$E(\bar{Y}_{(k)BRSS}) = \frac{1}{k} \sum_{i=1}^{k} \mu_{(i:k)} = \mu.$

Aliter:

$$E(\bar{Y}_{(k)BRSS}) = \frac{1}{km} \sum_{i=1}^{k} \sum_{j=1}^{m} E(Y_{(i:k)_j})$$
$$= \frac{1}{km} \sum_{i=1}^{k} \sum_{j=1}^{m} \int_{-\infty}^{\infty} y \frac{k!}{(i-1)!(k-i)!} [F(y)]^{i-1} [1 - F(y)]^{k-i} f(y)\, dy$$
$$= \int_{-\infty}^{\infty} y \left[\frac{1}{km} \sum_{i=1}^{k} \sum_{j=1}^{m} \binom{k-1}{i-1} \right] [F(y)]^{i-1} [1 - F(y)]^{k-i} f(y)\, dy$$
$$= \int_{-\infty}^{\infty} y f(y)\, dy = \mu.$$

Now,

$$V(\bar{Y}_{(k)BRSS}) = \frac{1}{k^2 m^2} \sum_{i=1}^{k} \sum_{j=1}^{m} V(Y_{(i:k)_j}) = \frac{1}{k^2 m^2} \sum_{i=1}^{k} \sum_{j=1}^{m} \sigma^2_{(i:k)}$$
$$= \frac{1}{k^2 m} \sum_{i=1}^{k} \sigma^2_{(i:k)}.$$

Hence the theorem. □

For SRS with $n = km$ measurements, the variance of the sample mean is $V(\bar{Y}_{SRS}) = \frac{\sigma^2}{km}$.

Therefore the relative precision (RP) of RSS with respect to SRS is

$$RP_{SRS:BRSS} = \frac{V(\bar{Y}_{SRS})}{V(\bar{Y}_{(k)BRSS})} = \frac{\sigma^2}{\bar{\sigma}^2}$$

where $\bar{\sigma}^2 = \frac{1}{k} \sum_{i=1}^{k} \sigma^2_{(i:k)}$ is the average of the within-rank variances.

It can also be seen that

$$V(\bar{Y}_{(k)BRSS}) = \frac{1}{k^2 m} \sum_{i=1}^{k} E\left(Y_{(i:k)} - \mu_{(i:k)}\right)^2$$
$$= \frac{\sigma^2}{mk} - \frac{1}{k^2 m} \sum_{i=1}^{k} \left(\mu_{(i:k)} - \mu\right)^2$$
$$= V(\bar{Y}_{SRS}) - \frac{1}{k^2 m} \sum_{i=1}^{k} \left(\mu_{(i:k)} - \mu\right)^2$$

Since $\frac{1}{k^2 m} \sum_{i=1}^{k} \left(\mu_{(i:k)} - \mu\right)^2 \geq 0$, therefore $V(\bar{Y}_{(k)BRSS}) \leq V(\bar{Y}_{SRS})$, where

$E\left(Y_{(i:k)}\right) = \mu_{(i:k)}.$

Takahasi and Wakimoto (1968) proved that under the perfect ranking, the $RP_{SRS:BRSS}$ lies between 1 and $\frac{k+1}{2}$ while estimating population mean. It was established by Dell

Table 19.1 $RP_{SRS:BRSS}$ for some distributions for $k = 2(1)8$

Distribution	μ	σ^2	Set size (k)						
			2	3	4	5	6	7	8
LN (0,1)	1.6487	4.6708	1.1872	1.3393	1.4711	1.5891	1.6971	1.7974	1.8914
Pareto (3)	1.5	0.75	1.1364	1.2422	1.3305	1.4072	1.4755	1.5373	1.5941
Pareto (5)	1.25	0.1042	1.2277	1.4179	1.5861	1.7390	1.8797	2.0126	2.1373
Weibull(0.5)	2	20	1.1268	1.2362	1.3345	1.4250	1.5094	1.5891	1.6648
Gamma (1)	1	1	1.3333	1.6364	1.9200	2.1898	2.4490	2.6997	2.9435
N(0,1)	0	1	1.467	1.914	2.347	2.770	3.186	3.595	3.999
U(0, 1)	0	1	1.500	2.000	2.500	3.000	3.500	4.000	4.500

and Clutter (1972) that the RSS performs better than SRS even in presence of the ranking errors.

Example 19.5 Based upon the computations on some of the skewed and symmetric distributions, the $RP_{SRS:BRSS}$ for the values of $k = 2, 3, \ldots 8$ are given in Table 19.1. For these distributions, the means and variances of order statistics are taken from Harter and Balakrishnan (1996). From Table 19.1, it is seen that $RP_{SRS:BRSS} > 1$ or all k of all the distributions and $RP_{SRS:BRSS}$ increases as k increases.

19.3   Unbalanced RSS

It is seen that BRSS is a more precise method than SRS. However, the gain in the performance of the RSS can be improved when an appropriate unequal allocation is made. The resulting RSS procedure is called unbalanced RSS (URSS) or RSS with unequal allocation. The measured units under URSS are presented in Fig. 19.2.

$$Y_{(1:k)1}, Y_{(1:k)2}, \ldots Y_{(1:k)m_1}$$
$$Y_{(2:k)1}, Y_{(2:k)2}, \ldots Y_{(2:k)m_2}$$
$$\vdots$$
$$Y_{(k:k)1}, Y_{(k:k)2}, \ldots Y_{(k:k)m_k}$$

where $Y_{(i:k)j}$ denotes the y-value of ith order statistics under jth cycle for set size k; $j = 1, 2, \ldots m_i$. It is to be noted that for different values of i, $Y_{(i:k)j}$'s are independent and for fixed i, $Y_{(i:k)j}$ are identically distributed.

McIntyre (1952) proposed that the sample size corresponding to each rank order should be proportional to the standard deviation. It is also known as unequal allocation based on Neyman's approach, gives an optimal allocation when the distribution under study is positively skew.

19.3.1   Skewed Underlying Distributions

In many environmental situations, the data obtained is positively skewed. For instance, verification data obtained after the remediation of a site is generally skewed with heavy right tail. For skewed distributions, Takahasi (1970), Yanagawa and Shirahata (1976), and Yanagawa and Chen (1980) used the random allocations in RSS.

Fig. 19.2 Representation of the measured $\sum\limits_{i=1}^{n} m_i$ units in URSS

$$Y_{(1:k)1}, Y_{(1:k)2}, \ldots Y_{(1:k)m_1}$$
$$Y_{(2:k)1}, Y_{(2:k)2}, \ldots Y_{(2:k)m_2}$$
$$\vdots$$
$$Y_{(k:k)1}, Y_{(k:k)2}, \ldots Y_{(k:k)m_k}$$

Utilizing the fact that the standard deviations usually tend to increase with increasing rank values for positively skewed distributions, Kaur et al. (1994, 1997) suggested two 'near' optimal approaches for positively skewed distributions and called them the 't-model' and '(s, t)-model'. The main objective of proposing these models was the unavailability of population parameters, particularly standard deviation of rank order statistics, as required for Neyman's method. They found that the performance of both models was better than the BRSS. They also studied the role played by skewness, kurtosis, and the coefficient of variation for obtaining the allocation factor(s) and in devising the rules-of-thumb. It was proved that the '(s, t)-model' performs better than the 't-model', although the performance of Neyman's method is the best. These allocation models do not provide the integer allocation values, and therefore Tiwari and Chandra (2011) proposed a systematic allocation procedure for the skewed distributions. Recently, Bhoj and Chandra (2019) proposed the unequal allocation model for heavy right tail distributions.

Theorem 19.2 *For measuring* $m_i (> 0)$ *units corresponding to the ith rank, $i = 1, 2, \ldots k$, and suppose* $T_i = \sum_{j=1}^{m_i} Y_{(i:k)_j}$, *an unbiased estimator* $\bar{Y}_{(k)URSS}$ *of μ and* $V(\bar{Y}_{(k)URSS})$ *are, respectively, given by* $\bar{Y}_{(k)URSS} = \frac{1}{k} \sum_{i=1}^{k} \frac{T_i}{m_i}$ *and*

$$V(\bar{Y}_{(k)URSS}) = \frac{1}{k^2} \sum_{i=1}^{k} \frac{\sigma_{(i:k)}^2}{m_i} \tag{19.3}$$

Proof We have given that ith rank order statistic is measured m_i times, $i = 1, 2, \ldots k$. This results in $n = m_1 + m_2 + \cdots + m_k$ actual measurements for the sample. We have

$$E(\bar{Y}_{(k)URSS}) = \frac{1}{k} \sum_{i=1}^{k} \frac{1}{m_i} \sum_{j=1}^{m_i} E(Y_{(i:k)j})$$

Since, for fixed i, $Y_{(i:k)j}$, $j = 1, 2, \ldots m_i$, are i.i.d. with mean $\mu_{(i:k)}$ and variance $\sigma_{(i:k)}^2$, therefore,

$$E(\bar{Y}_{(k)BRSS}) = \frac{1}{k} \sum_{i=1}^{k} \mu_{(i:k)} = \mu$$

and,

$$V(\bar{Y}_{(k)URSS}) = \frac{1}{k^2} \sum_{i=1}^{k} \frac{1}{m_i^2} \sum_{j=1}^{m_i} V(Y_{(i:k)j}) = \frac{1}{k^2} \sum_{i=1}^{k} \frac{\sigma_{(i:k)}^2}{m_i}$$

Hence the theorem. $\qquad \square$

The optimal allocation for RSS turns out be the same as Neyman's optimal allocation, in which the sampling units are allocated into ranks in proportion to the standard deviation of each rank and is given by

$$m_i = \frac{n\sigma_{(i:k)}}{\sum_{i=1}^{k} \sigma_{(i:k)}} \tag{19.4}$$

The corresponding variance by using (19.3) in (19.4) becomes

$$V(\bar{Y}_{(k)Ney} = \frac{1}{nk^2}\left(\sum_{i=1}^{k}\sigma_{(i:k)}\right)^2 = \frac{\bar{\sigma}^2}{n}$$

where $\bar{\sigma} = \frac{1}{k}\sum_{i=1}^{k}\sigma_{(i:k)}$ is the average of the within-rank standard deviations.

The RP of Neyman's optimum allocation relative to SRS and BRSS are, respectively, given by

$$RP_{SRS:Ney} = \frac{V(\bar{Y}_{SRS})}{V(\bar{Y}_{(k)Ney})} = \frac{\sigma^2}{\bar{\sigma}^2}; RP_{BRSS:Ney} = \frac{V(\bar{Y}_{(k)BRSS})}{V(\bar{Y}_{(k)Ney})} = \frac{\bar{\sigma}^2}{\bar{\sigma}^2}.$$

Now, we consider the near optimal allocation models (Kaur et al. 1997) in the following theorem:

Theorem 19.3 *Consider the following two models:*

(i) '*t-model': Highest (kth) order statistics is measured t times more frequently than the remaining $k-1$ order statistics, i.e.,*

$$m' \equiv m_1 = m_2 = \cdots = m_{k-1} = \frac{m_k}{t}, (t \geq 1) \qquad (19.5)$$

(ii) '*(s, t)-model': The highest (kth and $(k-1)$th) order statistics are measured t and s times, respectively, more than the remaining $(k-2)$ order statistics, i.e.,*

$$m'' \equiv m_1 = m_2 = \cdots = \frac{m_{k-1}}{s} = \frac{m_k}{t}, (1 \leq s \leq t) \qquad (19.6)$$

For the above models, the variances of the estimators $(\bar{Y}_{(k)t}, \bar{Y}_{(k)st})$ of μ are given respectively as:

$V(\bar{Y}_{(k)t}) = \frac{1}{k^2m'}(a' + b'/t)$ *and* $V(\bar{Y}_{(k)st}) = \frac{1}{k^2m''}(a + b/s + c/t)$

where $a' = \sum_{i=1}^{k-1}\sigma_{(i:k)}^2$, $b' = \sigma_{(k:k)}^2$, $a = \sum_{i=1}^{k-2}\sigma_{(i:k)}^2$, $b = \sigma_{(k-1:k)}^2$ *and* $c = \sigma_{(k:k)}^2$.

Proof For the models (19.5) and (19.6), the sample sizes are $(k-1+t)m'$ and $(k-2+s+t)m''$, respectively. Using (19.3), (19.5) and (19.6), we get RPs of $\bar{Y}_{(k)t}$ and $\bar{Y}_{(k)st}$ relative to SRS as

$$RP_{SRS:t} = \frac{k^2\sigma^2}{(k-1+t)(a'+b'/t)}; RP_{SRS:st} = \frac{k^2\sigma^2}{(k-2+s+t)(a+b/s+c/t)}$$

and the RPs of $\bar{Y}_{(k)t}$ and $\bar{Y}_{(k)st}$ with respect to BRSS are

$$RP_{BRSS:t} = \frac{k^2\bar{\sigma}^2}{(k-1+t)(a'+b'/t)} \qquad (19.7)$$

$$RP_{BRSS:st} = \frac{k^2\bar{\sigma}^2}{(k-2+s+t)(a+b/s+c/t)} \qquad (19.8)$$

On optimizing (19.7) with respect to t and optimizing (19.8) with respect of s and t using the method of maxima and minima, the optimal values in 't-model' (t_{opt}) and in '(s, t)-model' (s^*, t^*) are obtained as

Table 19.2 RPs of three models w.r.t. SRS for LN (0, 1), P(3), and W(0.5) for k = 2(1)8

k	LN (0, 1)			P(3)			W(0.5)		
	t	(s, t)	Neyman	t	(s, t)	Neyman	t	(s, t)	Neyman
2	1.482		1.482	1.449		1.449	1.6306		1.6306
3	2.039	2.039	2.039	2.055	2.055	2.055	2.0625	2.2105	2.2105
4	2.324	2.595	2.595	2.466	2.591	2.591	2.5008	2.7913	2.7913
5	2.714	3.030	3.088	2.632	3.099	3.099	2.7847	3.19	3.384
6	3.098	3.283	3.560	3.156	3.487	3.631	3.0754	3.6799	3.9393
7	3.266	3.815	4.067	3.463	3.800	4.114	3.3506	3.9897	4.4796
8	3.656	4.198	4.532	3.629	4.272	4.619	3.5747	4.403	5.032

$$t_{opt} = \sqrt{\frac{b'(k-1)}{a'}}$$

$$s^* = \sqrt{\frac{(k-2)b}{a}}$$

$$t^* = \sqrt{\frac{(k-2)c}{a}}$$

Now for the t model, we have

$$V(\bar{Y}_{(k)t}) = \frac{1}{k^2} \sum_{i=1}^{k} \frac{\sigma^2_{(i:k)}}{m_i} = \frac{1}{k^2}\left[\sum_{i=1}^{k-1} \frac{\sigma^2_{(i:k)}}{m} + \frac{\sigma^2_{(k:k)}}{tm} \right]$$

Now putting optimized value of t, i.e., $t_{opt} = \sqrt{\frac{b'(k-1)}{a'}}$ to the above, we get

$$V(\bar{Y}_{(k)t}) = \frac{1}{k^2 m'}\left(a' + \frac{b'}{t} \right)$$

Similarly, for the (s, t) model, we have

$$V(\bar{Y}_{(k)st}) = \frac{1}{k^2}\left[\sum_{i=1}^{k-2} \frac{\sigma^2_{(i:k)}}{m''} + \frac{\sigma^2_{(k-1:k)}}{sm''} + \frac{\sigma^2_{(k:k)}}{tm''} \right]$$

Using the optimized values of s and t i.e. $s^* = \sqrt{\frac{(k-2)b}{a}}$ and $t^* = \sqrt{\frac{(k-2)c}{a}}$

$$V(\bar{Y}_{(k)st}) = \frac{1}{k^2 m''}\left(a + \frac{b}{s} + \frac{c}{t} \right)$$

Hence the theorem. □

Example 19.6 RPs of Neyman's, 't-model' and '(s, t)-model' for three skewed distributions, LN (0, 1), Pareto (3), and Weibull (0.5) for the set size $k = 2, 3, \ldots 8$ are shown in Table 19.2.

It is to be noted that for $k = 2$, RPs are the same for all the distributions and (s,t) model is applicable only for $k > 2$. Further, all three RPs above increases as k increases.

Example 19.7 We wish to see the performance of the above three URSS models with the increasing values of skewness of family of distributions. For that purpose, we choose one skew distribution $LN(0, \sigma)$ with pdf:

$$f(x) = \frac{1}{x\,\sigma\,\sqrt{2\pi}}\, exp\left[\frac{-1}{2}\left(\frac{log x - \mu}{\sigma} \right)^2 \right], for\ x > 0, \mu > 0, \sigma > 0.$$

Table 19.3 RPs of URSS models w.r.t. SRS for $LN(0, \sigma)$ distributions for $k = 5$

p	σ	Sk	BRSS	t-model	(s, t) model	Neyman's model
1.8	0.77	3.40	1.8888	2.6862	2.9037	2.9626
1.9	0.80	3.70	1.8402	2.7043	2.9054	2.9858
2.0	0.83	4.00	1.7971	2.7197	2.9085	3.0031
2.1	0.86	4.30	1.7586	2.7328	2.9352	3.0216
2.2	0.89	4.60	1.7241	2.7441	2.9575	3.0373
2.3	0.91	4.90	1.6928	2.7539	2.9762	3.0586
2.4	0.94	5.21	1.6645	2.7625	2.9985	3.0816
2.5	0.96	5.51	1.6386	2.7700	3.0168	3.1012
2.6	0.98	5.82	1.6148	2.7766	3.0328	3.1180
2.7	1.00	6.13	1.5929	2.7824	3.0468	3.1323
2.8	1.01	6.44	1.5727	2.7875	3.0590	3.1445
2.9	1.03	6.75	1.5540	2.7920	3.0699	3.1549
3.0	1.05	7.07	1.5366	2.7961	3.0794	3.1639

The skewness(Sk) is given by

$$Sk = \sqrt{\beta_1} = \sqrt{exp(\sigma^2) - 1}(exp(\sigma^2) + 2)$$

The shape parameter (p) of this distribution is $p = Exp(\sigma^2)$.

The effect of the skewness on the performance of BRSS, t-model, (s, t) model, and Neyman's model for $k = 5$ is presented in Table 19.3. The values of σ and Sk are given for two decimal places only. It is shown that BRSS is not much performing as skewness increases. However, the gain in Neyman's, t, and (s,t) models are increasing with the increase of skewness.

19.3.2 Symmetric Underlying Distributions

For the skewed distributions, Neyman's model provides a substantial gain over BRSS. This is not true for the case of symmetric distributions. In fact, for symmetric distributions, the gain due to Neyman's allocation is marginal and remains very close to that of BRSS. Therefore, in such distributions, the strategy is precisely the reverse of Neyman's allocation which measures more times those order statistics that have the large standard deviations. In symmetric distributions, the strategy should be in such a way that to give the least preference to the order statistics having large standard deviations while making actual measurements. Hence, the performance improvement over Neyman's allocation can be quite high if the estimator is constructed from the order statistics having the smallest standard deviations.

Kaur et al. (2000) suggested an optimal allocation model for symmetric distributions and compared it with the BRSS and Neyman allocation models in terms of the RP of the estimator of μ. As in Neyman's criterion, one requires to measure all rank orders to ensure general unbiasedness. Kaur et al. (2000) used the concept of exploiting the symmetry to allow most of the order statistics to vanish and thereby to broaden the class of linear unbiased estimator of the population mean. Depending upon the class of the distribution, they suggested to measure either to the extreme ranks or to the middle rank(s) so as to attain the optimum allocation model. Yanagawa and Chen (1980) given a minimum variance linear unbiased estimator of μ for a group of symmetric distributions. Shirahata (1982) given general procedures that are unbiased for symmetric distributions.

Based on the growth patterns in the variance of the order statistics against the rank orders, the symmetric distributions can be classified into two classes. These classes are named as mound- and U-shaped distributions. In case of mound-shaped distributions, $\sigma^2_{(i:k)}$ is increasing in i for $1 \leq i \leq M$ and $\sigma^2_{(i:k)}$ is decreasing in i for $M \leq i \leq k$, where $M = \frac{k+1}{2}$ is the unique middle rank order when k is odd. While in the case of U-shaped distributions $\sigma^2_{(i:k)}$ is decreasing in i for $1 \leq i \leq M$ and $\sigma^2_{(i:k)}$ is increasing in i for $M \leq i \leq k$. If k is even, there are two middle rank orders $M - \frac{1}{2}$ and $M + \frac{1}{2}$.

Kaur et al. (2000) given the optimal allocation model for the mound- and U-shaped symmetric distributions without measuring all the rank orders as requires in the case of Neyman's optimal allocation model for skewed distributions. Only the rank order having the smallest variances and the rank orders having the largest variances are measured for the mound and U-shaped symmetric distributions respectively.

Theorem 19.4 *For any symmetric distribution, the unbiased estimator of μ based on the ith and $(k - i + 1)$th order statistics is given by*

$$\hat{\mu}_{(i:k)} = \check{\mu}_{(i:k)} = \begin{cases} \frac{1}{2}\left(\frac{T_i}{m_i} + \frac{T_{k-i+1}}{m_{k-i+1}}\right) & ; \text{for } 1 \leq i < M \\ \frac{T_M}{m_M} & ; for\ i = M\ and\ k\ is\ odd \end{cases} \quad (19.9)$$

Provided that m_i and $m_{(k-i+1)}$ are both positive. The hat and check in $\hat{\mu}_{(i:k)}$ and $\check{\mu}_{(i:k)}$ represents for mound- and U-shaped distributions, respectively.

The corresponding variance is given by

$$V(\hat{\mu}_{(i:k)}) = V(\check{\mu}_{(i:k)}) = \begin{cases} \frac{\sigma^2_{(i:k)}}{4}\left(\frac{1}{m_i} + \frac{1}{m_{k-i+1}}\right) & ; \text{for } 1 \leq i < M \\ \frac{\sigma^2_{(M:k)}}{m_M} & ; for\ i = M\ and\ k\ is\ odd. \end{cases} \quad (19.10)$$

Proof For any symmetric distributions, we have that

$$\mu = \begin{cases} \frac{1}{2}\left(\mu_{(i:k)} + \mu_{(k-i+1:k)}\right) & ; \text{for } 1 \leq i < \text{l}M \\ \mu_{(i:k)} & ; for\ i = M\ and\ k\ is\ odd \end{cases} \quad (19.11)$$

and

$$\sigma^2_{(i:k)} = \sigma^2_{(k-i+1:k)}, for\ 1 \leq i \leq M$$

Now taking expectation on Eq.(19.9), we have that

$$E\left(\hat{\mu}_{(i:k)}\right) = \begin{cases} \frac{1}{2}\left(\frac{\sum_{j=1}^{m_i} E(Y_{(i:k)j})}{m_i} + \frac{\sum_{j=1}^{m_{(k-i+1)}} E(Y_{(k-i+1:k)j})}{m_{k-i+1}}\right) & ; \text{for } 1 \leq i < M \\ \frac{\sum_{j=1}^{m_M} E(Y_{(M:k)j})}{m_M} & ; \text{for } i = M \text{ and } k \text{ is odd.} \end{cases}$$

Since, $Y_{(i:k)j}$ and $Y_{(k-i+1:k)j}$ for fixed i and $Y_{(M:k)j}$ are i.i.d with their respective means $\mu_{(i:k)}$, $\mu_{(k-i+1:k)}$ and $\mu_{(M:k)}$ and their respective variance are $\sigma_{(i:k)}^2$, $\sigma_{(k-i+1:k)}^2$ and $\sigma_{(M:k)}^2$. Therefore, using Eq. (19.11), we get
$$E(\hat{\mu}_{(i:k)}) = E(\breve{\mu}_{(i:k)}) = \mu$$
and

$$V\left(\hat{\mu}_{(i:k)}\right) = V\left(\breve{\mu}_{(i:k)}\right) = \begin{cases} \frac{1}{4}\left(\frac{\sum_{j=1}^{m_i} V(Y_{(i:k)j})}{m_i^2} + \frac{\sum_{j=1}^{m_i} V(Y_{(k-i+1:k)j})}{m_{k-i+1}^2}\right) & ; \text{for } 1 \leq i < M \\ \frac{\sum_{j=1}^{m_M} V(Y_{(M:k)j})}{m_M^2} & ; \text{for } i = M \text{ and } k \text{ is odd.} \end{cases}$$

Or

$$V\left(\hat{\mu}_{(i:k)}\right) = V\left(\breve{\mu}_{(i:k)}\right) = \begin{cases} \frac{\sigma_{(i:k)}^2}{4}\left(\frac{1}{m_i} + \frac{1}{m_{k-i+1}}\right) & ; \text{for } 1 \leq i < M \\ \frac{\sigma_{(M:k)}^2}{m_M} & ; \text{for } i = M \text{ and } k \text{ is odd.} \end{cases}$$

Hence the theorem.   □

It is to be noted that the variance given in Eq. (19.10) is minimized if one take the allocation as balanced as possible from both side of the middle-order statistics. That is for fixed $n_i = m_i + m_{k-i+1}$, $m_i = m_{k-i+1} = \frac{n_i}{2}$. While making the practical application, the allocation m_i or m_{k-i+1} may be odd. In this case one may take

$$m_i = floor\left(\frac{n_i}{2}\right), m_{k-i+1} = ceil\left(\frac{n_i}{2}\right) \tag{19.12}$$

For example, for $n_i = 15$, we have $m_i = 7$ and $m_{k-i+1} = 8$.
  In this case the variance is given by

$$V(\hat{\mu}_{(i:k)}) = V(\breve{\mu}_{(i:k)}) = \begin{cases} \frac{\sigma_{(i:k)}^2}{n_i} & ; \text{for } 1 \leq i < M \text{ and } n_i \text{ is even} \\ \frac{n_i}{n_i^2-1}\sigma_{(i:k)}^2 & ; \text{for } 1 \leq i < M \text{ and } n_i \text{ is odd} \\ \frac{\sigma_{(M:k)}^2}{m_M} & ; \text{for } i = M \text{ and } k \text{ is odd.} \end{cases}$$
$$\tag{19.13}$$

Theorem 19.5 *Depending upon the class of symmetric distributions, the optimal allocation measures all sampling units either to the extreme or middle-order statistics. For mound-shaped distribution,*

$$m_i = m_{k-i+1} = \begin{cases} \frac{n}{2} & ; for\ i = 1 \\ 0 & ; for\ any\ i\ other\ than\ 1. \end{cases}$$

For U-shaped distribution,

$$m_i = m_{k-i+1} = \begin{cases} \frac{n}{2} & ; for\ i = \frac{k}{2}\ and\ k\ is\ even \\ n & ; for\ i = M\ and\ k\ is\ odd \\ 0 & ; for\ any\ i\ other\ than\ above. \end{cases}$$

In case n is odd, two values of $\frac{n}{2}$ may be taken as $m_i = floor(\frac{n}{2})$, $m_{k-i+1} = ceil(\frac{n}{2})$ for the practical utility and variance of the estimator is given by Eq.(19.13).

Theorem 19.6 *The asymptotic variances of the best linear unbiased estimator (BLUE) of μ under mound-shaped distributions are given, respectively, by*

$$\sigma^2_{Mound} = \frac{\sigma^2_{(1:k)}}{n} \tag{19.14}$$

The asymptotic variance of the BLUE of μ under U-shaped distributions is

$$\sigma^2_U = \begin{cases} \frac{\sigma^2_{(\frac{k}{2}:k)}}{n} & ; if\ k\ is\ even \\ \frac{\sigma^2_{(M:k)}}{n} & ; if\ k\ is\ odd \end{cases} \tag{19.15}$$

After comparing with the variance under SRS, the asymptotic RP are

$$RP_{Mound} = \frac{\sigma^2}{\sigma^2_{(1:k)}} \tag{19.16}$$

and

$$RP_U = \begin{cases} \frac{\sigma^2}{\sigma^2_{(\frac{k}{2}:k)}} & ; if\ k\ is\ even \\ \frac{\sigma^2}{\sigma^2_{(\frac{k+1}{2}:k)}} & ; if\ k\ is\ odd \end{cases} \tag{19.17}$$

Proof Here the linear unbiased estimator of $\hat{\mu}_{(i:k)}(i = 1, 2, \ldots M)$ for μ can be written as
$a'\hat{\mu} = \sum_{i=1}^{M} a_i \hat{\mu}_{(i:k)}$ with $\sum_{i=1}^{M} a_i = 1$,
where $a' = (a_1, a_2, \ldots a_M)$ and $\hat{\mu} = (\hat{\mu}_{(1:k)}, \hat{\mu}_{(2:k)}, \ldots \hat{\mu}_{(M:k)}$. All $\hat{\mu}_{(i:k)}(i = 1, 2, \ldots M)$ are the independent random variables with common mean μ and different variances as given in equation (19.10). Now the following lemma completes the proof. □

Lemma 19.1 *Let $X_1, X_2, \ldots X_n$, are n independent random variables with a common mean μ and with variances $\sigma_1^2, \sigma_2^2, \ldots \sigma_n^2$. The linear combination $A_1 X_1 + A_2 X_2 + \cdots + A_n X_n$, with $A_1 + A_2 + \cdots + A_n = 1$, which has the smallest vari-*

Table 19.4 RPs of different models w.r.t. SRS for U (0, 1) and N(0, 1) for k = 2(1)8

k	U (0, 1)		N(0,1)	
	Optimal model	Neyman's model	Optimal model	Neyman's model
2	1.5	1.5	1.467	1.467
3	2.222	1.846	2.229	1.747
4	3.125	2.381	2.774	2.199
5	4.2	2.983	3.486	2.656
6	5.445	3.5	4.062	3.186
7	6.857	4.081	4.752	3.633
8	8.437	4.563	5.342	3.932

ance and is obtained by taking A_i inversely proportional to σ_i^2. The resulting minimum variance is

$$\frac{1}{\frac{1}{\sigma_1^2}+\frac{1}{\sigma_2^2}+\cdots+\frac{1}{\sigma_n^2}}$$

Example 19.8 RPs of Optimal and Neyman's models for mound- U(0, 1) and U-shaped N(0, 1) distributions, for the set size $k = 2, 3, \ldots, 8$ are shown (Table 19.4).

It is seen that the performance of Neyman's model is marginal for both types of symmetric distribution, however, both are increasing in k.

19.4   Some Important Applications

In this section, we discuss few applications of RSS done in the past in various sectors which shows the advantages of RSS over SRS. The first important application of RSS is observed for the estimation of dry bark yield and quinine contents from cinchona plants in Sengupta et al. (1951). They found that stripping the bark, and drying it until the weight reaches a steady state is rather easy than the measurement of dry bark yield, as it requires uprooting the plant. Halls and Dell (1966) applied RSS for estimating the weights of browse and herbage in a pine-hardwood forest of east Texas, USA. Martin et al. (1980) applied the RSS procedure for estimating shrub phytomass in Appalachian Oak forests. Cobby et al. (1985) investigated the performance of RSS relative to SRS for estimation of herbage mass in pure grass swards, and of herbage mass and clover content in mixed grass-clover swards. Similarly, Johnson et al. (1993) studied to estimate the mean of forest, grassland, and other vegetation resources using RSS. Nussbaum and Sinha (1997) successfully used RSS in estimating mean Reid vapor pressure. Mode et al. (1999) investigated the conditions for ecological and environmental field study under which RSS becomes a cost-effective method and the rough but cheap measurement has a cost. Al-Saleh and Al-Shrafat (2001) studied the performance of RSS in estimation milk yield based on 402 sheep in East Jordan. Al-Saleh and Al-Omari (2002) proposed the multistage RSS and used to estimate the average of Olive's yields in a field in West Jordan. Wang et al. (2009) used the RSS in fisheries research. Chandra et al. (2018) attempted a study in response estimation of the developmental programs implemented by the government and non-government organizations in successive phases by the use of RSS.

19.5 Imperfect Ranking in RSS

It is understood that RSS procedures perform better in the case of perfect ranking, i.e., in the absence of ranking errors of units. This situation has rarely been seen in the practical situation for the large values of k. Minimal uncertainty in the rankings may not cause an excessive increase in the mean square error, but if the ranking process is not done correctly, the precision of RSS estimators (particularly in UBRSS) may be reduced. Bohn and Wolfe (1992) discussed Mann–Whitney–Wilcoxon (MWW) procedures under imperfect ranking proposing a model for the probabilities of imperfect judgment rankings based on the concept of expected spacing and have used this model to study the properties of tests based on the ranked set analogue of the MWW statistic. Greater precision in the RSS estimator requires more accurate rankings in each set for each cycle (David and Levine 1972; Barnett and Moore 1997; Stark and Wolfe 2002). Al-Omari and Bouza (2014) discussed the impact of perfect and imperfect ranking on the performance of RSS. Much of the literature detailing the improvements from RSS estimation is based on the assumption of perfect rankings. This is the natural starting point for developing theory, but it is not realistic in practice. With imperfect rankings, the precision of the URSS estimator may actually be worse than that for the SRS estimator. Some of the situations have been discussed by Dell and Clutter (1972)), in which they showed that the estimator for population mean based on RSS is at least as efficient as the estimator based on SRS with the same number of measurements even when there are ranking errors. This suggests that URSS should not be used without first considering the setting of the study for ranking. If the amount of error in the ranking procedure is expected to be the only minor, then it is safe to use RSS, balanced or unbalanced, to improve estimation. When considerable uncertainty exists regarding the exactness of rankings in the procedure, BRSS may be the safest approach to take.

19.6 Exercises

1. Describe the possible situations of the survey in which the RSS would be used. Define BRSS and URSS for the set size k with one example of each.
2. If a person draws a sample of size 15 using BRSS. Obtain the value of m when the set size $k = 2, 5, 10$ and 20. Also obtain the number of sampling units actually to be drawn from the population in each of these k.

3. Obtain the unbiased estimator of population mean under BRSS. Derive its variance. Show that the BRSS is more precise than the SRS in terms of relative precision while estimating the population mean.

4. Under the perfect ranking, show that $1 < RP_{SRS:BRSS} < \frac{k+1}{2}$ by establishing the population mean.

5. For the Weibull (0.5) distribution, obtain $RP_{SRS:BRSS}$ for $k = 5$, when $\mu_{(i:5)}$ and $\sigma^2_{(i:5)}$ are given below:

$\mu_{(i:5)}$	0.08000	0.30500	0.82722	2.11056	6.67722
$\sigma^2_{(i:5)}$	0.03200	0.22212	1.10282	5.87228	62.94732

6. Suppose the population mean and variance of any population is given as 6.5760 and 7.7462, respectively. The variances corresponding k=3 and 5 with m=1 were calculated from the population as \

k\i	1	2	3	4	5
3	2.9368	3.6079	6.5763		
5	2.9368	3.6079	6.5763	3.6952	5.1143

Obtain $V(\bar{Y}_{(k)BRSS})$, the term σ^2 and hence $RP_{SRS:BRSS}$.

7. Show that $RP_{SRS:BRSS}$ increases as the set size k increases for any given distributions $F(y)$.

8. Find the unbiased estimator of population mean under URSS. Derive its variance. Show that the URSS is more precise than the BRSS in terms of relative precision while estimating the population mean.

9. Discuss different important allocation models in RSS. Show that Neyman's allocation provides the optimal allocation in URSS for positively skewed distributions.

10. Derive the optimum value of s and t in (s, t) model. Find these values for the Weibull (0.5) distribution given in exercise 5 above.

11. Determine the performance of t and (s, t) model for any one symmetric distribution. Discuss the results you have obtained.

12. Show that $RP_{SRS:st} \geq RP_{SRS:t}$ for positively skewed distributions.

13. Explain the reasons why the Neyman allocation does not perform well under UBRSS for the symmetric distributions. Discuss the optimal model for the symmetric distributions. Also, highlight the limitations of the optimal model for symmetric distribution.

14. Write a short note about the imperfect rankings in RSS.

References

Al-Omari, A.I., Bouza, C.N.: Review of ranked set sampling: modifications and applications. Investig. Oper. **35**(3), 215–240 (2014)

Al-Saleh, M.F., Al-Omari, A.I.: Multistage ranked set sampling. J. Stat. Plan. Inference **102**, 273–286 (2002)

Al-Saleh, M.F., Al-Shrafat, K.: Estimation of milk yield using ranked set sampling. Envirometrics **12**, 395–399 (2001)

Barabesi, L., Fattorini, L.: Kernel estimators of probability density functions by ranked set sampling. Commun. Stat. Theory Methods **31**(4), 597–610 (2002)

Barnett, V., Moore, K.: Best linear unbiased estimates in ranked-set sampling with particular reference to imperfect ordering. J. Appl. Stat. **24**, 697–710 (1997)

Bhoj, D.S., Chandra, G.: Simple unequal allocation procedure for ranked set sampling with skew distributions. J. Mod. Appl. Stat. Methods **18**(2), eP2811 (2019)

Bohn, L.L., Wolfe, D.A.: Nonparametric two-sample procedures for ranked set samples data. J. Am. Stat. Assoc. **87**(418), 552–561 (1992)

Chandra, G., Pandey, R., Bhoj, D.S., Nautiyal, R., Ashraf, J., Verma, M.R.: Ranked set sampling approach for response estimation of developmental programs with linear impacts under successive phases. Pak. J. Stat. **34**(2), 163–176 (2018)

Chen, Z.: The optimal ranked set sampling scheme for inference on population quantiles. Stat. Sin. **11**, 23–37 (2001)

Chen, M., Lim, J.: Estimating variances of strata in ranked set sampling. J. Stat. Plan. Inference **141**, 2513–2518 (2011)

Chen, H., Stasny, E.A., Wolfe, D.A.: Unbalanced ranked set sampling for estimating a population proportion. Biometrics **62**, 150–158 (2006)

Cobby, J.M., Ridout, M.S., Bassett, P.J., Large, R.V.: An investigation into the use of ranked set sampling on grass and grassclover swards. Grass Forage Sci. **40**, 257–63 (1985)

David, H.A., Levine, D.N.: Ranked set sampling in the presence of judgment error. Biometrics **28**, 553–555 (1972)

Dell, T. R. (1969). The theory of some applications of ranked set sampling. Ph.D. Thesis, University of Georgia, Athens, GA

Dell, T.R., Clutter, J.L.: Ranked set sampling theory with order statistics background. Biometrics **28**, 545–555 (1972)

Frey, J.: Bootstrap confidence bands for the CDF using ranked set sampling. J. Korean Stat. Soc. **43**(3), 453–461 (2014)

Halls, L.S., Dell, T.R.: Trial of ranked set sampling for forage yields. For. Sci. **12**(1), 22–26 (1966)

Harter, H.L., Balakrishnan, N.: CRC Handbook of Tables for the Use of Order Statistics in Estimation. CRC Press, Boca Raton (1996)

Johnson, G.D., Paul, G.P., Sinha, A.K.: Ranked set sampling for vegetation research. Abstr. Bot. **17**, 87–102 (1993)

Kaur, A., Patil, G.P., Taillie, C.: Unequal allocation models for ranked set sampling with skew distributions. Technical report 94-0930, Centre for Statistical Ecology and Environmental Statistics, Pennsylvania State University, University Park (1994)

Kaur, A., Patil, G.P., Taillie, C.: Unequal allocation models for ranked set sampling with skew distributions. Biometrics **53**, 123–130 (1997)

Kaur, A., Patil, G.P., Taillie, C.: Optimal allocation for symmetric distributions in ranked sampling. Ann. Inst. Stat. Math. **52**(2), 239–254 (2000)

Lam, K.F., Philip, L.H., Lee, C.F.: Kernel method for the estimation of the distribution function and the mean with auxiliary information in ranked set sampling. Environmetrics **13**, 397–406 (2002)

Lim, J., Chen, M., Park, S., Wang, X., Stokes, L.: Kernel density estimator from ranked set samples. Commun. Stat. Theory Methods **43**, 2156–2168 (2014)

MacEachern, S.N., Ozturk, O., Wolfe, D.A., Stark, G.V.: A new ranked set sample estimator of variance. J. R. Stat. Soc. **B64**, 177–188 (2002)

Martin, W., Sharik, T., Oderwald, R., Smith, D.: Evaluation of ranked set sampling for estimating shrub phytomass in Appalachian oak forests. Publication Number FWS-4-80, School of Forestry and Wildlife Resources, Virginia Polytechnic Institute and State University, Blacksburg, Virginia (1980)

McIntyre, G.A.: A method for unbiased selective sampling using ranked sets. Aust. J. Agric. Res. 3, 385–390 (1952)

Mode, N.A., Conquest, L.L., Marker, D.M.: Ranked set sampling for ecological research: accounting for the total costs of sampling. Environmetrics 10, 179–194 (1999)

Murff, E.J., Sager, T.W.: The relative efficiency of ranked set sampling in ordinary least squares regression. Environ. Ecol. Stat. 13, 41–51 (2006)

Nussbaum, B.D., Sinha, B.K.: Cost effective gasoline sampling using ranked set sampling. In: Proceedings of the Section on Statistics and the Environment, pp. 83–87. American Statistical Association, Alexandria, VA (1997)

Perron, F., Sinha, B.K.: Estimation of variance based on a ranked set sample. J. Stat. Plan. Inference 120, 21–28 (2004)

Samawi, H.M.: On Quantile Estimation With Application to Normal Ranges and Hodges-Lehmann Estimate Using a Variety of Ranked Set Sample. Department of Statistics, Yarmouk Univervesity, Irbid, Jordan (1999)

Sengupta, J.M., Chakravarti, I.M., Sarkar, D.: Experimental survey for the estimation of cinchona yield. Bull. Int. Stat. Inst. 33, 313–331 (1951)

Sengupta, S., Mukhuti, S.: Unbiased variance estimation in a simple exponential population using ranked set samples. J. Stat. Plan. Inference 139, 1526–1553 (2006)

Shirahata, S.: An extension of the ranked set sampling theory. J. Stat. Plan. Inference 6, 65–72 (1982)

Stark, G.V., Wolfe, D.A.: Evaluating ranked set sampling estimators with imperfect rankings. Journal of Statistical Studies, Special Volume in Honour of Professor Mir Masoom Ali?s 65th Birthday, 77–103 (2002)

Stokes, S.L.: Estimation of variance using judgment ordered ranked-set samples. Biometrics 36, 35–42 (1980)

Takahasi, K.: Practical note on estimation of population means based on samples stratified by means of ordering. Ann. Inst. Stat. Math. 22(1), 421–428 (1970)

Takahasi, K., Wakimoto, K.: On unbiased estimates of the population mean based on the sample stratified by means of ordering. Ann. Inst. Stat. Math. 20, 1–31 (1968)

Tiwari, N., Chandra, G.: A systematic procedure for unequal allocation for skewed distributions in ranked-set sampling. J. Indian Soc. Agric. Stat. 65(3), 331–338 (2011)

Wang, Y.G., Ye, Y., Milton, D.A.: Efficient designs for sampling and subsampling in fisheries research based on ranked sets. J. Mar. Sci. 66, 928–934 (2009)

Wolfe, D.A.: Ranked set sampling: an approach to more efficient data collection. Stat. Sci. 19(4), 636–643 (2004)

Yanagawa, T., Chen, S.: The MG-procedure in ranked-set sampling. J. Stat. Plan. Inference 4, 33–44 (1980)

Yanagawa, T., Shirahata, S.: Ranked set sampling theory with selective probability matrix. Aust. J. Stat. 18, 45–52 (1976)

Zhu, M., Wang, Y.: Quantile estimation from ranked set sampling data. Sankhya Indian J. Stat. 67(2), 295–304 (2005)

Chapter 20
RSS in Other Parameteric and Non-parametric Inference

20.1 Introduction

In Chap. 19, the method of RSS for estimating the population mean of the distribution has been discussed. The theory of RSS has been extended in this chapter to estimate location and scale parameters of distributions, population proportion, and quantiles. The original concept (McIntyre 1952) of RSS is completely non-parametric in nature and assumed that the population distribution is not known beforehand, therefore, the concept of RSS to estimate location, scale, and quantiles of the distributions is helpful. The application of RSS has also been attempted for the non-parametric inference. Some non-parametric tests based on one sample and two samples are discussed.

20.2 Estimation of Location and Scale Parameters

For the situation of unknown parameters of the distribution, one has a great challenge to estimate them with minimum cost and maximum precision. A lot of work has already been attempted by various researchers in the area of parameter estimation of the distributions using SRS. However, recent researchers started to explore the use of RSS in parameter estimation of the well-known distributions. Lam et al. (1994) worked on the estimation of parameters of two-parameter exponential distribution. They showed that the proper use of RSS results in much improved estimators compared to SRS. Similar other studies are also available in the literature. For example, generalized geometric distribution (Bhoj and Ahsanullah 1996), normal and exponential distribution (Sinha et al. 1996), extreme value distribution (Bhoj 1997), Gumbel distribution (Yousef and Al-Subh 2014), lognormal distribution (Chandra et al. 2016), etc. The parametric inference for the parameters of the location-scale family of distributions based on RSS is given by Ozturk (2011).

© The Author(s), under exclusive license to Springer Nature Singapore Pte Ltd. 2021
R. Latpate et al., *Advanced Sampling Methods*,
https://doi.org/10.1007/978-981-16-0622-9_20

It is to be known that the real-valued parameter θ under probability distribution function (pdf) $f_\theta(y)$ of a random variable Y is the location parameter if $Y - \theta$ has pdf $f(y)$ which is free of θ. The parameter θ is said to be a scale parameter if the pdf of Y/θ has pdf $f(y)$ which is free of θ. If $\theta = (\mu, \sigma)$, then θ is a location-scale parameter if the pdf of $\frac{Y-\mu}{\theta}$ is free from μ and σ. Lloyd (1952) first proposed the estimation technique of location and scale parameters, based on an ordered sample. After two years, Downton (1954) used the method of least squares for estimation of location and scale parameters based on ordered observations. The method of RSS seems to be a bit more easy and precise due to the easy use of the diagonal matrix of variance–covariance matrix of ordered observation. This method is described as follows.

Suppose the location and scale parameters of a variate Y are μ and σ, respectively, which are not necessarily the mean and standard deviation and the distribution of Y is depending on only these two parameters. Let $(Y_1, Y_2, \ldots, Y_n)$ be a simple random sample of size n on Y. After arranging these observation (Y_i's) in ascending order of magnitude, suppose $Y_{1n}, Y_{2n}, \ldots, Y_{nn}$ be the order statistics, so that $Y_{1n} \leq Y_{2n} \leq \cdots \leq Y_{nn}$.

The standardized variate $U_{in} = \frac{Y_{in}-\mu}{\sigma}$, $i = 1, 2, \ldots, n$, whose distribution is parameter free, may be regarded as independent observations on the standardized variable $U = \frac{Y-\mu}{\sigma}$.

Let

$$E(U_{in}) = c_{i:n}, \quad i = 1, 2, \ldots, n. \tag{20.1}$$

$$V(U_{in}) = d_{ii}, \quad Cov(U_{in}, U_{jn}) = d_{ij}, \quad i, j = 1, 2, \ldots, n. \tag{20.2}$$

The quantities $E(U_{in})$, $V(U_{in})$, and $Cov(U_{in}, U_{jn})$ have known values depending on the form of the parent distribution but not on the parameters.

Suppose, the ranked set sample $\left(Y_{(11)}, Y_{(22)}, \ldots, Y_{(nn)}\right)$ are actually measured accurately on Y, which is to be used in estimating μ and σ. The resulting estimates are unbiased, linear in the ordered observations, and have minimal variance. It is known that $Y_{(ii)}$ is distributed the same as Y_{in} and the sample $\left(Y_{(11)}, Y_{(22)}, \ldots, Y_{(nn)}\right)$ are independent but not identically distributed.

After using Eqs. (20.1) and (20.2) to the original ordered observations for RSS, we have

$$E(Y_{(ii)}) = \mu + \sigma E(U_{in}) = \mu + c_{i:n}\sigma, \quad i = 1, 2, \ldots, n. \tag{20.3}$$

$V(Y_{(ii)}) = \sigma^2 V((U_{in}) = \sigma^2 d_{ii}, \quad i = 1, 2, \ldots, n$
$Cov(Y_{(ii)}, Y_{(jj)}) = 0, \quad i \neq j, \quad i, j = 1, 2, \ldots, n.$

The ordered observations have expectations which are linear functions of the parameters μ and σ, with known coefficients, variances and covariance which are known up to a scale factor σ^2. Therefore, the Gauss and Markoff theorem of least squares can be applied to them.

The parameters are, therefore, estimable by unbiased linear functions of $Y_{(ii)}$, with minimal variance. Equation (20.3) can be rewritten in the matrix form as
$$E(Y) = \mu 1 + \sigma C = p\Theta.$$
where $Y' = (Y_{(11)}, Y_{(22)}, \ldots, Y_{(nn)})$, C is the vector of c_{in}, 1 is the vector of unit elements, p is the $(n \times 2)$ matrix $(1, C)$ and $\Theta' = (\mu, \sigma)$.

The variance–covariance matrix of $Y_{(ii)}$ is
$$V(Y) = \sigma^2 d,$$
where
$$d = \begin{bmatrix} d_{11} & 0 & 0 & \ldots & 0 \\ 0 & d_{22} & 0 & \ldots & 0 \\ \vdots & \vdots & \vdots & \ddots & \vdots \\ 0 & 0 & 0 & \ldots & d_{nn} \end{bmatrix} \quad \text{is the symmetric positive-definite matrix.}$$

The diagonal matrix is obtained in the case of RSS not in SRS which led to easy computations. Using Gauss and Markoff theorem of least squares, the required estimator of the vector Θ of parameters is given by

$$\hat{\Theta} = (p'Dp)^{-1} p'DY, \tag{20.4}$$

where $D = d^{-1} = \begin{bmatrix} \frac{1}{d_{11}} & 0 & 0 & \ldots & 0 \\ 0 & \frac{1}{d_{22}} & 0 & \ldots & 0 \\ \vdots & \vdots & \vdots & \ddots & \vdots \\ 0 & 0 & 0 & \ldots & \frac{1}{d_{nn}} \end{bmatrix}.$

The variance–covariance matrix of the estimator is $(p'Dp)^{-1} \sigma^2$, where
$$p'Dp = \begin{bmatrix} 1'D1 & 1'DC \\ 1'DC & c'DC \end{bmatrix}$$
The inverse of this matrix is

$$(p'Dp)^{-1} = \frac{1}{\Delta} \begin{bmatrix} C'DC & -1'DC \\ -1'DC & 1'D1 \end{bmatrix} \tag{20.5}$$

where Δ is the determinant of the matrix $p'Dp$.

While applying Eq. (20.5) in (20.4), the required estimates are written as
$$\hat{\mu}_{RSS} = -C'\Gamma Y, \quad \hat{\sigma}_{RSS} = 1'\Gamma Y,$$
where, Γ is the skew-symmetric matrix defined by
$$\Gamma = \frac{D(1C' - C1')D}{\Delta}$$
Using Eq. (20.5), the variances and covariance of these estimates are written as
$$V(\hat{\mu}_{RSS}) = \frac{C'DC\sigma^2}{\Delta}, \quad V(\hat{\sigma}_{RSS}) = \frac{1'D1\sigma^2}{\Delta}, \quad Cov(\hat{\mu}_{RSS}, \hat{\sigma}_{RSS}) = \frac{-1'DC\sigma^2}{\Delta}$$
The unique minimum variance linear estimators of μ and σ based on $(Y_{(11)}, Y_{(22)}, \ldots, Y_{(nn)})$ are obtained as
$$\hat{\mu}_{RSS} = \sum_{i=1}^{n} v_i Y_{(ii)}, \quad \hat{\sigma}_{RSS} = \sum_{i=1}^{n} w_i Y_{(ii)}$$
where
$$v_i = \frac{U_n}{d_{ii}} + V_n \frac{c_{i:n}}{d_{ii}}, \quad w_i = \frac{V_n}{d_{ii}} + W_n \frac{c_{i:n}}{d_{ii}}$$

$$U_n = \frac{1}{\rho}\sum_{i=1}^n \frac{c_{i:n}^2}{d_{ii}}, \quad V_n = -\frac{1}{\rho}\sum_{i=1}^n \frac{c_{i:n}}{d_{ii}}, \quad W_n = \frac{1}{\rho}\sum_{i=1}^n \frac{1}{d_{ii}}$$

$$\rho = \left(\sum_{i=1}^n \frac{1}{d_{ii}}\right)\left(\sum_{i=1}^n \frac{c_{i:n}^2}{d_{ii}}\right) - \left(\sum_{i=1}^n \frac{c_{i:n}}{d_{ii}}\right)^2$$

Finally, the simplified form of $\hat{\mu}_{RSS}$ and $\hat{\sigma}_{RSS}$ are given as

$$\hat{\mu}_{RSS} = \frac{\left(\sum_{i=1}^n \frac{Y_{(ii)}}{d_{ii}}\right)\left(\sum_{i=1}^n \frac{c_{i:n}^2}{d_{ii}}\right) - \left(\sum_{i=1}^n \frac{c_{i:n}}{d_{ii}}\right)\left(\sum_{i=1}^n \frac{c_{i:n}Y_{(ii)}}{d_{ii}}\right)}{\left(\sum_{i=1}^n \frac{1}{d_{ii}}\right)\left(\sum_{i=1}^n \frac{c_{i:n}^2}{d_{ii}}\right) - \left(\sum_{i=1}^n \frac{c_{i:n}}{d_{ii}}\right)^2}$$

$$\hat{\sigma}_{RSS} = \frac{\left(\sum_{i=1}^n \frac{1}{d_{ii}}\right)\left(\sum_{i=1}^n \frac{c_{i:n}Y_{(ii)}}{d_{ii}}\right) - \left(\sum_{i=1}^n \frac{c_{i:n}}{d_{ii}}\right)\left(\sum_{i=1}^n \frac{Y_{(ii)}}{d_{ii}}\right)}{\left(\sum_{i=1}^n \frac{1}{d_{ii}}\right)\left(\sum_{i=1}^n \frac{c_{i:n}^2}{d_{ii}}\right) - \left(\sum_{i=1}^n \frac{c_{i:n}}{d_{ii}}\right)^2}.$$

Similarly, the simplified form of variance of the estimators $\hat{\mu}_{RSS}$ and $\hat{\sigma}_{RSS}$ and $Cov\left(\hat{\mu}_{RSS}, \hat{\sigma}_{RSS}\right)$ are written as

$$V\left(\hat{\mu}_{RSS}\right) = \sigma^2 \frac{\left(\sum_{i=1}^n \frac{c_{i:n}^2}{d_{ii}}\right)}{\left(\sum_{i=1}^n \frac{1}{d_{ii}}\right)\left(\sum_{i=1}^n \frac{c_{i:n}^2}{d_{ii}}\right) - \left(\sum_{i=1}^n \frac{c_{i:n}}{d_{ii}}\right)^2},$$

$$V\left(\hat{\sigma}_{RSS}\right) = \sigma^2 \frac{\left(\sum_{i=1}^n \frac{1}{d_{ii}}\right)}{\left(\sum_{i=1}^n \frac{1}{d_{ii}}\right)\left(\sum_{i=1}^n \frac{c_{i:n}^2}{d_{ii}}\right) - \left(\sum_{i=1}^n \frac{c_{i:n}}{d_{ii}}\right)^2},$$

$$Cov\left(\hat{\mu}_{RSS}, \hat{\sigma}_{RSS}\right) = -\sigma^2 \frac{\left(\sum_{i=1}^n \frac{c_{i:n}}{d_{ii}}\right)}{\left(\sum_{i=1}^n \frac{1}{d_{ii}}\right)\left(\sum_{i=1}^n \frac{c_{i:n}^2}{d_{ii}}\right) - \left(\sum_{i=1}^n \frac{c_{i:n}}{d_{ii}}\right)^2}.$$

The values of $c_{i:n}$'s and d_{ii}'s for many of the well-known distribution are available in the literature (Harter and Balakrishnan 1996). In order to compute the values of $\hat{\mu}_{RSS}$ and $\hat{\sigma}_{RSS}$, the coefficients v_i and w_i can be easily calculated. For instance, for lognormal distribution $LN(0, 1)$, the values of v_i and w_i, for sample size $n = 2, 3, \ldots, 10$, are computed in Chandra and Tiwari (2012). For $LN(0, 1)$, the values of RP of RSS estimate of the parameters with respect to their SRS estimates are given in Table 20.1. Table 20.1 also gives the RP for comparing the generalized variances of estimators under RSS with respect to generalized variance under SRS. The generalized variance under RSS is given by

$$GV(\hat{\mu}_{RSS}, \hat{\sigma}_{RSS}) = V\left(\hat{\mu}_{RSS}\right)V\left(\hat{\sigma}_{RSS}\right) - \left(Cov\left(\hat{\mu}_{RSS}, \hat{\sigma}_{RSS}\right)\right)^2.$$

A similar definition of generalized variance is followed for the case of SRS.

Table 20.1 indicates that RSS seems to be a better method for estimating location and scale parameters for higher values of set size k. The better side is that RP increases as the set size k increases for all the cases. Further, the gain in precision in terms of the generalized variance is considerably higher than the individual gains in estimating μ and σ.

Table 20.1 Comparison of RPs of RSS with respect to SRS for LN(0, 1)

n	2	3	4	5	6	7	8	9	10
$RP\left(\hat{\mu}_{RSS}, \hat{\mu}_{SRS}\right)$	0.633	0.682	0.772	0.870	0.968	1.064	1.157	1.249	1.339
$RP\left(\hat{\sigma}_{RSS}, \hat{\sigma}_{SRS}\right)$	0.813	0.995	1.231	1.486	1.749	2.016	2.286	2.557	2.829
$RP\left(GV\left(\hat{\mu}_{RSS}, \hat{\sigma}_{RSS}; \hat{\mu}_{SRS}, \hat{\sigma}_{SRS}\right)\right)$	0.862	1.225	1.738	2.360	3.078	3.884	4.775	5.744	6.794

20.3 Estimating Population Proportion

Although BRSS leads to improvement in precision over SRS for estimating population proportion, the kit is not in general, optimal for reducing variance. Chen et al. (2005) discussed first time the use of BRSS in estimating a population proportion. Their result showed that BRSS is performing much better over SRS when the population proportion is close to 0.5. The case of estimating population proportion from a URSS under perfect ranking was considered by Chen et al. (2006). Here, the probabilities of success for order statistics are functions of the underlying population proportion. They have shown that for estimating population proportions, Neyman's allocation is optimal within the class of available RSS estimators, which are simple averages of the means of the order statistics.

Suppose the variable of interest Y follows the Bernoullie distribution with parameter p, the probability of success outcome. In fact p is the population proportion. The values of Y's are the dichotomous-type values (0 or 1). A sample of size $n = \sum_{i=1}^{k} m_i$ is drawn using the method of URSS. The optimal URSS may be chosen as detailed in Chap. 19. Let $Y_{(i:k)j}$ (either 0 or 1) denote the value of characteristic under study of the ith-order statistics in the jth cycle of URSS for set size k. The unbiased estimator of population proportion p under URSS can be written as

$\hat{p}_{URSS} = \frac{1}{k} \sum_{i=1}^{k} \hat{p}_{(i)}$

where $\hat{p}_{(i)} = \frac{1}{m_i} \sum_{j=1}^{m_i} Y_{(i:k)j}$ is the estimator of proportion of success (value 1) of ith-order statistic. It also denotes the estimated probability of success for ith judgment ordered statistic from the Bernoullie distribution with parameter p. The variance of $\hat{p}_{RSS}$ is given by

$V\left(\hat{p}_{URSS}\right) = \frac{1}{k^2} \sum_{i=1}^{k} \frac{\hat{p}_{(i)}\left(1-\hat{p}_{(i)}\right)}{m_i}$

This RSS variance can be compared with the SRS variance for the same number of observations n.

The variance of SRS estimator of p can be written as

$V\left(\hat{p}_{SRS}\right) = \frac{p(1-p)}{n}$.

The RP of RSS estimator with the SRS estimator is written as

$RP_{UBRSS} = \frac{V\left(\hat{p}_{SRS}\right)}{V\left(\hat{p}_{URSS}\right)} = \frac{\frac{p(1-p)}{n}}{\frac{1}{k^2} \sum_{i=1}^{k} \frac{\hat{p}_{(i)}\left(1-\hat{p}_{(i)}\right)}{m_i}}$.

When we have the BRSS with $m_1 = m_2 = \cdots = m_k$, the RP shall become

$RP_{BRSS} = \frac{p(1-p)}{\frac{1}{k} \sum_{i=1}^{k} \hat{p}_{(i)}\left(1-\hat{p}_{(i)}\right)}$.

Chen et al. (2006) showed that RP_{BRSS} achieves the lower bound 1 while RP under the Neyman's allocation achieved the upper bound k. However, the results $1 \le RP_{BRSS} \le \frac{k+1}{2}$ and $0 \le RP_{UBRSS} \le k$ of Takahasi and Wakimoto (1968) are true for population proportion as well.

20.4   Quantile Estimation

This section deals with the problem of quantile estimation for the distributions whose parent form of the distribution is known but the parameters are unavailable. A URSS approach for quantile estimation is summarized as follows:

Suppose the unknown cumulative distribution function (CDF) is denoted by $F(y)$. A random sample of size n is drawn from F using URSS such that m_i observations are taken for measurements for ith rank order statistic, $i = 1, 2, \ldots k$, $\sum_{i=1}^{k} m_i = n$. Let $Y_{(i:k)j}$, $i = 1, 2 \ldots k$; $j = 1, 2 \ldots m_i$ denotes the jth observation of ith -order statistic for set size. Note that the observations corresponding to the ith rank order statistic are i.i.d. with mean $\mu_{(i:k)}$ and variance $\sigma^2_{(i:k)}$.

Letting $F_{(i)}$ be the CDF of ith rank order statistic and corresponding pdf is $f_{(i)}$. The aim is to estimate pth quantile, ζ_p under perfect ranking, where

$$\zeta_p = inf(y : F(y) \geq p) \tag{20.6}$$

We know that
$f_{(i)}(y) = \frac{k!}{(i-1)!(k-i)!} F^{i-1}(y)(1 - F(y))^{k-i} f(y)$.
The simplified form of $f_{(i)}(y)$ is

$$f_{(i)}(y) = b\left(F(y); i, k + 1 - i\right), \tag{20.7}$$

where $b\,(y; a, b) = \frac{y^{a-1}(1-y)^{b-1}}{\beta(a,b)}$ denotes the pdf of the random variable which follows beta distribution with parameters a and b, where $\beta(a, b) = \int_0^1 y^{a-1}(1 - y)^{b-1} dy = \frac{\Gamma(a)\Gamma(b)}{\Gamma(a+b)}$, $\Gamma a = (a - 1)!$. The corresponding CDF is denoted by $B(y; a, b)$.

Using these notations, $F_{(i)}(y)$ may be written as
$F_{(i)}(y) = B((F(y); i, k + 1 - i)$.
As $F(\zeta_p) = P(Y \leq \zeta_p) = p$, therefore,
$F_{(i)}(\zeta_p) = B(p; i, k + 1 - i)$.
The quantile ζ_p can be estimated by the use of quantile of $F_{(i)}$ corresponding to the probability of $B(p; i, k + 1 - i)$, which is known for any p. Let $\hat{F}_{(i)}(y) = \frac{1}{m_i} \sum_{j=1}^{m_i} I(Y_{(i)j} \leq y)$, denotes the empirical distribution of $F_{(i)}(y)$ based on ith order statistic. That is,

$$\hat{\zeta}_i = inf\left\{y : \hat{F}_{(i)}(y) \geq B(p; i, k + 1 - i)\right\} \tag{20.8}$$

Chen (2001) showed that each $\hat{\zeta}_i$ is a consistent estimator of ζ_p and the asymptotic variance is given by

$$AV(\hat{\zeta}_i) = \frac{B(p; i, k + 1 - i)\left[1 - B(p; i, k + 1 - i)\right]}{m_i \left[b(p; i, k + 1 - i)\right]^2} \tag{20.9}$$

Chen (2001) proved that the method of URSS outperforms than the methods of BRSS and SRS in terms of asymptotic relative efficiency (ARE). He used a probability vector $\mathbf{q} = (q_1, q_2, \ldots q_k)'$ with $\sum_{i=1}^{k} q_i = 1$ and $0 \leq q_i \leq 1$ in constructing a new CDF
$F_q(y) = \sum_{i=1}^{k} q_i F_{(i)}(y)$ and corresponding $f_q(y) = \sum_{i=1}^{k} q_i f_{(i)}(y)$ which is useful to find the asymptotical unbiased estimator of ζ_p.

Zhu and Wang (2005) given a weighted estimator of ζ_p which overcomes certain shortcomings of Chen (2001). They suggested that as each $\hat{\zeta}_i, i = 1, 2, \ldots k$ is a consistent estimator of ζ_p, and therefore a weighted estimator can be constructed as $\hat{\zeta}_W = \frac{\sum_{i=1}^{k} w_i \hat{\zeta}_i}{\sum_{i=1}^{k} w_i}$, where

$$w_i = \frac{1}{AV(\hat{\zeta}_i)}. \tag{20.10}$$

The asymptotic variance of $\hat{\zeta}_W$ is given by
$$AV(\hat{\zeta}_W) = \frac{1}{n} \left\{ \sum_{i=1}^{k} \frac{\frac{m_i}{n}[b(p;i,k+1-i)]^2 \zeta_p}{B(p;i,k+1-i)[1-B(p;i,k+1-i)]} \right\}^{-1}, \text{ where, } n = \sum_{i=1}^{k} m_i, \text{ the sample size.}$$

The estimator (20.10) was shown to have a smaller asymptotic variance for all distributions. It has also been shown that the optimal strategy is to select observations with one fixed rank from different ranked sets. The optimal rank and the gain in relative efficiency w.r.t. SRS are distribution free and depend on the set size and given probability only.

20.5   RSS in Non-parametric Inference

In recent years, RSS is found an alternative sampling scheme to other random sampling methods in various non-parametric areas. In non-parametric inference, Stokes and Sager (1988) first considered the properties of empirical distribution function based on the ranked set sample and used this empirical distribution function in constructing improved confidence bands for the population distribution function. They also discussed the method of Kolmorgrov–Smirnov statistic based on RSS. The non-parametric statistics is one of the areas where considerable work needs to be done as the underlying assumptions of parametric analysis and estimations, are not always fulfilled. The application of parametric methods leads the researchers to incorrect inferences with the danger of not knowing. Since the applications have a lasting effect on the general health of environment, alternative methods to parametric play an equally important role in planning data collection and their subsequent analysis. Some work in the area of non-parametric inference using RSS is already initiated. Out of which the cases of one sample and two sample cases are discussed as under.

20.5.1   One Sample Case

Suppose $Y_{(i:k)j}, i = 1, 2 \ldots k; j = 1, 2 \ldots m$ represents the BRSS of size mk taken from the population having continuous CDF $F(y - \theta)$ where $F(y)$ is the CDF of a continuous random variable with $F(0) = 0.5$.

Hettmansperger (1992) introduced the non-parametric sign test based upon the RSS under the null hypothesis $H_0 : \theta = \theta_0$. This test of non-parametric is based upon the direction of the observations from a pre-specified value. This test is used for research in which quantitative measurement is infeasible, but the information about the greater or lesser of observations with respect to the specified value is available.

The RSS-based sign statistic is given by
$S_{RSS} = \sum_{i=1}^{k} \sum_{j=1}^{m} \phi_{ij}$, where

$$\phi_{ij} = \left\{ \begin{array}{ll} 1 & ; if \ Z_{(i:k)j} > 0 \\ 0 & ; if \ Z_{(i:k)j} < 0 \end{array} \right.$$

and

$$Z_{(i:k)j} = Y_{(i:k)j} - \theta_0. \tag{20.11}$$

We consider, in general, Y_i is the random variable having CDF $F(y - \theta)$. Suppose the non-parametric statistic based on SRS is denoted by S_{SRS}. The statistics S_{SRS} follows the binomial distribution with parameters n (sample size) and $p = P(Y - \theta_0 > 0) = P(Y > \theta_0)$, and therefore we have

$$E(S_{SRS}) = np \ \text{and} \ V(S_{SRS}) = np(1 - p). \tag{20.12}$$

Hettmansperger (1992) showed that the mean values of S_{RSS} equals that of S_{SRS}. We have for the case of RSS
$E(S_{RSS}) = E(S_{SRS}) = mk \ P(Y > \theta_0) = mk \ (1 - F(\theta_0 - \theta))$
Under the null hypothesis $\theta = \theta_0$, we have $P(Y > \theta_0) = 0.5$. Therefore, $E(S_{RSS}) = \frac{mk}{2}$. For the closer case of null hypothesis, i.e., $\theta \approx \theta_0$, we have that
$E(S_{RSS}) \approx \frac{mk}{2} + mkf(0)(\theta - \theta_0)$
The $V(S_{RSS})$ can be obtained in similar to the $V(S_{SRS})$ as follows:

It is to be noted that for fixed i, the variables $Y_{(i:k)j}, j = 1, 2 \ldots, m$ are independent and identically distributed this implies that of $Z_{(i:k)j}$. Therefore,
$V(S_{RSS}) = m \sum_{i=1}^{k} V(Z_{(i:k)1} > 0) = m \sum_{i=1}^{k} V(Z_{(i:k)1} < 0)$. Using Eq. (20.11) and property of indicator variable, we have
$V(S_{RSS}) = m \sum_{i=1}^{k} P(Y_{(i:k)1} < \theta_0) \left[1 - P(Y_{(i:k)1} < \theta_0) \right]$. Denoting the CDF of $Y_{(i:k)}$ by $F_{(i)}(y)$. Using the properties of order statistics and $\sum_{i=1}^{k} F_{(i)} = kF$, we may write
$V(S_{RSS}) = mk \ F(\theta_0 - \theta) \ (1 - F(\theta_0 - \theta)) + m \sum_{i=1}^{k} \left[F^2(\theta_0 - \theta) - F_{(i)}^2(\theta_0 - \theta) \right]$
$= mk \ F(\theta_0 - \theta) \ (1 - F(\theta_0 - \theta)) - m \sum_{i=1}^{k} \left[F(\theta_0 - \theta) - F_{(i)}(\theta_0 - \theta) \right]^2$

$$= mk \, F(\theta_0 - \theta) \, (1 - F(\theta_0 - \theta)) \left[1 - \frac{\sum_{i=1}^{k} [F(\theta_0 - \theta) - F_{(i)}(\theta_0 - \theta)]^2}{k F(\theta_0 - \theta)(1 - F(\theta_0 - \theta))} \right]$$

Let us take $\epsilon^2 = \left[1 - \frac{\sum_{i=1}^{k} [F(\theta_0 - \theta) - F_{(i)}(\theta_0 - \theta)]^2}{k F(\theta_0 - \theta)(1 - F(\theta_0 - \theta))} \right]$, we can write

$$V(S_{RSS}) = mk \, F(\theta_0 - \theta) \, (1 - F(\theta_0 - \theta)) \, \epsilon^2 \qquad (20.13)$$

It is to be known that
$V(S_{SRS}) = mk \, F(\theta_0 - \theta) \, (1 - F(\theta_0 - \theta))$ and $V(S_{SRS}) \geq V(S_{RSS})$.
Therefore, $0 < \epsilon^2 \leq 1$. When there exist at least two ranks i and j such that $F_{(i)} \neq F_{(j)}$ then the strict inequality $0 < \epsilon^2 < 1$ holds.

Under the null hypothesis, we have

$$V(S_{RSS}) = \frac{mk \epsilon_0^2}{4} \qquad (20.14)$$

where $\epsilon_0^2 = 1 - 4 \sum_{i=1}^{k} \frac{(F_{(i)}(0) - 0.5)^2}{k}$
It is to be clear that the value of ϵ_0^2 is depending upon set size k. Under the perfect ranking, $F_{(i)}(0)$ can be written as
$F_{(i)}(0) = B(i, k - i + 1, 0.5)$, where $B(i, k - i + 1, y)$ denotes the CDF of beta distribution with parameters $i, k - i + 1$.

Under perfect ranking, the Pitman asymptotic relative efficiency (ARE) is given by
$ARE(S_{RSS}, S_{SRS}) = \frac{1}{\epsilon_0^2}$
The tabulated values of ϵ_0^2 for $k = 2, 3, \ldots 8$ and corresponding ARE are given in the Table 20.2. From Table 20.2, it is seen that the ARE is increasing as k increases. The sign test based upon RSS is more efficient than that based on the SRS approach.

Based upon the above RSS statistic, the following decision rules is in hand.

(i) Under $H_0 : \theta = \theta_0 \, vs \, H_1 : \theta > \theta_0$, the H_0 is rejected at α level of significance, if $S_{RSS} \geq m_{\alpha(min)}$, where $m_{\alpha(min)}$ is the smallest integer such that $P(S_{RSS} \geq m) \leq \alpha$.

(ii) Under $H_0 : \theta = \theta_0 \, vs \, H_1 : \theta < \theta_0$, the H_0 is rejected at α level of significance, if $S_{RSS} \leq m_{\alpha(max)}$, where $m_{\alpha(max)}$ is the largest integer such that $P(S_{RSS} \leq m) \leq \alpha$.

(iii) Under $H_0 : \theta = \theta_0 \, vs \, H_1 : \theta \neq \theta_0$, the H_0 is rejected at α level of significance, if $S_{RSS} \leq m_{\alpha/2(max)}$ or $S_{RSS} \geq m_{\alpha/2(min)}$, where $m_{\alpha/2(min)}$ is the smallest integer such that $P(S_{RSS} \geq m) \leq \alpha/2$ and $m_{\alpha/2(max)}$ is the largest integer such that $P(S_{RSS} \leq m) \leq \alpha/2$.

Table 20.2 The values of ϵ_0^2 and Pitman's ARE for $k = 2(1)8$

k	2	3	4	5	6	7	8
ϵ_0^2	0.750	0.625	0.547	0.490	0.451	0.416	0.371
ARE	1.3333	1.6000	1.8282	2.0408	2.2173	2.4038	2.6954

The critical values of $m_{\alpha(min)}$ and $m_{\alpha(max)}$ can be determined by the normal approximation as follows:
$m_{\alpha(min)} \approx \frac{mk}{2} - \frac{z_\alpha \epsilon_0 \sqrt{mk}}{2} - 0.5$ and $m_{\alpha(max)} \approx \frac{mk}{2} + \frac{z_\alpha \epsilon_0 \sqrt{mk}}{2} + 0.5$, where z_α is the $(1-\alpha)$th quantile of the standard normal distribution. However, for the small values of k, the critical values of $m_{\alpha(min)}$ and $m_{\alpha(max)}$ are available in the literature.

Ozturk and Wolfe (2000) studied the effects of different RSS protocols on the sign test under the sampling protocols sequential, mid-range and fixed sampling designs. They have shown that the introduction of any correlation structure in quantified observation reduces the pitman efficiency of the design. When the sampling protocol is fixed, design optimality is achieved when only the middle observation is measured from each set. Koti and Babu (1996) computed the exact null distribution of the RSS sign test. They compared the power of this test with the SRS sign test for some continuous symmetric distributions and demonstrated the superiority if RSS over SRS.

When the relative magnitudes along with the direction of the observation are also available, the Wilcoxon-signed rank test is more appropriate. Bohn (1994) defined RSS-based Wilcoxon-signed rank statistic. RSS-based Wilcoxon-signed rank statistic is given by
$W_{RSS} = \sum_{i=1}^{k} \sum_{j=1}^{m} \phi_{ij} R_{ij(R)}$, where $R_{ij(R)}$ be the rank of $Z_{(i:k)j}$ among $Z_{(i:k)j}$, $j = 1, 2 \ldots m$.

It is proven that for $F(t) + F(-t) = 1$, for all t, the distribution of W_{RSS} under H_0 does not depend on the parent form of $F(y)$.

Similar to the sign test, Bohn (1994) showed that the mean value of W_{RSS} equals to that of statistic based on SRS, W_{SRS}. This mean value is given by
$E(W_{RSS}) = E(W_{SRS}) = \frac{mk(mk+1)}{4}$. He also calculated ARE of RSS with respect to the method of SRS as
$ARE(W_{RSS}, W_{SRS}) = \epsilon^2 \frac{k+1}{2} \times ARE(W_{SRS}, S_{SRS}) = 3\epsilon^2 \frac{k+1}{2} \frac{\left[\int_{-\infty}^{\infty} (f^2(x)dx)^2\right]^2}{f^2(0)}$
where, $f(x) = F'(x)$
$ARE(W_{RSS}, W_{SRS}) = \frac{k+1}{2}$
$ARE(W_{RSS}, W_{SRS})$ is the same for all continuous symmetric and it increases as k increases.

20.5.2 Two Sample Case

Under the two sample case, the Mann–Whitney–Wilcoxon (MWW) test based on RSS has been considered. MWW test is one of the important non-parametric tests and tests whether two independent samples are drawn from the same population or not. The first time, Bohn and Wolfe (1992) used RSS empirical distribution function in constructing distribution free estimation procedures in place of the standard MWW test based on SRS. The MWW test based on two independent BRSS is explained as follows:
Suppose the two independent BRSS of sizes mk and pq are, respectively, drawn from

$F(y)$ and $G(x) = F(y - \lambda)$ with $-\infty < \lambda < \infty$. These samples are denoted by $Y_{(i:k)j}, i = 1, 2 \ldots, k, j = 1, 2, \ldots, m$ and $X_{(l:p)m}, l=1, 2, \ldots, p; m = 1, 2, \ldots, q$. The null hypothesis is that both the sample has drawn from the same population, i.e., $H_0 : \lambda = 0$. The alternative hypothesis in this case may be taken as $H_1 : \lambda > 0$. Under the null hypothesis the RSS-based MWW statistic is defined as
$U_{RSS} = \sum_{l=1}^{p} \sum_{m=1}^{q} \sum_{i=1}^{k} \sum_{j=1}^{m} \phi(X_{(l:p)m} - Y_{(i:k)j})$.
Where the indicator function,

$$\phi_t = \left\{ \begin{array}{ll} 1 & ; \text{for } t \geq 0 \\ 0 & ; \text{otherwise} \end{array} \right.$$

It is to be noted that H_0 is rejected for sufficiently large values of U_{RSS}. This means that if there are a number of X's that are sufficiently higher than the number of Y's in the RSS data, the H_0 is rejected.

Bohn and Wolfe (1992) showed that the properties of U_{RSS} are same as those for SRS-based MWW statistic U_{SRS}. The important property is that the distribution of U_{RSS} under H_0 is free from the form of $F(y)$ and is symmetric about $E_0(U_{RSS}) = \frac{mkpq}{2}$. They gave the null distribution tables for perfect ranking and presented Pitman $ARE(U_{RSS}, U_{SRS})$. For the case of $m = k$, it was shown that $ARE(U_{RSS}, U_{SRS}) = ARE(W_{RSS}, W_{SRS}) = \frac{k+1}{2}$.

20.6 Exercises

1. Describe the Lloyd's estimation technique for estimating the location and scale parameters μ and σ respectively based on SRS and hence obtain the unique minimum variance linear unbiased estimators for μ and σ.
2. Obtain the unique minimum variance linear unbiased estimators of μ and σ based on RSS. Also calculate their variances.
3. Define the term generalized variance. Calculate the relative precision of $(\hat{\mu}_{RSS}, \hat{\sigma}_{SRS})$ over $(\hat{\mu}_{SRS}, \hat{\sigma}_{RSS})$ in terms of the generalized variance for standard normal distribution with $k = 5$. The values of $\mu_{(i:5)}$ and $\sigma_{(i:5)}^2$ are given as:

$\mu_{(i:5)}$	1.16296	0.49502	0	0.49502	1.16296
$\sigma_{(i:5)}^2$	0.44753	0.31152	0.28683	0.31152	0.44753

4. Show that the unbiased estimator of population proportion (p) under URSS is given by $\hat{p}_{RSS} = \frac{1}{k} \sum_{i=1}^{k} \hat{p}_{(i)}$, where $\hat{p}_{(i)} = \frac{1}{m_i} \sum_{j=1}^{m_i} Y_{(i:k)j}$. Further find $V(\hat{p}_{RSS})$.
5. Discuss the method of estimating pth quantile, ζ_p under perfect ranking of RSS. Obtain the consistent estimator of ζ_p and also find its asymptotic variance.

6. Consider the null hypothesis $H_0 : \theta = \theta_0$ of continuous CDF $F(y - \theta)$ with $F(0) = 0.5$. If S_{RSS} denotes the BRSS-based non-parametric sign statistic of one sample case. Under H_0, show that $E(S_{RSS}) = \frac{mk}{2}$. Further obtain $V(S_{RSS})$.

7. Under the similar notations of exercise 6, if $V(S_{SRS}) = mkF(\theta_0 - \theta)(1 - F(\theta_0 - \theta))$, show that $V(S_{SRS}) \geq V(S_{RSS})$. Also obtain the expression of Pitman asymptotic relative efficiency $ARE(S_{RSS}, S_{SRS})$.

8. Obtain the Mann–Whitney–Wilcoxon test statistic based on RSS. Describe the Mann–Whitney–Wilcoxon test for the null hypothesis that two samples under study has been drawn from the same population.

9. List the non-parametric areas where the RSS has been successfully used so far.

References

Bhoj, D.S.: Estimation of parameters of the extreme value distribution using ranked set sampling. Commun. Stat. Theory Methods **26**(3), 653–667 (1997)

Bhoj, D.S., Ahsanullah, M.: Estimation of parameters of the generalized geometric distributions using ranked set sampling. Biometrics **52**, 685–694 (1996)

Bohn, L.L.: A ranked-set sample signed-rank statistic. Technical report no. 426, Department of Statistics, The University of Florida, Gainevesville, Florida (1994)

Bohn, L.L., Wolfe, D.A.: Nonparametric two-sample procedures for ranked set samples data. J. Am. Stat. Assoc. **87**(418), 552–561 (1992)

Chandra, G., Tiwari, N.: Estimation of location and scale parameters of lognormal distribution using ranked set sampling. J. Stat. Appl. **7**(3–4), 139–152 (2012)

Chandra, G., Tiwari, N., Nautiyal, R., Gupta, D.S.: On partial ranked set sampling in parameter estimation of lognormal distribution. Int. J. Stat. Agric. Sci. **12**(2), 321–326 (2016)

Chen, Z.: The optimal ranked set sampling scheme for inference on population quantiles. Stat. Sin. **11**, 23–37 (2001)

Chen, H., Stasny, E.A., Wolfe, D.A.: Ranked set sampling for efficient estimation of a population proportion. Stat. Med. **24**, 3319–3329 (2005)

Chen, H., Stasny, E.A., Wolfe, D.A.: Unbalanced ranked set sampling for estimating a population proportion. Biometrics **62**, 150–158 (2006)

Downton, F.: Least-square estimates using ordered observations. Ann. Math. Stat. **25**, 303–316 (1954)

Harter, H.L., Balakrishnan, N.: CRC Handbook of Tables for the Use of Order Statistics in Estimation. CRC Press, Boca Raton (1996)

Koti, K.M., Babu, G.J.: Sign test for ranked-set sampling. Commun. Stat. Theory Methods **25**, 1617–1630 (1996)

Lam, K., Sinha, B.K., Wu, Z.: Estimation of parameters in a two parameter exponential distribution using ranked set sample. Ann. Inst. Stat. Math. **46**, 723–736 (1994)

Lloyd, E.H.: Least square estimation of location and scale parameters using order statistics. Biometrika **39**, 88–95 (1952)

McIntyre, G.A.: A method for unbiased selective sampling using ranked sets. Aust. J. Agric. Res. **3**, 385–390 (1952)

Ozturk, O.: Parametric estimation of location and scale parameters in ranked set sampling. Fuel Energy Abstr. **141**(4), 1616–1622 (2011)

Ozturk, O., Wolfe, D.A.: Alternative ranked set sampling protocols for sign test. Stat. Probab. Lett. **47**, 15–23 (2000)

Sinha, B.K., Sinha, B.K., Purakayastha, S.: On some aspects of ranked set sampling for estimation of normal and exponential parameters. Stat. Decis. **14**, 223–240 (1996)

Stokes, S.L., Sager, T.W.: Characterization of a ranked set sample with application to estimating distribution functions. J. Am. Stat. Assoc. **83**, 374–381 (1988)

Takahasi, K., Wakimoto, K.: On unbiased estimates of the population mean based on the sample stratified by means of ordering. Ann. Inst. Stat. Math. **20**, 1–31 (1968)

Yousef, O.M., Al-Subh, S.A.: Estimation of Gumbel parameters under ranked set sampling. J. Mod. Appl. Stat. Methods **13**(2), 432–443 (2014)

Zhu, M., Wang, Y.: Quantile estimation from ranked set sampling data. Sankhya Indian J. Stat. **67**(2), 295–304 (2005)

Chapter 21
Important Versions of RSS

21.1 Introduction

McIntyre's RSS has been studied and modified by several authors for different purposes and situations. Some of the well-known modifications of RSS, which are more practicable and less prone to problems resulting from imperfect ranking are extreme ranked set sampling (ERSS), median ranked set sampling (MRSS), and double-ranked set sampling (DRSS). These versions are discussed in this chapter. Besides, the researchers developed other editions by combining these methods, which are also listed. Samawi et al. (1996) introduced the method of ERSS which actually measures only extreme order statistics but all units may require ranking by any rough gauging method, which does not require actual measurements. The method of MRSS was proposed by Muttlak (1997) to reduce the ranking error, and increase the precision over RSS and SRS while estimating the population mean. In MRSS, a modification of RSS was done in which only the median rank order statistic from each of the drawn set is taken for measurement. Al-Saleh and Al-Kadiri (2000) given the DRSS by introducing the advantages of double sampling.

21.2 Extreme RSS (ERSS)

The purpose of introducing ERSS was to avoid difficulty in ranking a sample of moderate size and observing the ranked units other than extreme order statistics. In ERSS, the possibility of imperfect ranking is reduced because of the ranking of only first- and last-order statistics in case of even set size and additional ranking of mid-order statistics for odd set size. The original form of ERSS (Samawi et al. 1996) is explained as follows.

Suppose, n is the required sample size under ERSS. Firstly, a random sample of size n from a population under consideration is drawn. This drawing of random

© The Author(s), under exclusive license to Springer Nature Singapore Pte Ltd. 2021
R. Latpate et al., *Advanced Sampling Methods*,
https://doi.org/10.1007/978-981-16-0622-9_21

samples is repeated for n times. For even n, the smallest and largest ranked units are alternately taken from the first to the nth random sample for taking actual measurements and others are discarded. In this case, first- and last-order statistics are allocated equally, i.e., $\frac{n}{2}$ each for actual measurements and forms the extreme ranked set sample, denoted by ERSSE (Extreme Ranked Set Sample for even n). It is assumed that the lowest or largest units of each set can be easily detected by visual inspection. On the other hand, for odd n, the smallest and largest ranked units are alternately selected from the first random sample to the $(n-1)$th random sample. From the nth random sample, either the mean of the smallest and largest unit is chosen or the median of the whole set is taken for the actual measurement. In this case, either $\frac{n-1}{2}$ each to first- and last-order statistics and one for middle-rank order statistics or $\frac{n-1}{2} + \frac{1}{2}$ each to first- and last-order statistics are taken for the measurements and the sample is denoted by ERSSO (Extreme Ranked Set Sample for odd n). The present chapter deals with considering taking the mean of the smallest and largest ranked unit from the nth sample.

Suppose, the n independent random samples of size n are taken from the population with mean μ and variance σ^2. These unordered units are shown as

$y_{11}, y_{12}, \ldots, y_{1n}$

$y_{21}, y_{22}, \ldots, y_{2n}$

$\ldots$

$y_{n1}, y_{n2}, \ldots, y_{nn}$

Here, the rows are showing n independent random samples each of size n and y_{ij} denotes the value of jth unit of the ith sample, $i, j = 1, 2, \ldots, n$. From the first set of n units, largest ranked unit is selected for measurement. From the second set, lowest ranked unit is selected for measurement. From the third set, the largest ranked unit is selected for measurement. This process is continued until the n largest and n smallest ranked units are selected for measurements.

Let $y_{i(1)}, y_{i(2)}, \ldots, y_{i(n)}$ denote the value of ordered statistics of ith sample, $(i = 1, 2, \ldots, n)$. Then the RSS, ERSSE, and ERSSO shall be $y_{1(1)}, y_{2(2)}, \ldots, y_{n(n)}$; $y_{1(1)}, y_{2(n)}, y_{3(1)}, \ldots, y_{n-1(1)}, y_{n(n)}$ and $y_{1(1)}, y_{2(n)}, y_{3(1)}, \ldots, y_{n-1(n)}, \frac{1}{2}(y_{n(1)} + y_{n(n)})$, respectively.

As usual, based upon SRS in which first observation is measured from each set (say), unbiased estimator of μ is

$\bar{Y} = \frac{1}{n} \sum_{i=1}^{n} y_{i1}$ with $Var(\bar{Y}) = \frac{\sigma^2}{n}$.

Since the samples are independent, one may write

$E(y_{i(1)}) = \mu_{(1)}$ with $V(y_{i(1)}) = \sigma_{(1)}^2$ and

$E(y_{i(n)}) = \mu_{(n)}$ with $V(y_{i(n)}) = \sigma_{(n)}^2$

Theorem 21.1 *For the estimator* $\bar{Y}_{ERSSE} = \frac{T_1 + T_2}{2}$ *of* μ *using ERSSE, where*

$T_1 = \frac{2}{n} \sum_{i=1}^{\frac{n}{2}} y_{2i-1(1)}$ *and* $T_2 = \frac{2}{n} \sum_{i=1}^{\frac{n}{2}} y_{2i(n)}$, *the mean and variance is given by*

$E(\bar{Y}_{ERSSE}) = \frac{\mu_{(1)} + \mu_{(n)}}{2}$ *and* $V(\bar{Y}_{ERSSE}) = \frac{\sigma_{(1)}^2 + \sigma_{(n)}^2}{2n}$

Proof It is to be noted that, $y_{1(1)}, y_{3(1)}, \ldots, y_{n-1(1)}$ are independently and identically distributed with mean and variance $\mu_{(1)}$ and $\sigma_{(1)}^2$, respectively. Similarly, the units

$y_{2(n)}, y_{4(n)}, \ldots, y_{n(n)}$ are independently and identically distributed with mean $\mu_{(n)}$ and variance $\sigma^2_{(n)}$. Also the units in T_1 are independent with T_2. Therefore, the proof is easy.

Hence the theorem. $\square$

Theorem 21.2 *For the estimator* $\bar{Y}_{ERSSO} = \frac{T_3+T_4}{2} + T_5$ *of* μ *under ERSSO, where,*

$T_3 = \frac{2}{n} \sum_{i=1}^{\frac{n-1}{2}} y_{2i-1(1)}$, $T_4 = \frac{2}{n} \sum_{i=1}^{\frac{n-1}{2}} y_{2i(n)}$ *and* $T_5 = \frac{y_{n(1)}+y_{n(n)}}{2}$, *the mean and variance is given by*

$$E(\bar{Y}_{ERSSO}) = \frac{\mu_{(1)}+\mu_{(n)}}{2} \text{ and } V(\bar{Y}_{ERSSO}) = \frac{(2n-1)(\sigma^2_{(1)}+\sigma^2_{(n)})+2Cov(y_{n(1)},y_{n(n)})}{4n^2}.$$

Proof It is easy to see that $y_{1(1)}, y_{3(1)}, \ldots, y_{n-2(1)}$ and $y_{n(1)}$ are identically distributed with mean and variance $\mu_{(1)}$ and $\sigma^2_{(1)}$, respectively, and so are $y_{2(n)}, y_{4(n)}, \ldots, y_{n-1(n)}$ and $y_{n(n)}$ with mean and variance $\mu_{(n)}$ and $\sigma^2_{(n)}$ respectively. Therefore, it is easy to show that
$E(\bar{Y}_{ERSSO}) = \frac{\mu_{(1)}+\mu_{(n)}}{2}$.
Further, the units in T_3 and T_4 are independent random variables which are also independent with the units of T_5. However, the units in T_5, i.e., $y_{n(1)}$ and $y_{n(n)}$ are not independent. Therefore, using the simple derivations we get
$V(\bar{Y}_{ERSSO}) = \frac{(2n-1)(\sigma^2_{(1)}+\sigma^2_{(n)})+2Cov(y_{n(1)},y_{n(n)})}{4n^2}$.
Samawi et al. (1996) presented the results for uniform distribution $U(0, \theta)$. For this distribution, it was shown that
$E(\bar{Y}_{ERSSE}) = E(\bar{Y}_{ERSSO}) = \frac{\theta}{2}$ with $V(\bar{Y}_{ERSSE}) = \frac{\theta^2}{(n+1)^2(n+2)}$ and
$Var(\bar{Y}_{ERSSO}) = \frac{(2n^2-n+1)\theta^2}{2n^2(n+1)^2(n+2)}$.
For uniform distribution, it is known that
$E(\bar{Y}) = \frac{\theta}{2}$ with $V(\bar{Y}) = \frac{\theta^2}{12n}$.
On comparison, it is easily shown that $V(\bar{Y}) > V(\bar{Y}_{ERSSE})$ and $V(\bar{Y}) > V(\bar{Y}_{ERSSO})$ for $n \geq 2$. Similar results may also be observed for the other distributions.

Hence the theorem. $\square$

21.3 Median Ranked Set Sampling (MRSS)

With the same purpose to get difficulty in ranking of a sample and bearing much cost for measuring ranked units in usual RSS, the MRSS (Muttlak 1997) was introduced. The goal of MRSS is to collect observations from a population that are more likely to spread the full range of values in the population and, therefore it is more representative of it through taking the median values only, under the minimal errors in ranking, than the same number of observations obtained via SRS.

The MRSS consists of selecting k simple random samples each of size k and quantifying the median of each set. The procedure of MRSS for even and odd k is explained as follows:

For even k: First identify k^2 sampling units (k sets, each of size k) from the targeted population $F(x)$ with mean μ and variance σ^2. Each set is ranked based on a perception of relative values for the variable under interest, say Y. Select the $\frac{k}{2}$th smallest rank from the first $\frac{k}{2}$ sets and select the $\left(\frac{k+2}{2}\right)$th smallest rank from the other $\frac{k}{2}$ sets for the actual measurement. By repeating the process for m cycles, one gets the desired sample of size $n = mk$.

The estimator under MRSS when k is even is given by

$$\bar{Y}_{(k)MRSSE} = \frac{1}{km} \sum_{j=1}^{m} \left(\sum_{i=1}^{\frac{k}{2}} Y_{i\left(\frac{k}{2}:k\right)j} + \sum_{i=\frac{k}{2}+1}^{k} Y_{i\left(\frac{k+2}{2}:k\right)j} \right).$$

With its variance by

$$V(\bar{Y}_{(k)MRSSE}) = \frac{1}{k^2m} \sum_{i=1}^{\frac{k}{2}} \sigma^2_{i\left(\frac{k}{2}:k\right)} + \sum_{i=\frac{k}{2}+1}^{k} \sigma^2_{i\left(\frac{k+2}{2}:k\right)}.$$

where $Y_{i(p:k)j}$ represents the measured value of the pth rank order statistic of ith set under jth cycle and $\sigma^2_{i(p:k)}$ represents variance of pth order statistic of ith set.

For odd k: In this case, identify k^2 sampling units (k sets, each of size k) and each set is ranked similar to the case of even k as above. Select the $\left(\frac{k+1}{2}\right)$th smallest rank from all sets for the actual measurement. The process is repeated for m cycles to get the desired $n = mk$. The estimator under MRSS when k is odd is given by

$$\bar{Y}_{(k)MRSSO} = \frac{1}{km} \sum_{j=1}^{m} \left(\sum_{i=1}^{k} Y_{i\left(\frac{k+1}{2}:k\right)j} \right)$$

with

$$V(\bar{Y}_{(k)MRSSO}) = \frac{1}{k^2m} \left[\sum_{i=1}^{k} \sigma^2_{i\left(\frac{k+1}{2}:k\right)} \right].$$

Muttlak (1997) showed the MRSS provides an unbiased estimator of mean when the underlying distribution is symmetric, otherwise, it is biased. In any of cases, this estimator has been shown better than the RSS method.

21.4   Double-Ranked Set Sampling (DRSS)

Al-Saleh and Al-Kadiri (2000) given the method of DRSS for estimating the population mean. The method starts with selecting k^3 units from the target population $F(x)$. Divide these units randomly into k sets each of size k^2. In the first stage, from each set, select a sample of size k using the ordinary RSS method. This provides a ranked set sample of size k^2. In the second stage, the RSS procedure is applied to this sample of size k^2. From these two stages, the double-ranked set sample of size k is obtained. The cycle may be repeated m times to obtain the desired sample of size $n = mk$.

The unbiased estimator of μ based on DRSS is given by

$$\bar{Y}_{DRSS} = \frac{1}{mn} \sum_{j=1}^{m} \sum_{i=1}^{k} y_{(i)j}.$$

where $y_{(i)j}$ denotes the ith samples unit selected under jth cycle of DRSS, $i = 1, 2, \ldots, k$; $j = 1, 2, \ldots, m$.

Suppose, for all j,

$E(y_{(i)j}) = \mu_{(i)}$

$V(y_{(i)j}) = \sigma^2_{(i)}$

Then it is easily shown that

$$V(\bar{Y}_{DRSS})=\frac{1}{mk}\left\{\sigma^2 - \frac{1}{k}\sum_{i=1}^{k}\left(\mu_{(i)} - \mu\right)^2\right\} = \frac{1}{mk^2}\sum_{i=1}^{k}\sigma_{(i)}^2.$$

Al-Saleh and Al-Kadiri (2000) showed that the sample mean based on DRSS is more efficient than the sample mean with RSS.

21.5 Some Other Versions

Researchers introduced a number of other variations of RSS; some of them are listed below.

21.5.1 Double ERSS (Samawi 2002)

This is an extension to the ERSS procedure and results in an increase of efficiency for the estimation of population mean when the underlying distribution of interest is symmetric. This procedure improves the efficiency when the regression estimator is used.

21.5.2 Quartile RSS (Muttlak 2003a) and Percentile RSS (Muttlak 2003b)

The Quartile RSS and Percentile RSS were compared to the SRS, RSS, and MRSS methods while estimating population mean. It turns out that QRSS is more efficient than its counterpart SRS method for the symmetric distributions.

21.5.3 LRSS (Al-Nasser 2007)

The main idea behind LRSS was to remove extreme observations and replace them with the next largest values. In this method, we first set the coefficient $p = [k\alpha]$ such that $0 \le \alpha \le 0.5$.

Then from the first $p + 1$ sample sets, select the $(p + 1)$th ranked unit and from the last $k - p$ sample sets select the $(k - p)$th ranked unit and select the ith ranked unit from the ith sample sets, where $\alpha \in (p + 2, k - p - 1)$.

21.5.4 Balanced Groups RSS (Jemain et al. 2008)

In this method, $p = 3k$ $(k = 1, 2, \ldots)$ sets each of size p from the target population are selected, and rank the units within each set with respect to the variable of interest.

Then divide the $3k$ selected sets randomly into three groups, each of size k. For each group, select the lowest ranked unit from each set in the first group, the median ranked unit from each set in the second group, and the largest ranked unit from each set in the third group for actual measurements. This process may be repeated m times to get the balanced group ranked set sample of size $3\,km$.

21.5.5 Selective Order RSS (Al-Subh et al. 2009)

In this method, a random sample of size k is drawn and rank the units in the sample as usual. Select the ith ranked unit as of the unit of interest. Repeat this m times to obtain a selective order ranked set sample of size m. Essentially, this is a random sample found based on the RSS involving a set size m via the ith-order statistic.

21.5.6 Robust ERSS (Al-Nasser and Bani-Mustafa 2009)

Robust ERSS method comes into existence by slight change in the process of RSS. When k (set size) is even, one may select the $(p + 1)$th ranked unit from the first $p/2$ sample sets and select the $(k - p)$th ranked unit from the last $p/2$ sample sets as the final sample for measurement, where $p = [k\beta]$, $0 < \beta < 0.5$.

21.5.7 Quartile DRSS (Al-Saleh and Al-Omari 2002)

By combining the Quartile RSS and DRSS, a new method for estimating the population median was given. This compared with the SRS, RSS, and quartile RSS.

21.5.8 Record RSS (Salehi and Ahmadi 2014)

This particular method is helpful when the aim is to generate record-breaking data. A distribution-free two-sided prediction interval for future order statistics has also been derived.

21.5.9 Mixed RSS (Haq et al. 2014)

A suitable mixture of SRS and RSS schemes was developed for estimation of the population mean and median. It turns out that the mean and median estimates under the mixed RSS scheme are more precise than those based on the SRS scheme. When estimating the mean of symmetric and some skewed populations, the mean estimates under mixed RSS scheme are found to be more efficient than the mean estimates with the Partial RSS scheme.

21.5.10 Neoteric RSS (Zamanzade and Al-Omari 2016)

This method was derived from the RSS scheme for estimating the population mean and variance. NRSS provides more efficient estimators as compared to RSS. This provides a uniform improvement over RSS without any additional costs. Taconeli and Cabral (2019) given two-stage sampling designs based on Neoteric RSS. Five different sampling schemes were given by them which outperform RSS, Neoteric RSS, and the DRSS designs.

21.5.11 Robust Extreme-DRSS (Majd and Saba 2018)

More efficient estimators of population mean with smaller variances than the usual SRS, RSS, DRSS, ERSS have been shown for both the symmetric and skewed distributions, mostly based upon the simulation results.

21.5.12 Extreme-cum-Median RSS (Ahmed and Shabbir 2019)

This method is a mixture of ERSS and DRSS for obtaining a more representative sample using three out of five number of summary statistics [i.e., Minimum, Median, and Maximum]. This provides an unbiased estimator of mean for symmetric population and gives moderate efficiency for skewed distribution as well under perfect and imperfect rankings.

21.6 Exercises

1. Describe the advantages of ERSS over RSS.
2. Obtain the estimator of μ under ERSS for both the cases of even and odd sample size. Also, obtain the MSE of these estimators.
3. Consider the standard normal distribution. Show that the Bias of the estimator of the population mean under ERSS is zero for the even sample size $n = 2, 4$ and 6.

4. Obtain the estimator of μ under MRSS for both the cases of even and odd sample sizes. Also, find the MSE of these estimators.
5. Explain the advantages and limitations of DRSS over RSS. Discuss one case in which the use of DRSS is appropriate to apply.

References

Samawi, H.M.: On Quantile Estimation With Application to Normal Ranges and Hodges–Lehmann Estimate Using a Variety of Ranked Set Sample. Department of Statistics, Yarmouk University, Irbid, Jordan (1999)

Samawi, H.M., Ahmed, M.S., Abu-Dayyeh, W.: Estimating the population mean using extreme ranked set sampling. Biom. J. **38**, 577–586 (1996)

Muttlak, H.A.: Median ranked set sampling. J. Appl. Stat. Sci. **6**(4), 245–255 (1997)

Al-Saleh, M.F., Al-Kadiri, M.A.: Double ranked set sampling. Stat. Probab. Lett. **48**, 205–212 (2000)

Samawi, H.M.: On double extreme ranked set sample with application to regression estimator. Metron **60**(1–2), 53–66 (2002)

Muttlak, H.A.: Investigating the use of quartile ranked set samples for estimating the population mean. J. Appl. Math. Comput. **146**, 437–443 (2003a)

Muttlak, H.A.: Modified ranked set sampling methods. Pak. J. Stat. **19**(3), 315–323 (2003b)

Al-Nasser, D.A.: L-ranked Set sampling: a generalization procedure for robust visual sampling. Commun. Stat. Simul. Comput. **36**, 33–43 (2007)

Jemain, A.A., Al-Omari, A., Ibrahim, K.: Some variations of ranked set sampling. Electron. J. Appl. Stat. Anal. **1**, 1–15 (2008)

Al-Subh, S.A., Al-Odat, M.T., Ibrahim, K., Jemain, A.A.: EDF goodness of fit tests of logistic distribution under selective order statistics. Pak. J. Stat. **25**(3), 265–274 (2009)

Al-Nasser, D.A., Bani-Mustafa, A.: Robust extreme ranked set sampling. J. Stat. Comput. Simul. **79**(7), 859–867 (2009)

Al-Saleh, M.F., Al-Omari, A.I.: Multistage ranked set sampling. J. Stat. Plan. Inference **102**, 273–286 (2002)

Salehi, M., Ahmadi, J.: Record ranked set sampling scheme. Metron **72**, 351–365 (2014)

Haq, A., Brown, J., Moltchanova, E., Al-Omari, A.I.: Mixed ranked set sampling design. J. Appl. Stat. **41**(10), 2141–2156 (2014)

Zamanzade, E., Al-Omari, A.I.: New ranked set sampling for estimating the population mean and variance. Hacet. J. Math. Stat. **45**(6), 1891–1905 (2016)

Taconeli, C.A., Cabral, A.S.: New two-stage sampling designs based on neoteric ranked set sampling. J. Stat. Comput. Simul. **89**(2), 232–248 (2019)

Majd, M.H.H., Saba, R.A.: Robust extreme double ranked set sampling. J. Stat. Comput. Simul. **88**(9), 1749–1758 (2018)

Ahmed, S., Shabbir, J.: Extreme-cum-median ranked set sampling. Braz. J. Probab. Stat. **33**(1), 24–38 (2019)

Chapter 22
Non-sampling Errors

22.1 Introduction

In the previous chapters, we have seen that despite using the appropriate sampling methods and other measures like estimators for estimating the population parameters viz. mean, total, proportions, etc., there is a difference between the estimate and the true value of the parameter. This difference is termed as *sampling errors*. Sampling error is varied from sample to sample. In general, it is measured by the standard error of the estimator. For instance, it is known that under SRSWR, sample mean $\bar{y}_n$ is an unbiased estimator of the population mean $\bar{Y}$ with $V(\bar{y}_n) = \frac{\sigma^2}{n}$. Therefore, sampling error, in general, is measured by $\frac{\sigma}{\sqrt{n}}$. This means that the sampling error decreases as the sample size increases. The estimator will be free from the sampling error when the sample and population size becomes the same, i.e., $n = N$. In this case, the estimated value is exactly the same as the population value. Besides sampling errors, there is another kind of error present which comes through different sources including measurement of units, inadequately determination of the sample size, respondent's non-response/poor response/incorrect response, processing of data. The precision of an estimate due to errors arising from incomplete coverage and faulty procedures of estimation, observational errors and non-response, etc., are termed as non-sampling errors. This means that non-sampling errors have basically arisen when one deviates from the prescribed rule of the sampling and not getting the information from the sampling units to the extent it is essentially required for the survey. Many times, the non-sampling error is more serious than the sampling errors and this affects substantially the survey estimates. In this chapter, we discuss the non-sampling errors.

22.2 Types of Non-sampling Errors

The non-sampling errors may arise broadly from any one of the following sources.

© The Author(s), under exclusive license to Springer Nature Singapore Pte Ltd. 2021
R. Latpate et al., *Advanced Sampling Methods*,
https://doi.org/10.1007/978-981-16-0622-9_22

22.2.1   Measurement Error

The objective of the sample survey is to obtain information on the true population value. But we don't get a close estimate due to the faulty measuring instrument and sometimes it is too costly and surveyors don't have the resources. The eye estimate in the agricultural field may far away from reality and it requires extensive training from the surveyor. This error may be reduced by some of the following ways:

(i) Adequate scrutiny of the data before including them in the estimation. The method of trimmed mean, box plots, etc., may be useful to reduce the outliers.

(ii) Use of proper coding, listing of the units.

(iii) Use accurate unit of measurements transformation to make the same unit of measurements for the sampled observations.

(iv) Accurate tabulation of the data for the analysis. Any single error in tabulation may seriously affect the outputs.

22.2.2   Absence of Pilot Study

A pilot study is a pretest of the sampling methods. It is conducted on a small group of typical people as a dress rehearsal. It helps to reduce the sample size and gives the idea about the appropriateness of the sampling method. Also, it warrants the change of sampling methods and revision of the questionnaire. In some of the situations, it is required to conduct and hence some extent of non-sampling error may be reduced.

22.2.3   Inadequate Sample Size

The sample size plays a pivotal role while conducting the sample survey. It depends on the cost, time, and availability of resources. The sample size does not depend on the population size but it depends on the variation in the population, i.e., the lesser the variation in the population, the smaller the population size and vice versa. The sampling errors decrease as sample size increases and non-sampling errors increase as sample size increases. This is so happening due to a large amount of measurement errors and other kinds of errors. Therefore, an adequate sample size is required for the survey.

22.2.4 Response Errors

In some of the situations, respondents do not provide the correct information due to some unknown and personal reasons, and therefore the deviation from the actual individual value is known as the response errors. For example, in a socioeconomic survey, a respondent reports the total income of his family is $500 per month, which was actually $1000 per month. Therefore, this survey contains response errors. This error may be reduced if (i) there is adequate training to the field staff for communication with respondents and techniques to extract the information from the respondents, (ii) Supervision and inspection of the field staff from time to time, (iii) Identifications of the respondents not providing the correct data for omission purpose, and (iii) Solving the problems during data collection and reducing errors on the part of the respondents.

22.2.5 Non-response Error

This kind of error arises due to refusals, not available at the time of call and the questionnaire is too difficult to understand. Low response will make the sample unrepresentative to the population.

The problem of non-response can be reduced by using the following techniques.

1. Designing the short questionnaire.
2. Sponsorship by bodies which are highly regarded.
3. Assuring confidentiality.
4. Start the questionnaire with non-threatening questions.
5. Providing small rewards.
6. Offer to provide final result later on.
7. Callbacks.

Hansen and Hurwitz Pioneer Model:
Hansen and Hurwitz (1946) proposed a model for mail survey to minimize the cost of sample survey. It involves a large non-response than the personal interview. This model is a combination of mail surveys and personal interviews. It provides unbiased estimators of population mean or total.

The Hansen and Hurwitz model consist of the following steps.

1. Mail a questionnaire to all the selected individuals and give them a deadline.
2. Collect all the filled questionnaires from the individuals who send up to a deadline are called respondents and those who do not send are called non-respondents.
3. Take subsample from the non-respondents and conduct their personal interview.
4. Combine the data from respondents and non-respondents to estimate the population parameters of the interest.

The population of size N is partitioned into N_1 number of respondents and N_2 number of non-respondents. The sample of size n_1 is selected from the class of respondents

N_1 and sample of size n_2 is selected from the class of non-respondents. Again, h_2 denotes the size of subsample from the n_2 non-respondents to be interviewed so that $n_2 = h_2 g$ where $g > 1$ and constant.

The unbiased estimators of N_1 and N_2 are respectively, $\hat{N}_1 = \frac{n_1}{n} N$ and $\hat{N}_2 = \frac{n_2}{n} N$. Where n is total sample size i.e. number of questionnaires is sent.

Theorem 22.1 *An unbiased estimator of the population mean $\bar{Y}$ is given by*
$$\bar{y}_{..} = \frac{n_1 \bar{y}_1 + n_2 \bar{y}_{h_2}}{n}$$
where $\bar{y}_{h_2}$ is subsample mean of h_2 observations from non-respondents class.

Proof By using the definition of conditional expectation,
$$E(\bar{y}_{..}) = E_1\big(E_2(\bar{y}_{..}|n_1, n_2)\big)$$
$$= E_1\Big(\frac{n_1 \bar{y}_1 + n_2 \bar{y}_2}{n}\Big)$$
$$= E_1(\bar{y}_{.}) = \bar{Y}$$
Hence the theorem. □

Theorem 22.2 *The variance of the estimator $\bar{y}_{..}$ is given by*
$$V(\bar{y}_{..}) = (\tfrac{1}{n} - \tfrac{1}{N})S_y^2 + (\tfrac{g-1}{n})(\tfrac{N}{N_2})S_{hy}^2$$
where S_{hy}^2 is the population mean square for the non-respondents class.

Proof By using the definition of conditional variance, we have
$$V(\bar{y}_{..}) = V_1\big(E_2(\bar{y}_{..}|n_1, n_2)\big) + E_1\big(V_2(\bar{y}_{..}|n_1, n_2)\big)$$
$$= V_1\big(E_2(\tfrac{n_1 \bar{y}_1 + n_2 \bar{y}_{h_2}}{n}|n_1, n_2)\big) + E_1\big(V_2(\tfrac{n_1 \bar{y}_1 + n_2 \bar{y}_{h_2}}{n}|n_1, n_2)\big)$$
$$= V_1(\bar{y}_{.}) + E_1\big(\tfrac{n_2^2}{n^2} V(\bar{y}_{h_2})\big)$$
$$= V_1(\bar{y}_{.}) + E_1\big(\tfrac{n_2^2}{n^2}(\tfrac{1}{h_2} - \tfrac{1}{n_2})s_{hy}^2\big)$$
$$= V_1(\bar{y}_{.}) + (\tfrac{g-1}{n})E_1\big(\tfrac{n_2}{n} s_{hy}^2\big)$$
$$= (\tfrac{1}{n} - \tfrac{1}{N})S_y^2 + \tfrac{g-1}{n}\tfrac{N_2}{N} S_{hy}^2$$
Hence the theorem. □

Example 22.1 Pune City in India consists of 50 lakh persons. We want to to estimate the monthly average expenditure on their daily need. We selected an SRSWOR sample of 10,000 persons and mailed a questionnaire to each of them regarding their monthly average expenditure on their daily needs. We received 8000 questionnaires and their monthly average expenditure was $800. After the deadline of questionnaire submission is over, we select 200 people out of 2000 by using SRSWOR from no-respondent. and take a personal interview. The monthly average expenditure was $700. Estimate the monthly average expenditure of persons living in Pune City.

The Hansen–Hurwitz estimator is given by
$$\bar{y}_{..} = \frac{n_1 \bar{y}_1 + n_2 \bar{y}_{h_2}}{n}$$
$$= \frac{80000 \times 800 + 2000 \times 700}{10000} = 780$$
The average monthly expenditure the peoples on their daily needs is $780.

22.3 Exercises

1. Define the term Sampling errors.
2. Describe the problem of non-response.
3. Describe the problem of non-response and explain Hansen and Hurwitz Pioneer Model.
4. Define the term non-sampling errors.
5. Explain the problem of non-response and Hansen and Hurwitz Pioneer Model. Also, show that Hansen and Hurwitz's estimator is unbiased for the population mean and obtain its variance.

Reference

Hansen, M.H., Hurwitz, W.N.: The problem of non-response in sample surveys. J. Am. Stat. Assoc. **41**, 517–529 (1946)

Printed in the United States
by Baker & Taylor Publisher Services